CIA Cold War Records

CORONA: America's First Satellite Program

CIA Cold War Records Series

Editor in Chief
J. Kenneth McDonald

CIA Documents on the Cuban Missile Crisis, 1962,
Mary S. McAuliffe, editor (1992)

Selected Estimates on the Soviet Union, 1950–1959,
Scott A. Koch, editor (1993)

The CIA under Harry Truman,
Michael S. Warner, editor (1994)

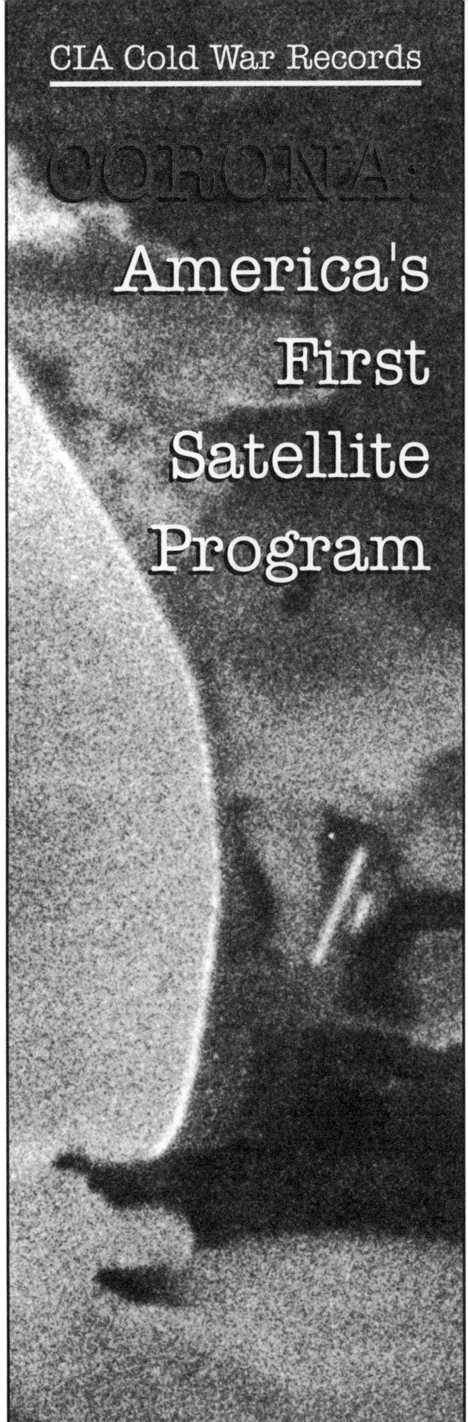

Kevin C. Ruffner
Editor

History Staff
Center for the Study of Intelligence
Central Intelligence Agency
Washington, D.C.
1995

These documents have been approved for release through the Historical Review Program of the Central Intelligence Agency.

US Government officials may obtain additional copies of this document directly through liaison channels from the Central Intelligence Agency.

Requesters outside the US Government may obtain subscriptions to CIA publications similar to this one by addressing inquiries to:

Document Expedition (DOCEX) Project
Exchange and Gift Division
Library of Congress
Washington, DC 20540

or: **National Technical Information Service**
5285 Port Royal Road
Springfield, VA 22161

Requesters outside the US Government not interested in subscription service may purchase specific publications either in paper copy or microform from:

Photoduplication Service
Library of Congress
Washington, DC 20540

or: **National Technical Information Service**
5285 Port Royal Road
Springfield, VA 22161
(to expedite service call the
NTIS Order Desk 703-487-4650)

Comments and queries on this paper may be directed to the DOCEX Project at the above address or by phone (202-707-9527), or the NTIS Office of Customer Service at the above address or by phone (703-487-4660). Publications are not available to the public from the Central Intelligence Agency.

Requesters wishing to obtain copies of these and other declassified documents may contact:

Military Reference Branch (NNRM)
Textual Reference Division
National Archives and Records Administration
Washington, DC 20408

Contents

CORONA: America's First Satellite Program

Foreword

Since the CORONA satellite made its first successful flight in August 1960, the Intelligence Community's overhead reconnaissance programs have been among the nation's most closely guarded secrets. The end of the Cold War, however, has at last made it possible to declassify both information and imagery from the first American satellite systems of the 1960s. To do this, President William Clinton in February of this year ordered the declassification within 18 months of historical intelligence imagery from the early satellite systems known as the CORONA, ARGON, and LANYARD. Because the President's Executive Order 12951 (see appendix) envisions scientific and environmental uses for this satellite imagery, the declassified photographs will be transferred to the National Archives with a copy sent to the US Geological Survey. Vice President Albert Gore, who first urged the Intelligence Community to open up its early imagery for environmental studies, unveiled the first CORONA satellite photographs for the American press and public at CIA Headquarters on 24 February 1995.

To mark this new initiative, CIA's Center for the Study of Intelligence and the Space Policy Institute at George Washington University are cosponsoring a conference, "Piercing the Curtain: CORONA and the Revolution in Intelligence," in Washington on 23–24 May 1995. On the occasion of this conference, the CIA History Staff is publishing this collection of newly declassified documents and imagery from the CORONA program. This is the fourth volume in the CIA Cold War Records Series, which began in 1992 when Director of Central Intelligence Robert Gates launched CIA's Openness Policy and reorganized the Center for the Study of Intelligence to include both the History Staff and a new Historical Review Group to declassify historically important CIA records.

The editor of this new volume, Dr. Kevin C. Ruffner, has an A.B. from the College of William and Mary and an M.A. in history from the University of Virginia. He joined the CIA History Staff in 1991, soon after he received his Ph.D. in American Studies from George Washington University.

The documents and imagery in this volume were reviewed and declassified with unusual dispatch by a special working group of declassification officers from the National Reconnaissance Office, the Central Imagery Office, CIA's Directorate of Science and Technology, and its National Photographic Interpretation Center. The group's prompt work is especially notable since many documents required consultation with the US Air Force, National Security Agency, Defense Intelligence Agency, Department of Energy, Department of State, and CIA's Collection Requirements and Evaluation Staff.

This volume's appearance just three months after President Clinton's declassification order is yet another tribute to the skill and speed that the History Staff of the Center for the Study of Intelligence has come to expect from the Design Center and Publications Center in the Directorate of Intelligence, and from the Directorate of Administration's Printing and Photography Group.

J. Kenneth McDonald
Editor in Chief

CORONA: America's First Satellite Program

Preface

The CORONA reconnaissance satellites revolutionized the collection of intelligence in the 1960s. This was a time when it was still extraordinarily difficult to gather information by any other means from "denied areas" including the Soviet Union, Communist China, and their allies. The need for intelligence about Soviet strategic weapon systems and bases dramatically increased after 1 May 1960, when the Soviets shot down an American U-2 aircraft and captured its CIA pilot, Francis Gary Powers. Within a few months, however, on 18 August the United States launched its first successful reconnaissance satellite, which in one mission provided more photographic coverage of the Soviet Union than all previous U-2 missions. On 19 August 1960, the recovery of Mission 9009 with a KH-1 camera marked the beginning of the CORONA reconnaissance satellite program's long and valuable service. The story of this program's success is one of the most remarkable in the annals of American science and intelligence.

The US Government did not acknowledge that it used satellite systems and imagery for intelligence purposes until 1978. Although President Jimmy Carter then announced that the United States used satellites to verify arms control treaties, it has only been the past few years that officials have talked openly about these systems and their intelligence uses.

CORONA, the program name for a series of satellites with increasingly more accurate cameras, provided coverage of the Soviet Union, China, and other areas from the Middle East to Southeast Asia. From its start in the late 1950s until its retirement in 1972, CORONA (in its several versions) both proved valuable in itself and set the stage for the satellite programs that followed it. For the first time US policymakers had encompassing coverage of the Soviet Union and China that was both timely and accurate. Since the 1960s a significant percentage of finished intelligence—intelligence reports sent to policymakers—has been largely derived from reconnaissance satellites. Satellite imagery is used for a variety of analytical purposes from assessing military strength to estimating the size of grain production. Far and away its greatest utility, however, has been to monitor the deployment of Soviet strategic forces and to verify compliance with arms control agreements.

While orbiting the earth, CORONA concentrated principally on photographing the USSR and China. One intelligence community study summarized CORONA's efforts over the Soviet Union:

> CORONA's initial major accomplishment was imaging all Soviet medium-range, intermediate-range, and intercontinental ballistic missile launching complexes. CORONA also identified the Plesetsk Missile Test Range, north of Moscow. Repetitive coverage of centers like Plesetsk provided information as to what missiles were being developed, tested, and/or deployed. Also, the unequivocal fact of observation gave the United States freedom from concern over many areas and locations which had been suspect in the past.
>
> Severodvinsk, the main Soviet construction site for ballistic-missile-carrying submarines was first seen by CORONA. Now it was possible to monitor the launching of each new class of submarines and follow it through deployment to operational bases. Similarly, one could observe Soviet construction and deployment of the ocean-going surface fleet. Coverage of aircraft factories and airbases provided an inventory of bomber and fighter forces. Great strides were also made in compiling an improved Soviet ground order of battle.
>
> It was CORONA imagery which uncovered Soviet antiballistic missile activity. Construction of the GALOSH sites around Moscow and the GRIFFON site near Leningrad, together with construction of sites around Tallinn for the Soviet surface-to-air missile known as the SA-5, were first observed in CORONA imagery. HEN HOUSE, DOG HOUSE, and the Soviet Union's first phased-array radars—all associated with the Soviet ABM program—were also identified in CORONA imagery.
>
> CORONA "take" was also used to locate Soviet SA-1 and SA-2 installations; later its imagery was used to find SA-3 and SA-5 batteries. The precise location of these defenses provided Strategic Air Command planners with the information needed to determine good entry and egress routes for US strategic bombers.
>
> CORONA imagery was also adapted extensively to serve the needs of the Army Map Service and its successor, the Defense Mapping Agency (DMA). Enhanced by improvements in system attitude control and ephemeris data plus the addition of a stellar-index camera, CORONA eventually became almost the sole source of DMA's military mapping data.

Some explanation of the terms used in the CORONA program may be helpful. The imagery acquired from the satellites and cameras that composed the CORONA program had a specific security system called TALENT-KEYHOLE. This added the codeword KEYHOLE, for satellite collection, to the codeword TALENT, which was originally used for imagery collected by aircraft.

The first four versions of CORONA were designated KH-1 through KH-4 (KH denoted KEYHOLE); KH-4 went through three versions. The camera in KH-1—public cover name DISCOVERER—had a nominal ground resolution of 40 feet. (Ground resolution is the ground size equivalent of the smallest visible imagery and its associated space.) By 1963 improvements to the original CORONA had produced the KH-2 and KH-3, with cameras that achieved resolutions of 10 feet.

The first KH-4 mission was launched in 1962 and brought a major breakthrough in technology by using the MURAL camera to provide stereoscopic imagery. This meant that two cameras photographed each target from different angles, which allowed imagery analysts to look at KH-4 stereoscopic photos as three-dimensional. In the KH-4, the workhorse of the CORONA system, three camera models with different resolutions were the principal difference between the versions, KH-4, KH-4A, and KH-4B. By 1967, the J-3 camera of KH-4B had entered service with a resolution of 5 feet. This final version of CORONA continued overflights until 1972.

Two other systems, separate but closely allied with CORONA, also operated during this time with less success. The KH-5, or ARGON, performed mapping services for the Army in a few missions in the early 1960s with mediocre results. The same disappointing performance afflicted the LANYARD system, or KH-6, which was both begun and abandoned in 1963.

The following outlines the CORONA versions from 1959 to 1972:

	Camera	Units Launched	Time Period
KH-1		10	1959–60
KH-2	C' (C Prime)	10	1960–61
KH-3	C''' (C Triple Prime)	6	1961–62
KH-4	M (Mural)	26	1962–63
KH-4A	J (J-1)	52	1964–69
KH-4B	J-3	17	1967–72

This volume of newly declassified documents and photos is organized in four parts. Part 1 presents the first history of the CORONA program, an article published in 1973 in a classified special supplement to CIA's professional quarterly, *Studies in Intelligence*. Part 2 provides a brief look at how the interdepartmental Committee on Overhead Reconnaissance, formed in 1960 to coordinate satellite collection, implemented the new system. Part 3 includes a number of National Photographic Interpretation Center and other

CIA reports on the analysis of CORONA imagery, while Part 4 concludes with an example of a nonmilitary use of satellite imagery. In each part, a brief introduction is followed by the relevant documents in chronological order.

CORONA was the United States' response to a growing need in the 1960s for detailed photographic coverage of countries behind the Iron Curtain. The introduction of newer—and still classified—satellite systems after 1964 further improved the program's utility and performance. The sheer volume of documents and imagery associated with CORONA—its imagery alone is estimated at over 2 million linear shelf feet—made it both important and difficult to select representative samples for this volume.

In the spring of 1992, Robert Gates, then Director of Central Intelligence (DCI), formed the Environmental Task Force to determine how the Intelligence Community could use its technology to assist scientists in studying the environment. Spurred by then Senator Albert Gore, the CIA also formed a DCI Classification Review Task Force to examine the declassification of satellite imagery collected by obsolete, broad-area-search satellite systems. Both the Environmental Task Force and the DCI Classification Review Task Force determined that imagery produced from KH-1 through KH-6 systems offered unusual information for scientists, scholars, and historians. The declassification of this imagery, both panels concluded, presented no threat to national security.

DCI R. James Woolsey approved the recommendations of the two task forces and on 22 February 1995, President William Clinton signed an Executive order directing the declassification of more than 800,000 early satellite images. These images, collected by the CORONA, ARGON, and LANYARD systems, provide extensive coverage of the earth's surface.

This book of documents is but the first installment of information on America's first satellite system. In the years ahead the American public can look forward to a wealth of declassified reports and imagery from the CORONA program.

Part I

History of CORONA Program

Part I: History of the CORONA Program

After the CORONA program drew to a close in 1972, the CIA published a survey account of the program in a special Spring 1973 supplement to its classified professional journal, *Studies in Intelligence*. Kenneth E. Greer's article focuses on the program's early years, its uncertainties and frustrations. CIA manager, Richard M. Bissell, Jr., commented after the second mission—DISCOVERER I—failed in 1959:

> It was a most heartbreaking business. If an airplane goes on a test flight and something malfunctions, and it gets back, the pilot can tell you about the malfunction, or you can look it over and find out. But in the case of a recce [reconnaissance] satellite, you fire the damn thing off and you've got some telemetry and you never get it back. There is no pilot, of course, and you've got no hardware, you never see it again. So you have to infer from telemetry what went wrong. Then you make a fix, and if it fails again you know you've inferred wrong. In the case of CORONA it went on and on.[1]

In its first years CORONA encountered considerable difficulties, which did not immediately diminish even after the first successful mission in August 1960. Indeed, of the first 30 missions from 1960 through 1962, only 12 were considered productive. The description of the recovery of Mission 1005 in South America illustrates some of the problems that the intelligence community confronted and overcame in developing and employing CORONA.

The *Studies* article also highlights CORONA's considerable achievements. When *The New York Times* on 12 August 1960 reported the safe return of DISCOVERER XIII and its triumphant procession from the Pacific Ocean to President Eisenhower at the White House, the paper immediately recognized that this startling reentry signaled a new era:

> The technological feat marks an important step toward the development of reconnaissance satellites that will be able to spy from space. The same ejection and recovery techniques eventually will be used for returning photographs taken by reconnaissance satellites. Indirectly the technique will also contribute to the eventual return of manned spacecraft.

Within a week, Air Force Capt. Harold E. Mitchell and his crew conducted the first aerial recovery when DISCOVERER XIV (or Mission 9009), the first satellite with film, returned to earth on 19 August 1960. Six days later,

[1]Quoted in Leonard Mosley, *Dulles: A Biography of Eleanor, Allen, and John Foster Dulles and Their Family Network* (New York: The Dial Press/James Wade, 1978), p. 432.

President Eisenhower and Director of Central Intelligence Allen Dulles inspected the mission's photographs. In films "good to very good," the camera had photographed 1.5 million square miles of the Soviet Union and East European countries. From this imagery 64 Soviet airfields and 26 new surface-to-air missile (SAM) sites were identified. That the first satellite mission could produce such results stunned knowledgeable observers from imagery analysts to the President.

1. Kenneth E. Greer, "Corona," *Studies in Intelligence*, Supplement, 17 (Spring 1973): 1-37.

The first photographic reconnaissance satellite

CORONA

Kenneth E. Greer

When the U-2 began operating in the summer of 1956, it was expected to have a relatively short operational life in overflying the Soviet Union—perhaps no more than a year or two. That expectation was based not so much on the likelihood that the Soviets could develop the means of shooting it down, as on their ability to develop a radar surveillance network capable of tracking the U-2 reliably. With accurate tracking data in hand, the Soviets could file diplomatic protests with enough supporting evidence to generate political pressures to discontinue the overflights. As it turned out, the United States had underestimated the Soviet radars, which promptly acquired and continuously tracked the very first U-2 flight over Soviet territory. The Soviets filed a formal protest within days of the incident, and a standdown was ordered.

For nearly four years, the U-2 ranged over much of the world, but only sporadically over the Soviet Union. Soviet radar was so effective that each flight risked another protest, and another standdown. Clearly, some means had to be found to accelerate the initial operational capability for a less vulnerable successor to the U-2. Fortunately, by the time Francis Gary Powers was shot down near Sverdlovsk on 1 May 1960 (fortunate for the intelligence community, that is—not for Powers), an alternative means of carrying out photographic reconnaissance over the Soviet Union was approaching operational readiness. On 19 August 1960, just 110 days after the downing of the last U-2 overflight of the Soviet Union, the first successful air catch was made near Hawaii of a capsule of exposed film ejected from a photographic reconnaissance satellite that had completed seven passes over denied territory and 17 orbits of the earth. The feat was the culmination of four years of intensive and often frustrating effort to build, launch, orbit, and recover an intelligence product from a camera-carrying satellite.

At about the time the U-2 first began overflying the Soviet Union in 1956, the U.S. Air Force was embarking on the development of a strategic reconnaissance weapons system employing orbiting satellites in a variety of collection configurations. The program, which was designated WS-117L, had its origins in 1946 when a requirement was placed on the RAND Corporation for a study of the technical feasibility of orbiting artificial satellites. The first real breakthrough had come in 1953 when the USAF Scientific Advisory Board reported to the Air Staff that it was feasible to produce relatively small and light-weight thermonuclear warheads. As a result of that report, the ATLAS ICBM program was accorded the highest priority in the Air Force.

1. *(Continued)*

Since the propulsion required to place a satellite in orbit is of the same general order of magnitude as that required to launch an ICBM, the achievement of an ICBM-level of propulsion made it possible to begin thinking seriously of launching orbital satellites. Accordingly, General Operational Requirement No. 80 was levied in 1955 with the stated objective of providing continuous surveillance of pre-selected areas of the world to determine the status of a potential enemy's war-making capacity.

The Air Research and Development Command, which had inherited the RAND study program in 1953, assigned the satellite project to its Ballistic Missile Division. The development plan for WS-117L was approved in July 1956, and the program got under way in October 1956 with the awarding of a contract to the Lockheed Aircraft Corporation for the development and testing of the system under the program name [redacted]

The planning for WS-117L contemplated a family of separate systems and subsystems employing satellites for the collection of photographic, [redacted] and infrared intelligence. The program, which was scheduled to extend beyond 1965, was divided into three phases. Phase I, the THOR-boosted test series, was to begin in November 1958. Phase II, the ATLAS-boosted test series, was scheduled to begin in June 1959 with the objective of completing the transition from the testing phase to the operational phase and of proving the capability of the ATLAS booster to launch heavy loads into space. Phase III, the operational series, was to begin in March 1960 and was to consist of three progressively more sophisticated systems: the Pioneer version (photographic and [redacted]), the Advanced version (photographic and [redacted]), and the Survellance version (photographic, [redacted] and infrared). It was expected that operational control of WS-117L would be transferred to the Strategic Air Command with the initiation of Phase III.

It was an ambitious and complex program that was pioneering in technical fields about which little was known. Not surprisingly, it had become apparent by the end of 1957 that the program was running behind schedule. It also was in trouble from the standpoint of security. The U-2 program was carried out in secret from 1956 until May 1960. Its existence was no secret to the Soviets, of course, but they chose to let it remain a secret to the general public (and to most of the official community) rather than publicize it and thereby admit that they lacked the means of defending their air space against the high-flying U-2. WS-117L was undertaken as a classified project, but its very size and the number of people involved made it impossible to conceal the existence of the program for long. The press soon began speculating on the nature of the program, correctly identifying it as involving military reconnaissance satellites, and referring to it as BIG BROTHER and SPY IN THE SKY. The publicity was of concern, because the development of WS-117L was begun in a period when the international political climate was hostile to any form of overflight reconnaissance.

It was against this background that the President's Board of Consultants on Foreign Intelligence Activities submitted its semi-annual report to the President on 24 October 1957. The Board noted in its report that it was aware of two

1. *(Continued)*

advanced reconnaissance systems that were under consideration. One was a study then in progress in the Central Intelligence Agency concerning the feasibility of a manned reconnaissance aircraft designed for greatly increased performance and reduced radar cross-section; the other was WS-117L. However, there appeared little prospect that either of these could produce operational systems earlier than mid-1959. The Board emphasized the need for an interim photo reconnaissance system and recommended that an early review be made of new developments in advanced reconnaissance systems to ensure that they were given adequate consideration and received proper handling in the light of then-existing and future intelligence requirements. The Executive Secretary of the National Security Council on 28 October notified the Secretary of Defense and the Director of Central Intelligence that the President had asked for a joint report from them on the status of the advanced systems. Secretary Quarles responded on behalf of himself and Mr. Dulles on 5 December with a recommendation that, because of the extreme sensitivity of the subject, details on the new systems be furnished through oral briefings.

As a consequence, there are no official records in CIA's Project CORONA files bearing dates between 5 December 1957 and 21 March 1958, but it is clear that major decisions were made and that important actions were undertaken during the period. In brief, it was decided that the photographic subsystem of WS-117L offering the best prospect of early success would be separated from WS-117L, designated Project CORONA, and placed under a joint CIA-Air Force management team—an approach that had been so successful in covertly developing and operating the U-2.

The nucleus of such a team was then constituted as the Development Projects Staff under the direction of Richard Bissell, who was Special Assistant to the DCI for Planning and Development. Bissell was designated as the senior CIA representative on the new venture, and his Air Force counterpart was Brigadier General Osmond Ritland, who, as Colonel Ritland, had served as Bissell's first deputy in the early days of the Development Projects Staff and later became Vice Commander of the Air Force Ballistic Missile Division.

Bissell recalls that he first learned of the new program and of the role intended for him in it "in an odd and informal way" from Dr. Edwin Land. Dr. Land had been deeply involved in the planning and development of the U-2 as a member of the Technological Capabilities Panel of the Office of Defense Mobilization. He continued an active interest in overhead reconnaissance and later headed the Land Panel, which was formed in May 1958 to advise on the development of OXCART, the aircraft planned as the successor to the U-2. Bissell also recalls that his early instructions were extremely vague: that the subsystem was to be split off from WS-117L, that it was to be placed under separate covert management, and that the pattern established for the development of the U-2 was to be followed. One of the instructions, however, was firm and precise: none of the funds for the new program were to come from monies authorized for already approved Air Force programs. This restriction, although seemingly clear at first glance, later led to disagreement over its interpretation. CORONA mangement expected that the boosters already approved

1. *(Continued)*

for the THOR test series of WS-117L would simply be diverted to the CORONA program; this proved not to be so. As a consequence, CIA had to go back to the President with an admission that the original project proposal had understated the estimated cost and with a request for more money.

Roughly concurrent with the decision to place one of the WS-117L subsystems under covert management, the Department of Defense realigned its structure for the management of space activities. The Advanced Research Projects Agency (ARPA) was established on 7 February 1958 and was granted authority over all military space projects. The splitting off of CORONA from WS-117L was accomplished by a directive from ARPA on 28 February 1958, assigning responsibility for the WS-117L program to the Air Force and ordering that the proposed WS-117L interim reconnaissance system employing THOR boost be dropped.

The ARPA directive ostensibly cancelling the THOR-boosted interim reconnaissance satellite was followed by all of the notifications that would normally accompany the cancellation of a military program. The word was passed officially within the Air Force, and formal contract cancellations were sent out to the prospective suppliers. There was much furore when the cancellations went out: contractors were furious over the suddenness of the action; Air Force personnel were thunderstruck at the abandonment of the WS-117L photographic subsystem that seemed to have the best chance of early success. After the cancellation, very limited numbers of individuals in the Air Force and in the participating companies were cleared for Project CORONA and were informed of the procedures to be followed in the covert reactivation of the cancelled program.

After Bissell and Ritland had worked out the arrangements for the [redacted] they then began tackling the technical problems associated with the design configuration they had inherited from WS-117L. The subsystem in point contemplated the use of the THOR IRBM as the first stage booster and, as a second stage, Lockheed's modification of a rocket engine that had been developed by Bell Aircraft for take-off assist and auxiliary power applications in the B-58 HUSTLER bomber. It was referred to as the HUSTLER engine during the development phase of WS-117L but soon came to be known as the AGENA—the name it bears today.

One of the very early CORONA plans called for spin stabilization of the payload, with the camera scanning as the payload rotated. The contractors working on this subsystem design were Lockheed on the space vehicle, and Fairchild on the camera. The camera was to have a focal length of six inches, without image motion compensation. Ground resolution was expected to be poor with this short focal length, particularly if combined with the readout techniques envisaged by WS-117L.

Several important design decisions were implemented in this organizational period of CORONA. Recognizing the need for resolution to meet the intelligence objectives, it was concluded that physical film recovery offered the most promising approach for a usable photographic return in the interim time period. This resulted in the addition to the design of a recovery pod or capsule with General Electric selected as the recovery vehicle contractor. In retrospect, the decision on film recovery would prove to be one of the most important made in U.S. reconnaissance activities, in that all photo reconnaissance systems developed up to the current time have relied on physical recovery of film.

Another major decision for the new CORONA Program came in late March 1958, following a three-day conference in San Mateo, California, among representatives of CIA, Air Force Ballistic Missile Division, Lockheed, General Electric, and Fairchild. The discussion revealed that, while work was going forward, the design was far from complete. The senior Lockheed representative reported that they had investigated the possibility of building a satellite vehicle shaped like a football, a cigar, or a sphere. They had finally decided, for the original drawings at least, on a football-shaped pod slightly elongated at each end to correct the center of gravity. There was discussion of the need for immediate contractual arrangements with the various suppliers. Bissell remarked that he was "faced with the problem at present of being broke" and would need estimates from all the suppliers as soon as possible in order to obtain the necessary financing to get the program under way. The suppliers agreed to furnish the required estimates by the following week.

The project quickly began taking formal shape following that meeting. Within a span of about three weeks, approval of the program and of its financing was obtained, and the design of the payload configuration evolved into a concept quite different from the spin-stabilized pod. It was at this point in late March and early April 1958 that major complications had arisen in the technical design of the Fairchild camera. Interest shifted to a competitive design submitted by the Itek Corporation, a spin-off of Boston University. Itek proposed a longer focal length camera scanning within an earth-center stabilized pod. The Itek design was based on the principle of the Boston University Hyac camera. Bissell recalls that he personally decided in favor of the Itek design, but only after much agonizing evaluation. The decision was a difficult one to make because it involved moving from a proven method of space vehicle stabilization to one that was technically more difficult to accomplish. It did, however, standardize on the 3-axis stabilization being pursued on the WS-117L AGENA development, and which has been a part of all subsequent photo reconnaissance systems.

Bissell's first project proposal, which was completed on 9 April 1958, requested approval for concurrent development of both the Fairchild and the Itek systems, with the Fairchild configuration becoming operational first and the Itek configuration being developed as a follow-on system. Within two days, however, Bissell had made the final decision to abandon the Fairchild spin-stabilized configuration entirely. He rewrote the project proposal, taking note of the earlier configuration and giving his reasons for favoring the Itek approach (principally the better resolution attainable, the lower overall cost, and the greater potential for growth). The proposal was rewritten a second time, retaining the Itek configuration but raising the cost estimate from [redacted] to [redacted]. Of the total estimated cost, [redacted] represented "a rather arbitrary allowance" for 12 each THOR boosters and Lockheed second stage vehicles, and was to be financed by ARPA through the Air Force. The remaining [redacted] was for [redacted] by CIA of the pods containing the reconnaissance equipment and the recoverable film cassettes.

The final project proposal was forwarded to Brigadier General Andrew J. Goodpaster, the President's Staff Secretary, on 16 April 1958 after having been reviewed by Mr. Roy Johnson and Admiral John Clark of ARPA; Mr. Richard Horner, Assistant Secretary of the Air Force for Research and Development; Brigadier General Osmond Ritland, Vice Commander, Air Force Ballistic Missile Division; and Dr. James Killian, Special Assistant to the President for Science and Technology. The proposal was approved, although not in writing. The only original record of the President's approval reportedly was in the form of a handwritten note on the back of an envelope by General C. P. Cabell, the Deputy Director of Central Intelligence.

Although it may have been the original intent that CORONA would be administered in a manner essentially the same as that of the U-2 program, it actually began and evolved quite differently. It was a joint CIA-ARPA-Air Force effort, much as the U-2 was a joint CIA-Air Force effort, but it lacked the central direction that characterized the U-2 program. The project proposal described the anticipated administrative arrangements, but it fell short of clarifying the delineation of authorities. It noted that CORONA was being carried out under the authority of ARPA and CIA with the support and participation of the Air Force. CIA's role was further explained in terms of participating in supervision of the technical development, especially as regards the actual reconnaissance equipment, handling all [redacted] The work statement prepared for Lockheed, the prime contractor, on 25 April 1958 noted merely that technical direction of the program was the joint responsibility of several agencies of the Government.

The imprecise statements of who was to do what in connection with CORONA allowed for a range of interpretation. The vague assignments of responsibilities caused no appreciable difficulties in the early years of CORONA when the joint concern was primarily one of producing as promised, but they later (1963) became a source of severe friction between CIA and the Air Force over responsibility for conducting the program.

Bissell, the recognized leader of the early CORONA program, gave this description of how the early program was managed:

> The program was started in a marvelously informal manner. Ritland and I worked out the division of labor between the two organizations as we went along. Decisions were made jointly. There were so few people involved and their relations were so close that decisions could be and were made quickly and cleanly. We did not have the problem of having to make

compromises or of endless delays awaiting agreement. After we got fully organized and the contracts had been let, we began a system of management through monthly suppliers' meetings—as we had done with the U-2. Ritland and I sat at the end of the table, and I acted as chairman. The group included two or three people from each of the suppliers. We heard reports of progress and ventilated problems—especially those involving interfaces among contractors. The program was handled in an extraordinarily cooperative manner between the Air Force and CIA. Almost all of the people involved on the Government side were more interested in getting the job done than in claiming credit or gaining control.

The schedule of the program, as it had been presented to the CORONA group at its meeting in San Mateo in late March 1958, called for a "count-down" beginning about the first of July 1958 and extending for a period of 19 weeks. It was anticipated that the equipment would be assembled, tested, and the first vehicle launched during that 19-week period, which meant that the fabrication of the individual components would have had to be completed by 1 July 1958. By the time Bissell submitted his project proposal some three weeks later, it had become apparent that the earlier tentative scheduling was unrealistic. Bissell noted in his project proposal that it was not yet possible to establish a firm schedule of delivery dates, but that it appeared probable that the first firing could be attempted no later than June 1959.

It is pertinent to note here that there was no expectation in 1958 that CORONA would still be operating over a decade later. The CORONA program got under way initially as an interim, short-term, high-risk development to meet the intelligence community's requirements for area search photographic reconnaissance pending successful development of other, more sophisticated systems planned for WS-117L. The original CORONA proposal anticipated the acquisition of only 12 vehicles, noting that at a later date it might be desirable to consider whether the program should be extended—with or without further technological improvement.

Having settled on the desired configuration and having received Presidential approval of the program and its financing, the CORONA management team moved forward rapidly with the contractual arrangements. The team of contractors for CORONA differed from the team on the WS-117L subsystem as a consequence of selecting Itek's earth-center stabilized approach. Itek was brought in as one of the two major subcontractors to Lockheed (General Electric being the other). However, to soften the financial blow to Fairchild, Itek was made responsible for the design and development of the camera subsystem with Fairchild producing the camera under subcontract to Itek. This contractor team continued throughout the CORONA program, although later in the program, the relationship was changed to that of associate contractors. The contractor relationships on the CORONA program were as friendly and cooperative as any that could have been set up, and this team dedication to the success of the program is one of the primary reasons for the success the program enjoyed. The final contractors were selected on 25 April 1958 and a work statement was issued to Lockheed on that date. The contractors began systems design on 28 April

1. *(Continued)*

~~Handle via TALENT KEYHOLE Controls~~

~~TOP SECRET~~

Corona

and completed them and submitted them for first review on 14 May. The designs were frozen on 26 July.

Thus, by mid-1958, the program was well down the road—on the contractors' side—toward meeting the goal of a first launch no later than mid-1959. The Government side, however, was running into difficulties. The first problem was money, the second was cover, and the two were inextricably intertwined. The [redacted] cost estimate for the 12-vehicle program had assumed that the cost of the THOR boosters would be absorbed by the Air Force by diverting them from the cancelled WS-117L subsystem. That assumption proved to be incorrect. An additional [redacted] had to be found to pay for the 12 THORs. Further, it had been decided that an additional four launch vehicles would be required for testing of launch, orbit, and recovery procedures and that an additional three would be required for biomedical launches in support of the CORONA cover story. ARPA could not see its way clear to making Defense Department funds available merely for testing or for cover support when there were other DoD space programs with pressing needs for money. Consequently, CORONA management had to go back to the President for approval of a revised estimate.

By August 1958, it had also become apparent to the project's managers that the original, but as yet unannounced, cover story conceived for the future CORONA launchings (an experimental program within the first phase of WS-117L) was becoming increasingly untenable. WS-117L had by then become the subject of fairly widespread public speculation identifying it as a military reconnaissance program. It was feared that linking CORONA to WS-117L in any way would inevitably place the reconnaissance label on CORONA, and—given the hostility of the international political climate to overflight reconnaissance—there was the risk that the policy level of government might cancel the program if it should be so identified. Some other story would have to be contrived that would dissociate CORONA from WS-117L and at the same time account for multiple launchings of stabilized vehicles in low polar orbits and with payloads being recovered from orbit.

It was decided, therefore, to separate the WS-117L photo reconnaissance program into two distinct and ostensibly unrelated series: one identified as

~~TOP SECRET~~

DISCOVERER (CORONA – THOR boost) and the other as SENTRY (later known as SAMOS – ATLAS boost). A press release announcing the initiation of the DISCOVERER series was issued on 3 December 1958 identifying the initial launchings as tests of the vehicle itself and later launchings as explorations of environmental conditions in space. Biomedical specimens, including live animals, were to be carried into space and their recovery from orbit attempted.

The new CORONA cover concept, from which the press release stemmed, called for a total of five biomedical vehicles, and three of the five were committed to the schedule under launchings three, four, and seven. The first two were to carry mice and the third a primate. The two uncommitted vehicles were to be held in reserve in event of failure of the heavier primate vehicle. In further support of the cover plan, ARPA was to develop two radiometric payload packages designed specifically to study navigation of space vehicles and to obtain data useful in the development of an early warning system (the planned [redacted]). It might be noted here that only one of the three planned animal-carrying missions was actually attempted (as DISCOVERER III), and it was a failure. ARPA did develop the radiometric payload packages, and they were launched as DISCOVERERs XIX and XXI in late 1960 and early 1961.

The photo reconnaissance mission of CORONA necessitated a near-polar orbit, by launching either to the north or to the south. There are few otherwise suitable areas in the continental United States where this can be done without danger that debris from an early in-flight failure could fall into populated areas. Cooke Air Force Base* near California's Point Arguello met the requirement for downrange safety, because the trajectory of a southward launch from there would be over the Santa Barbara channel and the Pacific Ocean beyond. Cooke was a natural choice, because it was the site of the first Air Force operational missile training base and also housed the 672nd Strategic Missile Squadron (THOR). Two additional factors favored this as the launch area: the manufacturing facilities and skilled personnel required were in the near vicinity, and a southward launch would permit recovery in the Hawaii area by initiating the ejection/recovery sequence as the satellite passed over the Alaskan tracking facility.

Unlike the U-2 flights, launchings of satellites from U.S. soil simply could not be concealed from the public. Even a booster as small as the THOR (small, that is, in comparison with present-day space boosters) launches with a thunderous roar that can be heard for miles; the space vehicle transmits telemetry that can be intercepted; and the vehicle can be detected in orbit by radar skin-track. The fact of a launch could not be concealed, but maintenance of the cover story for the DISCOVERER series required that the launchings of the uniquely configured photographic payloads be closed to observation by unwitting personnel. Vandenberg was excellent as a launch site from many standpoints, but it had one feature that posed a severe handicap to screening the actual launches from unwanted observation: the heavily traveled Southern Pacific railroad passes through it. The early launches from Vandenberg had to

*Cooke AFB was renamed Vandenberg AFB in October 1958.

be timed for early afternoon,* and the Southern Pacific schedule broke this period into a series of launch windows, some of which were no more than a few minutes between trains. Throughout its existence, the CORONA program at Vandenberg was plagued by having to time the launches to occur during one of the intervals between passing trains.

The planned recovery sequence involved a series of maneuvers, each of which had to be executed to near-perfection or recovery would fail. Immediately after injection into orbit, the AGENA vehicle was yawed 180 degrees so that the recovery vehicle faced to the rear. This maneuver minimized the control gas which would be required for re-entry orientation at the end of the mission, and protected the heat shield from molecular heating, a subject of considerable concern at that time. (Later in the J-3 design when these concerns had diminished, the vehicle would be flown forward until re-entry.) When re-entry was to take place, the AGENA would then be pitched down through 60 degrees to position the satellite recovery vehicle (SRV) for retro-firing. Then the SRV would be separated from the AGENA and be spin-stabilized by firing the spin rockets to maintain it in the attitude given it by the AGENA. Next the retro-rocket would be fired, slowing down the SRV into a descent trajectory. Then the spin of the SRV would be slowed by firing the de-spin rockets. Next would come the separation of the retro-rocket thrust cone followed by the heat shield and the parachute cover. The drogue (or deceleration) chute would then deploy, and finally the main chute would open to lower the capsule gently into the recovery area. The primary recovery technique involved flying an airplane across the top of the descending parachute, catching the chute or its shrouds in a trapeze-like hook suspended beneath the airplane and then winching the recovery vehicle aboard. C-119 Aircraft were initially used with C-130 aircraft replacing them later in the program. The recovery vehicle was designed to float long enough, if the air catch failed, for a water recovery by helicopter launched from a surface ship.

While the vehicle was still in the construction stage, tests of the air recovery technique were conducted by the 6593rd Test Squadron—with disheartening results. Of 74 drops using personnel-type chutes, only 49 were recovered. Using one type of operational drop chute, only four were recovered out of 15 dropped, and an average of 1.5 aircraft passes were required for the hook-up. Eleven drops with another type of operational chute resulted in five recoveries and an average of two aircraft passes for the snatch. Part of the difficulty lay in weak chutes and rigging, and in crew inexperience. The most serious problem, however, was the fast drop rate of the chutes. Parachutes that were available to support the planned weight of the recovery vehicle had a sink rate of about 33 feet per second. What was required was a sink rate approaching 20 feet per second so that the aircraft would have time to make three or four passes if necessary for hook-up. Fortunately, by the time space hardware was ready for launching,

*The early THOR-AGENA combination limited film to enough for a 24-hour mission of 17 orbits, seven of which would cross denied territory. Requirements for daylight recovery and for daylight passage over denied areas with acceptable sun angles dictated the afternoon launch time.

1. *(Continued)*

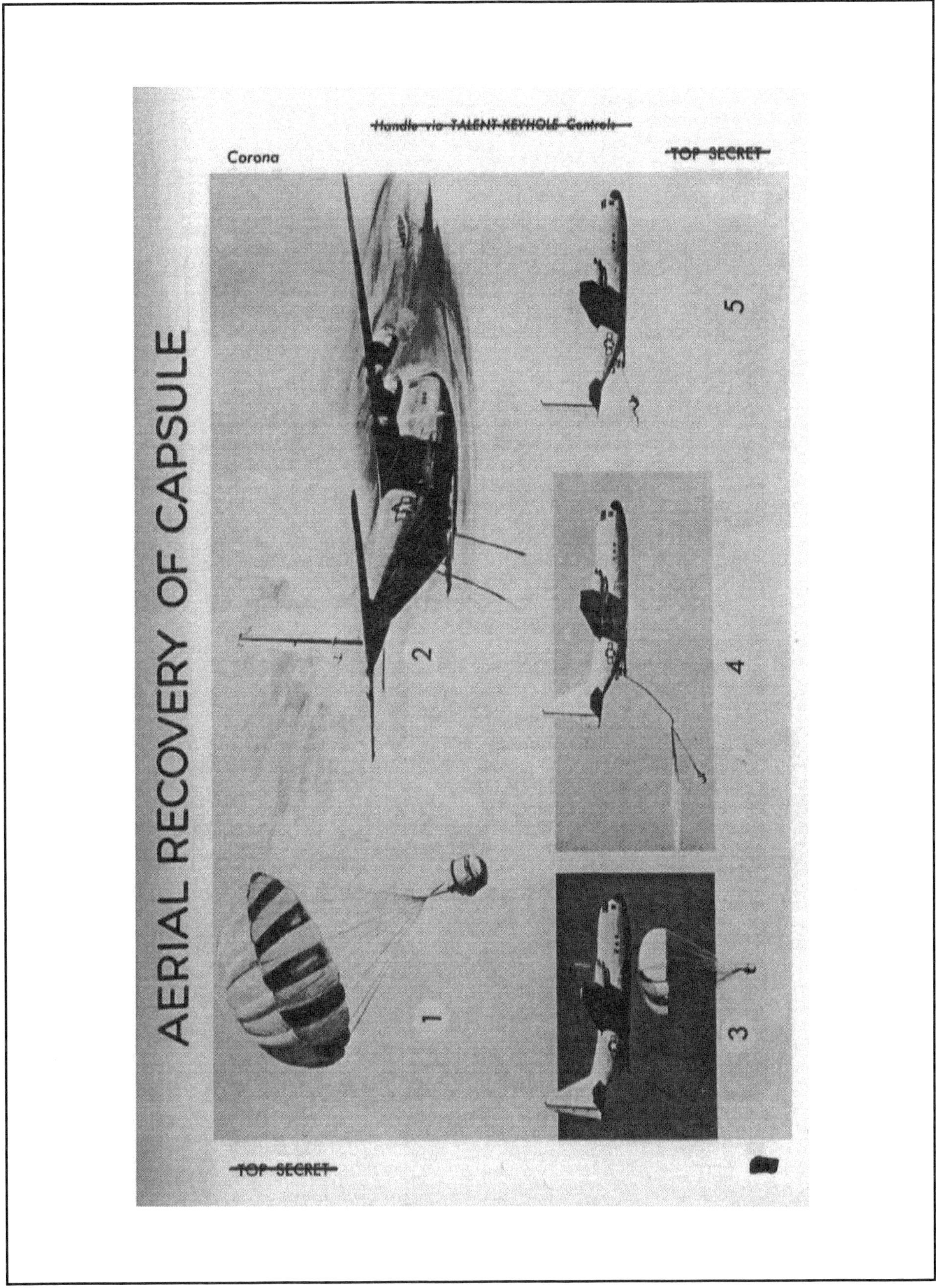
Handle via TALENT-KEYHOLE Controls
Corona
TOP SECRET
AERIAL RECOVERY OF CAPSULE
1
2
3
4
5
TOP SECRET

a parachute had been developed with a sink rate slow enough to offer a reasonable chance of air recovery.

The launch facilities at Vandenberg AFB were complete, and the remote tracking and control facilities which had been developed for WS-117L were ready for the first flight test of a THOR-AGENA combination in January 1959. The count-down was started for a launch on the 21st; however, the attempt aborted at launch minus 60 minutes. When power was applied to test the AGENA hydraulic system, certain events took place that were supposed to occur in flight but not while the vehicle was still sitting on the launch pad. The explosive bolts connecting the AGENA to the THOR detonated, and the ullage rockets* fired. The AGENA settled into the fairing attaching it to the THOR and did not fall to the ground, but appreciable damage was done.

A program review conference was held in Palo Alto two days after the launch failure to examine the possible causes of the abort and to assess its impact on the planned CORONA launch schedule. Fortunately, the problem was quickly identified and easily corrected, and it was felt that the system was ready for test launches at the rate of about one per month.

At the review conference, General Electric surfaced a new problem having to do with the stability of the nose cone during re-entry. The cone was designed for a film load of 40 pounds, but the first missions would be able to carry only 20 pounds. GE reported that about three pounds of ballast would have to be carried in the forward end of the cone to restore stability. The program officers decided to add an instrument package as ballast, either for diagnostic purposes or for support of the biomedical cover story, thus converting what could have been dead weight into a net plus for the test program.

The test plan contemplated arriving at full operational capability at a relatively early date through sequential testing of the major components of the system—beginning with the THOR-AGENA combination alone, then adding the nose cone to test the ejection/re-entry/recovery sequence, and finally installing a camera for a full CORONA systems test. Just how much confidence the project planners had in the imminence of success cannot now be discovered; however, if the confidence factor was very high at the start, it must soon have begun to wane. Beginning in February 1959 and extending through June 1960 an even dozen launches were attempted, with eight of the vehicles carrying cameras, and all of them were failures; no film capsules were recovered from orbit. Of the eight camera-carrying vehicles, four failed to achieve orbit, three experienced camera or film failures, and the eighth was not recovered because of a malfunction of the re-entry body spin rockets. These summaries of the initial launch attempts illustrate the nature and dimensions of the problems for which solutions had to be found.

---

*Ullage rockets are small solid propellant rockets attached to the AGENA. These rockets are fired just prior to ignition of the AGENA engine after its separation from the THOR to insure that the liquid AGENA propellants are pushed against the bottom of the tanks so that proper flow into the pumps will occur.

1. *(Continued)*

Handle via TALENT-KEYHOLE Controls
Corona
TOP SECRET
TOP SECRET

1. *(Continued)*

~~Handle via TALENT-KEYHOLE Controls~~

~~TOP SECRET~~ Corona

DISCOVERER I

The on-pad failure of 21 January was not assigned a number in the DISCOVERER series. DISCOVERER I was launched on 28 February 1959 with a light engineering payload as a test of THOR-AGENA performance. No recovery was planned. For a time there was uncertainty as to what had happened to it because no radio signals were received. At the time, it was believed to have obtained orbit with speculation that the protective nose cone over the antennas was ejected just before the AGENA fired and that the AGENA then rammed into the nose cone, damaging the antennas. Today, most people believe the DISCOVERER I landed somewhere near the South Pole.

DISCOVERER II

The second vehicle was launched on 13 April 1959. Orbit was officially announced about two hours later, along with a statement that the capsule carried a lightweight biomedical payload (as indeed it did). The Air Force reported on 15 April that plans to recover the capsule near Hawaii had been abandoned and that the capsule might descend somewhere in the Arctic. The announcement slightly understated the known facts. The capsule had ejected on the 17th orbit as planned, but a timing malfunction (actually a human programming error) had caused the ejection sequence to be initiated too early. The capsule was down, probably somewhere in the near-vicinity of the Spitsbergen Islands north of Norway. In fact, there were later reports that the falling capsule had actually been seen by Spitsbergen residents. The Air Force announced on the 16th that the Norwegian government had authorized a search for the capsule which would begin the following day. Planes scoured the area, and helicopters joined the search on the 20th. Nothing was found, however, and the search was abandoned on the 23rd.

DISCOVERER III

Much publicity attended the launching of DISCOVERER III: some of it planned and some uplanned (and unwanted). This was the first (and only) DISCOVERER flight to carry animals: four live black mice. Black mice were chosen in order to ascertain the possible hair-bleaching effects of cosmic rays. The mice were members of the C-57 strain, a particularly rugged breed. They had been "trained," along with 60 other mice, at the Air Force's Aeromedical Field Laboratory at Holloman AFB. They were seven to ten weeks old and

~~TOP SECRET~~

1. *(Continued)*

weighed slightly over an ounce apiece. A three-day food supply was provided, which consisted of a special formula containing peanuts, oatmeal, gelatin, orange juice, and water. Each mouse was placed in a small individual cage about twice its size, and each had a minuscule radio strapped to its back to monitor the effects of the space trip on heart action, respiration, and muscular activity.

The lift-off on 3 June 1959 was uneventful, but, instead of injecting approximately horizontally into orbit, the AGENA apparently fired downward, driving the vehicle into the Pacific Ocean and killing the mice. Looking back on the mission, the attempt to orbit the mice seems to have been jinxed from the very beginning.

Just before the first try at launch, telemetry indicated a lack of mouse activity. It was thought at first that the little fellows were merely asleep, so a technician was sent up in a cherry-picker to arouse them. He banged on the side of the vehicle and tried catcalls, but to no avail. When the capsule was opened, the mice were found to be dead. The cages had been sprayed with krylon to cover rough edges; the mice had found it tastier than their formula; and that was that.

"The Mouse That Poured"

The second try at launch several days later, with a back-up mouse "crew," was a near-abort when the capsule life cell humidity sensor suddenly indicated 100 percent relative humidity. The panic button was pushed, and troubleshooters were sent up to check. They found that when the vehicle was in a vertical position the humidity sensor was directly beneath the cages, and it did not distinguish between plain water and urine. The wetness dried out after a while, all was forgiven, and the vehicle was launched—unhappily into the permanent 100 percent moisture environment of the Pacific Ocean.

Also, the timing of the launch was unfortunate. The monkeys, Able and Baker, had survived a 300-mile flight in a JUPITER nose cone on 29 May in connection with another, unrelated test program. However, Able died during minor surgery on 3 June to remove an electrode that had been implanted under his skin. (This was the date of the DISCOVERER III launch.) The British Society Against Cruel Sports made a formal protest to the U.S. Ambassador, and the press raised quite a stink about the fatal mice flight—comparing it unfavorably with the Russians' successful launching of the dog, Laika, in SPUTNIK II back in November 1957, and demanding that orbit and recovery procedures be perfected before attempting further launches of mice or monkeys.

DISCOVERERS IV-VIII

DISCOVERER IV on 25 June 1959 was the first to carry a camera and thus the first true CORONA test, but the payload did not go into orbit. DISCOVERER V, again with a camera, attained orbit but the temperature inside the spacecraft was abnormally low and the camera failed on the first orbit. The recovery capsule was ejected at the proper time, but never showed up; early in 1960 it was discovered in a high near-polar orbit with an apogee of 1,058 miles. Failure of the spin rocket had caused the retro-rocket to accelerate rather than de-boost the package. DISCOVERER VI went into orbit six days later, but the camera failed on the second revolution, and the retro-rocket failed on the recovery attempt.*

DISCOVERER VII on 7 November did not go into orbit. DISCOVERER VIII on 20 November went into an eccentric orbit with an apogee of 913 miles, and the camera failed again. The recovery vehicle was ejected successfully, but the parachute failed to open.

It had become plain by the end of November 1959 that something (or, to be more precise, many things) had to be done to correct the multiple failures that were plaguing the CORONA system. Eight THOR-AGENA combinations and five cameras had been expended with nothing to show for the effort except accumulated knowledge of the system's weaknesses. The project technicians knew what was going wrong, but not always why. Through DISCOVERER VIII, the system had experienced these major failures:

One misfired on the launch pad.
Three failed to achieve orbit.
Two went into highly eccentric orbits.
One capsule ejected prematurely.
Two cameras operated briefly and then failed.
One camera failed entirely.
One experienced a retro-rocket malfunction.
One had very low spacecraft temperature.

A panel of consultants reviewed the various failures and their probable causes and concluded that what was needed most was "qualification, requalification, and multiple testing of component parts" before assembling them and sending them aloft. This called for more money. Accordingly, Bissell submitted a project amendment to the DDCI on 22 January 1960 asking approval of nearly [redacted] additional to cover the costs of the testing program. He apologized to General Cabell for submitting a request for funds to pay for work that was already under way: "Although such a sequence is regrettable, there has been con-

*One of these early launches tested a system for concealing the tell-tale payload doors from inquisitive eyes near the launch pad. The scheme was to cover them with paper, fastened over two lengths of piano wire with pingpong balls at the front end. The air flow at launch would use the pingpong balls and wire as "ripcords" to strip away the paper. The idea was tested on the side of a sports car simulating launch velocity as nearly as possible on the Bayshore Freeway late one evening. The test proved that the ripcords worked, and that Freeway patrolmen could overhaul a vehicle going only 90 m.p.h. Unfortunately, the ripcords malfunctioned on the next actual launch, and there was no consensus for another test round with the Freeway police.

siderable confusion in this program as to what the amount of the overruns would be and this has made it difficult to obtain approvals in an orderly fashion in advance."

As of the fall of 1959, major problems remained to be solved in achieving an acceptable orbit, in camera functioning, and in recovering the film capsule. These were the more serious of the specific failures that were occupying the attention of the technicians:

> The AGENA vehicle was designed for use with both the THOR and the ATLAS boosters. The ascent technique used by the AGENA vehicle was essentially the same in both combinations, but there were significant differences in the method of employing the booster. In the CORONA program, in order to conserve weight, the THOR booster followed a programmed trajectory using only its autopilot. Also, the THOR thrust was not cut off by command at a predetermined velocity (as in the ATLAS); instead, its fuel burned to near-exhaustion. This relatively inaccurate boosting profile, coupled with the low altitude of CORONA orbits, required great precision in the orbital injection. At a typical injection altitude of 120 miles, an angular error of plus or minus 1.1 degrees or a velocity deficit of as little as 100 feet per second would result in failure to complete the first orbit. This had happened repeatedly. Lasting relief from this problem lay some distance in the future: a more powerful AGENA was being developed, and the weight of instrumentation for measuring in-flight performance on the early flights would be reduced on later operational missions. The short-term remedy lay in a drastic weight-reduction program. This was carried out in part (literally, it is said) by attacking surplus metal with tin snips and files.
>
> The system was designed to operate without pressurization (again to conserve weight), and the acetate base film being used was tearing or breaking in the high vacuum existing in space and causing the camera to jam. A solution for this problem was found in substituting polyester for acetate base film. The importance to the reconnaissance programs of this achievement by Eastman Kodak in film technology cannot be overemphasized. It ranks on a level with the development of the film recovery capsule itself. The first orbital flight in which the camera was operated with polyester film was DISCOVERER XI (Mission 9008) in April 1960. Although recovery was not successful, one of the major space reconnaissance problems had been solved.
>
> The equipment was built to work best at an even and predetermined temperature. To save weight, only passive thermal control was provided. The spacecraft's internal temperature had varied on the flights thus far, and it was much lower than desired on one flight. An interim solution for this problem was found in varying the thermal painting of the vehicle skin.
>
> The spin and de-spin rockets used to stabilize the recovery vehicle during re-entry had a tendency to explode rather than merely to fire. Several had blown up in ground tests. A solution was found in substituting cold gas spin and de-spin rockets.

One of the most intractable problems, which was to persist for many months, was that of placing the satellite recovery vehicle (SRV) into a descent trajectory that would terminate in the recovery zone. This required ejecting the SRV from the AGENA at precisely the right time, and decelerating it by retro-rocket firing to the correct velocity and at a suitable angle. There was very little margin for error in this phase: each one-second error in ejection timing could shift the recovery point five miles; a retro-velocity vector error of more than ten degrees would cause the capsule to miss the recovery zone completely.

One might ask why the CORONA program officers persisted in the face of such adversity. The answer lay in the overwhelming intelligence needs of the period. The initial planning of CORONA began at a time when we did not know how many BEAR and BISON aircraft the Soviets had, whether they were introducing a new and far more advanced long range bomber than the BISON, or whether they had largely skipped the build-up of a manned bomber force in favor of missiles. There had been major changes in intelligence estimates of Soviet nuclear capabilities and of the scope of the Soviet missile program on the basis of the results of the relatively small number of U-2 missions approved for the summer of 1957. However, by 1959, the great "missile gap" controversy was very much in the fore. The Soviets had tested ICBM's at ranges of 5,000 miles, proving they had a capability of building and operating them. What was not known was where they were deploying them operationally, and in what numbers. In the preparation of the National Intelligence Estimate on guided missiles in the fall of 1959, the various intelligence agencies held widely diverse views on Soviet missile strength. Nineteen Sixty ushered in an election year in which the missile gap had become a grave political issue, and the President was scheduled to meet with Soviet leaders that spring without—it appeared—the benefit of hard intelligence data. The U-2 had improved our knowledge of the Soviet Union, but it could not provide area coverage and the answers to the critical questions, and it was increasingly becoming less an intelligence asset than a political liability. It was judged to be only a matter of time until one was shot down—with the program coming to an end as an almost certain consequence.

DISCOVERERS IX-XII

A standdown was in effect in CORONA from 20 November 1959 until 4 February 1960 to allow time for intensive R&D efforts to identify and eliminate the causes of failure. On 4 February, DISCOVERER IX was launched and failed to achieve orbit.

The first recovery of film from a CORONA vehicle occurred with the launching of DISCOVERER X on 19 February 1960, but in a manner such that no one boasted of it. The THOR booster rocket began to fishtail not long after it left the launch pad and was destroyed by the range safety officer at 52 seconds after lift-off. The payload came down about a mile from Pad 5, was located by helicopter, and the recovery was made by a crew that rode to the scene by Jeep.*

DISCOVERERs VII through X carried only a quarter of a load of film (10 pounds) to permit the carrying of additional instrumentation for testing vehicle performance. DISCOVERER XI was launched on 15 April 1960 carrying a camera and 16 pounds of film. A reasonably good orbit was achieved (380 miles at apogee and 109.5 miles at perigee), and the camera operated satisfactorily.** All of the film was exposed and transferred into the recovery capsule. Unfortunately, the problem of the exploding spin rockets, which had been observed in ground tests, occurred during the recovery sequence, and the payload was lost.

Another standdown—a major one—was imposed following the failure of DISCOVERER XI. As of mid-April 1960, there had been 11 launches and one abort on pad. Seven of the launches achieved orbit, but no capsules had been recovered. DISCOVERER XII was planned as a diagnostic flight—without camera payload—heavily instrumented to determine precisely why recovery of capsules had failed previously. The vehicle was launched on 29 June 1960, but the AGENA failed to go into orbit.

DISCOVERER XIII—Partial Success

The next flight, on August 1960, was launched as a repeat of the no-orbit DISCOVERER XII diagnostic flight, without camera and film. The vehicle was launched and successfully inserted into orbit. The recovery package was ejected on the 17th orbit, and retro-firing and descent were normal—except that the capsule came down well away from the planned impact point. The nominal impact area was approximately 250 miles south of Honolulu where C-119 and C-130 aircraft circled awaiting the capsule's descent. The splash-down occurred about 330 miles northwest of Hawaii. The airplanes were backed up by surface ships deployed in a recovery zone with a north-south axis of some 250 miles and an east-west axis extending about 550 miles to either side of the expected impact point. Although beyond the range of the airborne recovery aircraft, the DISCOVERER XIII capsule descended near enough to the staked-out zone to permit an attempt at water recovery. A ship reached the scene before the capsule sank

*This was one of the few launch failures for the remarkable Douglas team which prepared the THOR boosters at Vandenberg Air Force Base. The early CORONA launches provided many exciting moments for the Douglas crew, however. Several of the crew were holdovers from the V-2 "broomlighters," who on V-2 launch days would actually ignite reluctant rocket engines with kerosene-soaked brooms. At Vandenberg AFB they did not have to resort to this tactic, but they were required on numerous occasions to return to the launch pad as late as T minus 15 seconds to unfreeze valves with the touch of a sledgehammer. Other members of the blockhouse crew would marvel as the "Douglas Daredevils" would race their vehicles in reverse the entire way from the launch pad to the blockhouse, arriving just as ignition would begin.

**This was the first mission on which the camera operated successfully throughout the mission, primarily because of the change from acetate base to polyester base film.

and fished it out of the ocean. Much of the credit for the success was attributed to the inauguration (on the unsuccessful DISCOVERER XII launch) of the cold gas spin and de-spin system.

For the first time ever, man had orbited an object in space and recovered it. This American space "first" beat the Russians by just nine days. The Soviets had tried to recover SPUTNIK IV the previous May but failed when the recovery capsule ejected into a higher orbit. They did succeed in de-orbiting and recovering SPUTNIK V carrying the dogs, Belka and Strelka, on 20 August 1960.

Arrangements were made for extensive publicity concerning this success in recovering an object from orbit—in large measure to support the cover story of DISCOVERER/CORONA as being an experimental space series. News photos were released of the lift-off from Vandenberg, of the capsule floating in the ocean, and of the recovery ship *Haiti Victory*. President Eisenhower displayed the capsule and the flag it had carried to the press, and it was later placed on exhibit in the Smithsonian Institution for public viewing.

In anticipation of the first recovery being a reconnaissance mission, a plan had been developed under which the capsule would be switched in transit through Sunnyvale. Since DISCOVERER XIII was a diagnostic flight, the project office was spared the necessity of executing a clandestine switch of capsules prior to shipment to Washington, and the President and Smithsonian received the actual hardware from the first recovery.

We have all watched television coverage of the U.S. man-in-space programs with the recovery of astronauts and capsules after splash-down in the ocean. A helicopter flies from the recovery ship to the floating capsule and drops swimmers to attach a line to the capsule. After the astronauts are removed, the helicopter hoists the capsule from the water and carries it to the recovery ship. What most of us don't realize is that the recovery technique was developed for and perfected by the CORONA program as a back-up in event of failure of the air catch.

DISCOVERER XIV—Full Success

Success! ! ! DISCOVERER XIV was launched on 18 August 1960, one week after the successful water recovery of the DISCOVERER XIII capsule. The vehicle carried a camera and a 20-pound load of film. The camera operated satisfactorily, and the full load of film was exposed and transferred to the recovery capsule. The AGENA did not initially position itself in orbit so as to permit the recovery sequence to occur. It was on the verge of tumbling during the first few orbits, and an excessive quantity of gas had to be used in correcting the situation. Fortunately, vehicle attitude became stabilized about midway through the scheduled flight period, thus relieving the earlier fear that recovery would be impossible. The satellite recovery vehicle was ejected on the 17th pass, and the film capsule was recovered by air snatch.

Captain Harold E. Mitchell of the 6593rd Test Squadron, piloting a C-119 (flying boxcar) called Pelican 9, successfully hooked the descending capsule on

1. *(Continued)*

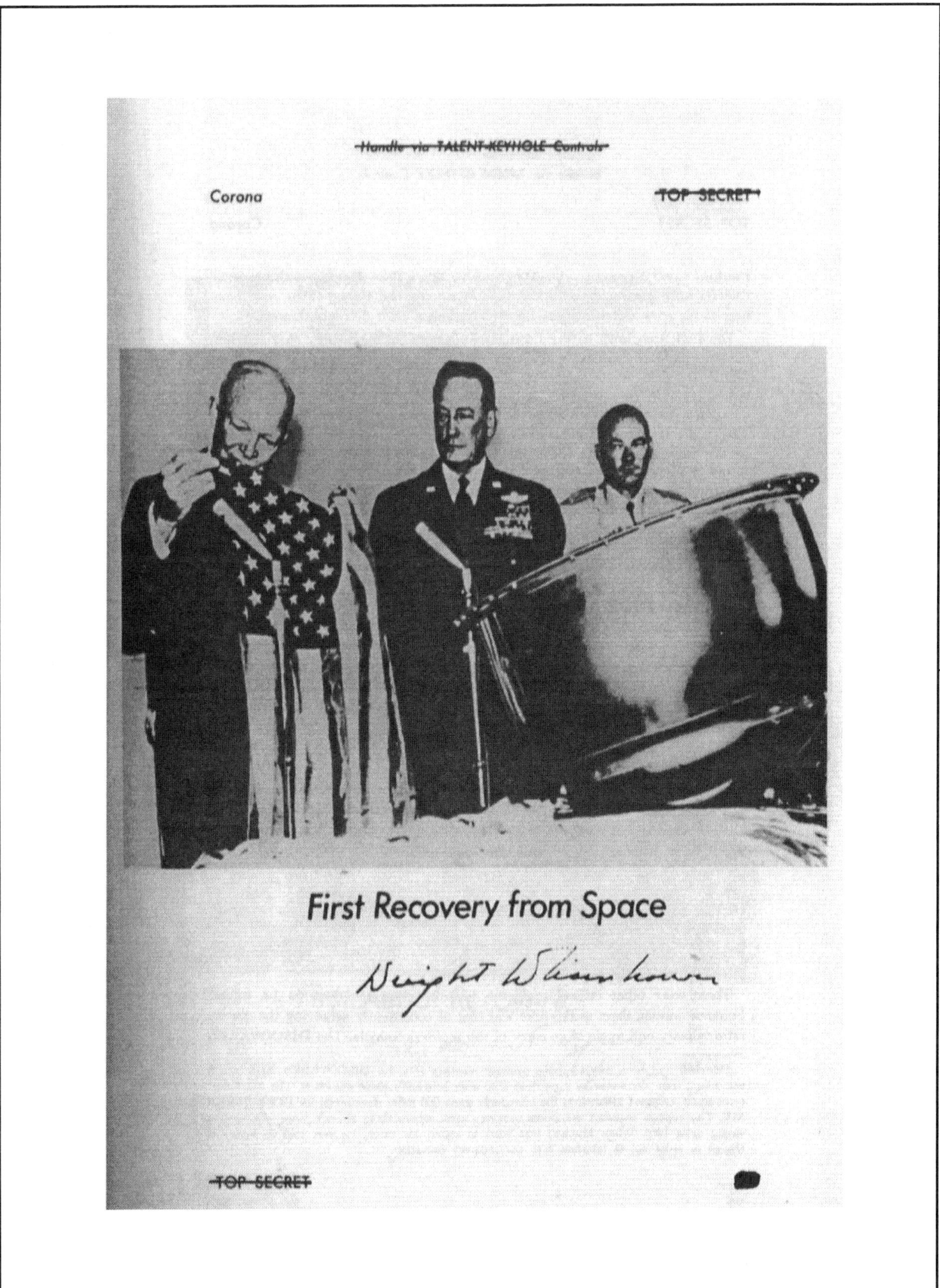
~~Handle via TALENT-KEYHOLE Controls~~

Corona ~~TOP SECRET~~

~~TOP SECRET~~

his third pass.* Upon arrival at Hickham Air Force Base, Hawaii, with his prize, Captain Mitchell was decorated with the Distinguished Flying Cross, and members of his crew were awarded the Air Medal for their accomplishments.

The film was flown to the [redacted] [redacted] for development and was then delivered to PIC (now known as NPIC) for readout and reporting. The resolution was substantially lower than that obtainable from the U-2, but the photography had intelligence value, and it covered areas of the USSR which the U-2 had never reached. This one satellite mission, in fact, yielded photo coverage of a greater area than the total produced by all of the U-2 missions over the Soviet Union. The only major deficiencies in the photography were plus and minus density bars running diagonally across the format. Some were due to minor light leaks, and others were the result of electrostatic discharge known as corona. These marks showed that the program security officer had had great insight when he named the program. There are two types of corona markings, a glow which caused the most difficulty, and a dendritic discharge which is more spectacular in appearance.

A press release announced the success of the mission but naturally made no mention of the *real* success: the delivery of photographic intelligence. The announcement noted that the satellite had been placed into an orbit with a 77.6 degree of inclination, an apogee of 502 miles, a perigee of 116 miles, and an orbital period of 94.5 minutes. A retro-rocket had slowed the capsule to re-entry velocity, and a parachute had been released at 60,000 feet. The capsule, which weighed 84 pounds at recovery, was caught at 8,500 feet by a C-119 airplane on its third pass over the falling parachute.

Progress and Problems

The program officers did not take the success of DISCOVERER XIV to mean that their problems with the system were at an end, but many of the earlier difficulties had been surmounted. The orbital injection technique had been improved to a level at which vehicles were repeatedly put into orbit with injection angle errors of less than four-tenths of a degree. The timing of the initiation of the recovery sequence had been so refined that ejection of the DISCOVERER XI SRV occurred within five seconds of the planned time. Parachute deceleration and air catch of the capsule had been accomplished repeatedly with test capsules dropped from high-altitude balloons. The last two cameras placed in orbit had operated well.

There were other critical problems, however, that remained to be solved. Foremost among them at the time was that of consistently achieving the correct retro-velocity and angle of re-entry of the recovery vehicle. The DISCOVERER

*Mitchell had been patrolling the primary recovery zone for DISCOVERER XIII, which was fished from the water by a recovery ship after Mitchell's plane missed it. The Air Force, pride stung, assigned Mitchell to the boondocks some 500 miles downrange for DISCOVERER XIV. The capsule overshot the prime recovery area, where three aircraft were chasing the wrong radar blip. When Mitchell first tried to report his catch, he was told to keep off the air in order not to interfere with the recovery operation.

1. *(Continued)*

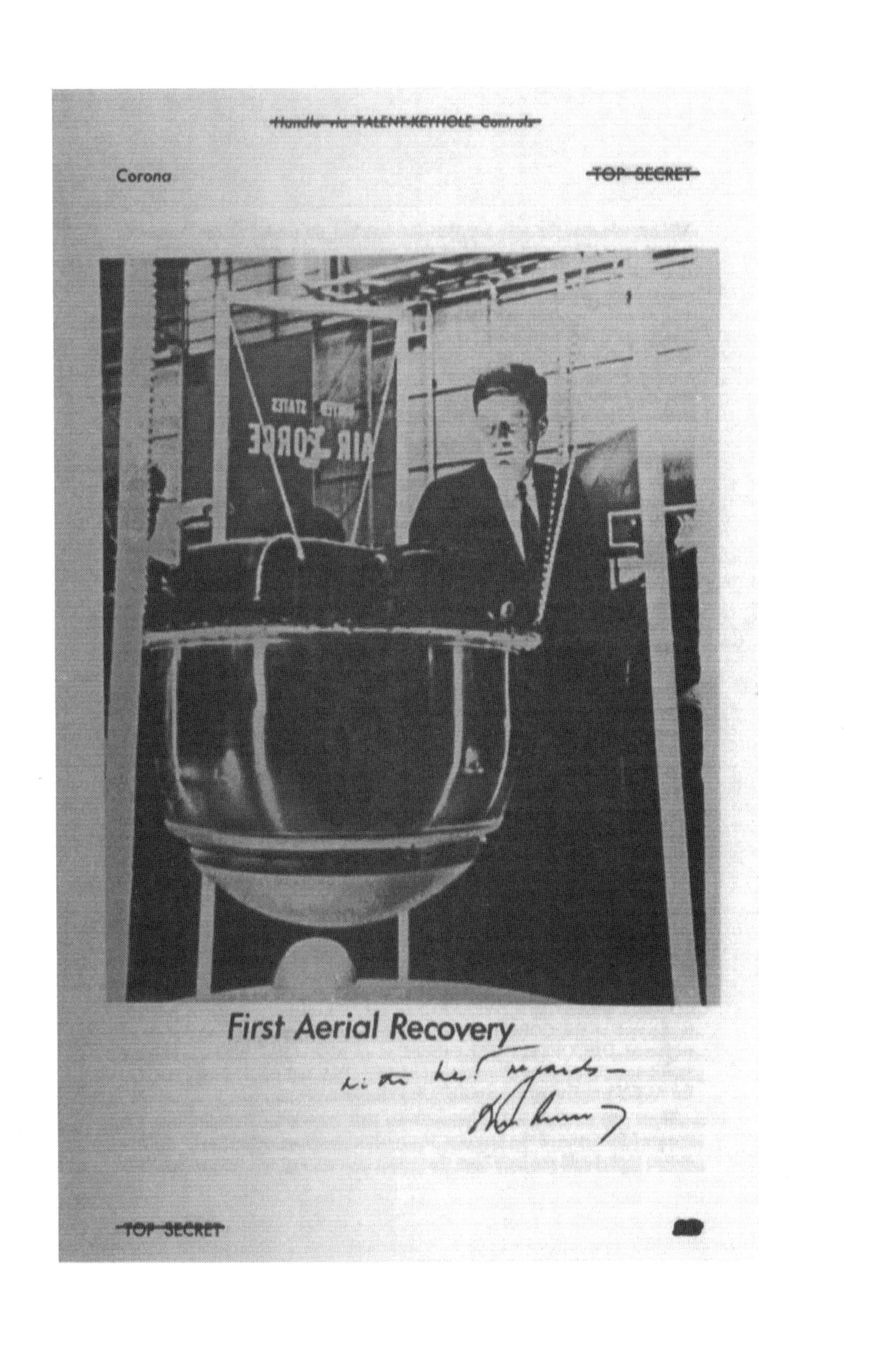

~~Handle via TALENT-KEYHOLE Controls~~

Corona ~~TOP SECRET~~

First Aerial Recovery

with best regards —

~~TOP SECRET~~

1. *(Continued)*

XIV capsule was the only one thus far that had descended in the designated impact zone. This was a problem that was to receive major attention during the next few weeks.

Four more cameras were launched within the next four months, with one success and three failures. DISCOVERER XV was sent aloft on 13 September. The vehicle was successfully inserted into orbit, and the camera functioned properly. However, the recovery vehicle re-entered at the wrong pitch attitude, causing the capsule to come down outside the recovery zone and demonstrating that the technicians' concern over the retro-firing sequence was well founded. The capsule was located, but it sank before a recovery ship could reach it. DISCOVERER XVI was launched on 26 October, but the AGENA failed to go into orbit because of a malfunction of a timing device.

The first ten camera-equipped vehicles carried what was known as the C camera: a single, vertical-looking, reciprocating, panoramic camera that exposed the film by scanning at a right angle to the line of flight. DISCOVERER XVI carried the first of a new series of cameras known as the C Prime (C′). The C′ differed only slightly from the original C configuration and was essentially little more than a follow-on procurement of the C camera.

The DISCOVERER XVII mission was launched on 12 November and went the full route through successful air catch—except for one mishap: the film broke after 1.7 feet of the acetate base leader had fed through the camera. There is an inconsistency in the records on this and the succeeding mission. The press release concerning this mission announced that the AGENA B, a more powerful second-stage engine, was used for the first time; the project files record the first use of the B vehicle on the following mission. In either event, it was the first of the two-day missions. The capsule was recovered on the 31st orbit.

DISCOVERER XVIII was launched on 10 December 1960 carrying 39 pounds of film. Orbit was achieved, and the camera worked well, exposing the entire film load. The recovery vehicle was ejected on revolution number 48 after three days in orbit, and the capsule was retrieved by air snatch. This was the first successful mission employing the C′ camera and the AGENA B second stage. There was fogging on the first, second, and last frame of each photo pass due to mirror light leaks, but image quality was otherwise as good as the best from DISCOVERER XIV.

CORONA in 1961

Of the next ten launches, extending through 3 August 1961, only four were CORONA missions. DISCOVERERs XIX and XXI carried radiometric payloads in support of the CORONA cover story, and they were not intended to be recovered. DISCOVERER XXI included an experiment that was to be of major significance in the later development of CORONA and other space programs: the AGENA engine was successfully restarted in space.

There was another "first" during these 1961 launches. When the film was removed from one of the capsules, the quality assurance inspector found three objects that should not have been there: two quarters and a buffalo nickel. Early capsules had contained a flag, so that there would be one to present to President Eisenhower after the first successful recovery. This had apparently inspired program personnel at Vandenberg to make their own payload additions during flight preparation. The Washington program office sent a sharply worded message to the West Coast project office charging it with responsibility for ensuring that the practice of souvenir-launching be stopped. (Years later NASA would find itself in the same position after the Apollo moon flights.)

DISCOVERER XX was the first of a dozen launches extending over a period of three years carrying mapping cameras, a program sponsored by the U.S. Army, which the President had approved for inclusion within the CORONA project. The purpose of the mapping program, which was known as ARGON, was to obtain precise geodetic fixes and an extension of existing datum planes within the Soviet Union. DISCOVERER XX was a bust on a number of counts: the camera failed; there were no shutter firings; and the orbital programmer malfunctioned. This last-named failure led to an important change in control procedures for CORONA. On this and all prior flights the recovery sequence was initiated automatically by an ejection command cut into the program tape. The program timer failed temporarily on orbit 31 of this mission, causing the entire sequence to be about one-half cycle out of phase. The automatic initiation of the recovery sequence was eliminated from the program tape on subsequent missions. Thereafter, the positive issuance of an injection command was required.

Of the four CORONA missions attempted between December 1960 and August 1961, two did not go into orbit as a consequence of AGENA failures, and two were qualified successes. DISCOVERER XXV was launched on 16 June and exposed its full load of film. The air catch failed, but the back-up water recovery was successful. The camera failed on revolution 22 of DISCOVERER XXVI, which was launched on 7 July, but about three-quarters of the film was exposed and was recovered by air catch.

Going into August 1961, a total of 17 camera-carrying CORONA missions had been attempted, and usable photography had been recovered from only four of them. These four successful missions, however, had yielded plottable coverage of some 13 million square miles, or nearly half of the total area of interest.

Camera Improvements

The first substantial upgrading of the CORONA camera system came with the introduction in August 1961 of the C Triple Prime (C′′′) camera. The original C camera was a scanning panoramic camera in which the camera cycling rate and the velocity-over-height ratio were constant and were selected before launching. Image motion compensation was fixed mechanically to the velocity-over-height ratio. A brief explanation of these terms may be helpful in understanding the nature of the problems with which the camera designers had to cope.

> A means must be provided for matching the number of film exposures in a given period of time (camera cycling rate) with the varying ratio between vehicle altitude and velocity on orbit (velocity-over-height) so that

the ground area is photographed in a series of swaths with neither gaps nor excessive overlapping in the coverage.

If the subject moves just as a snapshot is taken with a hand-held camera, and if the camera shutter speed is not fast enough to "stop" the motion, the photographic image will be smeared. To a camera peering down from an orbiting CORONA space vehicle, the earth's surface appears to be passing beneath the camera at a speed of roughly five miles per second. A camera photographing the earth's surface from a satellite moving at that speed would yield smeared photography if some means were not provided for stopping the relative motion. The technique used in accomplishing this is known as image motion compensation.

The C Triple Prime was the first camera built totally by the Itek Corporation. The C''' was also a reciprocating camera with a rotating lens cell, which exposed the film during a segment of its rotation. The new camera had a larger aperture lens, an improved film transport mechanism, and a greater flexibility in command of camera and vehicle operations—especially as regards control of the velocity-over-height factor. The larger aperture lens permitted use of slower film emulsions, which, combined with the improved resolving power of the lens itself, offered the prospect of resolution approximately twice as good as the C and C' cameras.

The first C''' camera system with a 39-pound film load was launched on 30 August 1961. The mission was a success, with the full film load being transferred and with ejection and recovery occurring on the 32nd orbit. All frames of the photography however, were out of focus. The cause was identified and was corrected by redesigning the scan head. Seven more missions were launched during the last four months of 1961, three with the C' camera and four with the C'''. Six of them attained orbit, and the cameras operated satisfactorily on all six. Film was recovered from four of the missions. The last of the four, which carried a C''' camera system, was rated the best mission to date. It also had a cover assignment to carry out: the injection of a secondary satellite, dubbed OSCAR (orbital satellite carrying amateur radio), into a separate orbit. OSCAR was a small radio satellite broadcasting a signal on 145 megacycles for pick-up by amateurs as an aid in the study of radio propagation phenomena.

Slowly but surely the bugs were being worked out, but it seemed that just as one was laid to rest another arose to take its place. Perhaps what was actually happening was that various sets of problems existed simultaneously, but the importance of some of them was masked by others. The elimination of a particular problem made it possible to recognize the significance of another. The recent successes had resulted largely from correcting weaknesses in the payload portion of the system. At the same time, difficulties in the AGENA vehicle began to surface. Of the last seven missions in 1961, four experienced on-orbit difficulties with the AGENA power supply or control gas system.

Power system components for general use in satellite systems were designed, developed, and tested in the CORONA program. Foremost among those components were the static electronic inverters used to convert direct current

1. *(Continued)*

battery energy into the various alternating current voltages required by the other subsystems. Static inverters, which were first flown aboard CORONA vehicles, were considered essential, because they had half the weight and double the efficiency of their rotary counterparts. Unfortunately, they are rather temperamental gadgets. The history of inverter development had been marked by high failure rates in system checkouts on the ground. Despite the lessons that had been learned and the improvements in circuit design that resulted from them, the recent on-orbit power failures demonstrated a need for further research and development.

The Last DISCOVERER

The AGENA failed on DISCOVERER XXXVII, launched on 13 January 1962, and the payload did not go into orbit. It was the last mission to carry the C''' camera system, and with it the DISCOVERER series came to an end. After 37 launches or launch attempts, the cover story for DISCOVERER had simply worn out. With the improved record of success and the near-certainty of an even better record in the future, it seemed likely that there would be as many as a dozen and a half to two dozen launches per year for perhaps years to come.

CORONA Goes Stereo

The 1961 R&D effort was not confined to improving the performance of the existing system. A major development program was concurrently under way on a much better camera subsystem. A contract was awarded on 9 August 1961, retroactively effective to 20 March, for a new camera configuration to be known as MURAL. The MURAL camera system consisted essentially of two C''' cameras mounted with one pointing slightly forward and the other slightly backward. Two 40-pound rolls of film were carried in a double-spool film supply cassette. The two film webs were fed separately to the two cameras where they were panoramically exposed during segments of the lens cells' rotations and then were fed to a double-spool take-up cassette in the satellite recovery vehicle. The system was designed for a mission duration of up to four days.

The vertical-looking C, C', and C''' cameras had photographed the target area by sweeping across it in successive overlapping swaths. The MURAL concept involved photographing each swath area twice. The forward-looking camera first photographed the swath at an angle 15 degrees from the vertical. About a half-dozen frames later, the backward-looking camera photographed the same swatch at an angle also 15 degrees from the vertical. When the two resulting photographs of the same area or object were properly aligned in a stereo-microscope, the photography would appear to be three-dimensional. Simultaneous operation of both instruments was required for stereo photography. If either camera failed, photography could still be obtained from the other, but it could be viewed in only two dimensions.

The first MURAL camera system was launched as program flight number 38 on 27 February 1962. On the first M flight, an anomaly occurred during re-entry. The RV heat shield failed to separate and was recovered by the aircraft along with the capsule. This anomaly provided valuable diagnostic data on the re-entry effects, which served the program well in later years, when program stretchouts caused shelf life of the heat shields to be a major concern. The twenty-sixth and last in the MURAL series was launched on 21 December 1963. Twenty of the SRV's were recovered, 19 of them by air snatch. The one water recovery was of a capsule that splashed down a thousand miles from the nominal impact point. An interesting aspect of this recovery was that the capsule turned upside down in the water, causing loss of the beacon signals. It was located during the search by an alert observer who spotted the sun shining on the gold capsule. Of the six vehicles that failed, two malfunctioned in the launch sequence, one SRV failed to eject properly, and three capsules came down in the ocean and sank before they could be recovered. Twenty successes out of 26 tries appeared to be a remarkable record when viewed against the difficulties experienced only two years earlier.

The three capsules that sank came down in or near the recovery zone, indicating that the problems previously encountered in the reentry sequence had been solved. They were not supposed to sink so quickly, however. (One of them floated for less than three minutes.) To minimize the chance that a capsule might be retrieved by persons other than the American recovery crew, the capsules were designed to float for a period ranging originally from one to three days and then to sink. The duration of the flotation period was controlled by a capsule sink valve containing compressed salt, which would dissolve in sea water at a rate that could be predicted within rather broad limits. When the salt plug had dissolved, water entered the capsule, and it sank—ingenious but simple.

More Problems, More Answers

Other significant improvements in the CORONA program were inaugurated during the lifetime of the MURAL system. One of them was an aid to photointerpretation. In order to read out the photography, the photointerpreter must be able to determine for each frame the portion of the earth's surface that is imaged, the scale of the photography, and its geometry. In simplest terms, he must know where the vehicle was and how it was oriented in space at the precise time the picture was taken. Until 1962, the ground area covered by a particular frame of photography was identified by combining data provided on the orbital path of the vehicle with the time of camera firing. The orientation or attitude of the vehicle on orbit was determined from horizon photographs recorded at each end of every other frame from a pair of horizon cameras that were included in the CORONA camera system.

Beginning with the first of the MURAL flights, an index camera was incorporated into the photographic system, and a stellar camera was added a few missions later. The short focal length index camera took a small scale photograph of the area being covered on a much larger scale by successive sweeps of the pan cameras. The small scale photograph, used in conjunction with orbital data, simplified the problem of matching the pan photographs with the terrain. Photographs taken of stars by the stellar camera, in combination with those taken of the horizons by the horizon cameras, provided a more precise means of determining vehicle attitude on orbit.

The photography from program flight number 47, a MURAL mission launched on 27 July 1962, was marred by heavy corona and radiation fogging. The corona problem was a persistent one—disappearing for a time only to reappear later—and had become even more severe with the advent of the complicated film transport mechanisms of the MURAL camera. Corona marking was caused by sparking of static electricity from moving parts of the system, especially from the film rollers. The problem was eventually solved by modifications of the parts themselves and by rigid qualification testing of them.

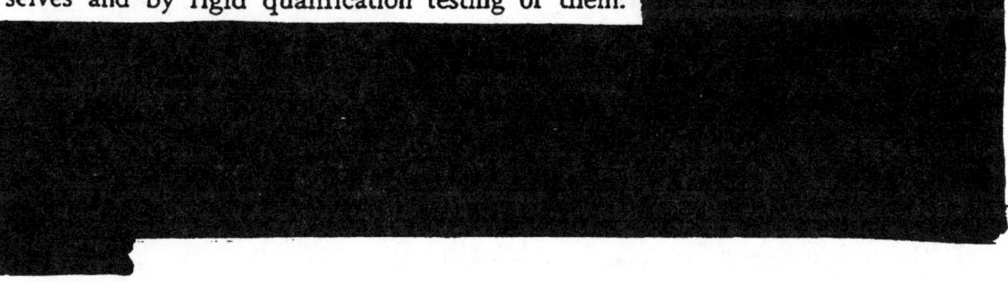

The boosting capacity of the first-stage THOR was substantially increased in early 1963 by strapping to the THOR a cluster of small solid-propellant rockets, which were jettisoned after firing. This Thrust Augmented THOR, or TAT as it came to be known, was first used for the launching of the heavier LANYARD camera system. LANYARD was developed within the CORONA program as a film recovery modification of one of the cameras designed for the SAMOS system and, with its longer focal length, was expected to yield better resolution than the CORONA cameras. It had a single lens cell capable of stereoscopic coverage by swinging a mirror through a 30-degree angle. Three flights were attempted, only one of which was partially successful. The camera had a serious lens focus problem, which was later traced to thermal factors and corrected. The LANYARD program was initiated as an interim system pending the completion of a high-resolution spotting system then under development. It was cancelled upon the success of the spotting system. The TAT booster itself was a significant success, permitting the later launching of heavier, more versatile CORONA systems.

The Two-Bucket System

Program flight number 69, launched on 24 August 1963, introduced the first two-bucket configuration—the next major upgrading of the CORONA system.

(The film recovery capsule is commonly referred to as a bucket, although it more nearly resembles a round-bottomed kettle.) The new modification, which was known as the J-1 system, retained the MURAL stereoscopic camera concept but added a second film capsule and recovery vehicle. With two SRV's in the system, film capacity was increased to 160 pounds (versus the 20-pound capacity of the first few CORONA missions). The two-bucket system was designed to be deactivated or stored in orbit in a passive (zombie) mode for up to 21 days. This permitted the recovery of the first bucket after half of the film supply was exposed. The second bucket could begin filling immediately thereafter, or its start could be delayed for a few days. A major redesign of the command and control mechanisms was required to accommodate the more complicated mission profile of the two-bucket system.

As with each of the major modifications of CORONA, the J-1 program had a few early bugs. On the first mission, the shutter on the master horizon camera remained open about 1,000 times seriously fogging the adjacent panoramic photography, and the AGENA current inverter failed in mid-flight, making it impossible to recover the second bucket. Also, the J-1 system initially experienced a rather severe heat problem, which was solved by reducing the thermal sensitivity of the camera and by better control of vehicle skin temperature through shielding and varying the paint pattern.

Back in 1960 and 1961, the successful recovery of a CORONA film bucket was an "event." A mere two years later, with the advent of the J-1 system, success had become routine and a failure was an "event." By the end of 1966, 37 J-1 systems had been launched; 35 of them were put into orbit; and 64 buckets of film were recovered. There were *no* failures at recovery in the three years following 1966: 28 buckets were launched, and 28 buckets were recovered. Also, mission duration was greatly expanded during the lifetime of the J-1 system. A mission in June 1964 yielded four full days over target for each of the two buckets. Five full days of operation with each bucket was attained in January 1965. In April 1966, the first bucket was recovered after seven days on orbit. A 13-day mission life was achieved in August 1966, and this was increased to 15 days in June 1967.

The increased mission life and excellent record of recovery resulted from a number of successive improvements that were incorporated into the J-1 time period. Among them was a subsystem known as LIFEBOAT, a completely redundant and self-contained apparatus built into the AGENA that could be activated for recovering the SRV in event of an AGENA power failure (which still happened occasionally). Another improvement was the introduction of the new and more powerful THORAD booster. A third was the addition of a rocket orbit-adjust system. The CORONA vehicles were necessarily flown over the target areas with quite a low perigee in order to increase the scale of the photography, and this led to a relatively rapid decay of the orbit. The orbit-adjust system compensated for the decay. It consisted of a cluster of small rockets, known as drag make-up units, which were fired individually and at selected intervals. Each firing accelerated the vehicle slightly, boosting it back into approximately its original orbit.

A Maverick

The CORONA camera system was to undergo one more major upgrading but we cannot leave the J-1 program without giving an account of one mission failure of truly magnificent proportions. Program flight number 78 (CORONA Mission Number 1005), a two-bucket J-1 system, was launched on 27 April 1964. Launch and insertion into orbit were uneventful. The master panoramic camera operated satisfactorily through the first bucket, but the slave panoramic camera failed after 350 cycles when the film broke. Then the AGENA power supply failed. Vandenberg transmitted a normal recovery enable command on southbound revolution number 47 on 30 April. The vehicle verified receipt of the command, but nothing happened. The recovery command was repeated from various control stations—in both the normal and back-up LIFEBOAT recovery modes—on 26 subsequent passes extending through 20 May. The space vehicle repeatedly verified that it had received the commands, but the ejection sequence did not occur. After 19 May, the vehicle no longer acknowledged receipt, and from 20 May on it was assumed that the space hardware of Mission 1005 was doomed to total incineration as the orbit decayed.

But Mission 1005, it later developed, had staged its own partial re-entry, stubborn to the end. At six minutes past midnight on 26 May, coinciding with northbound revolution No. 452 of Mission 1005, observers in Maracaibo, Venezuela saw five burning objects in the sky.

On 7 July, two farm workers found a battered golden object on a farm in lonely mountain terrain near La Fria in Tachira State, southwestern Venezuela, a couple of miles from the Colombian border. They reported it to their employer, Facundo Albarracin, who had them move it some 100 yards onto his own farm and then spread the news of his find in hopes of selling it. Albarracin got no offers from the limited market in Tachira, however—not even from the smugglers with access to Colombia—so he hacked and pried loose the radio transmitter and various pieces of the take-up assembly to use as household utensils or toys for the children.

Ultimately word of the find reached San Cristobal, the nearest town of any size. Among the curious who visited La Fria was a commercial photographer, Leonardo Davila, who telephoned the U.S. Embassy in Caracas on 1 August that he had photographed a space object. It was the first bucket from Mission 1005, with one full spool of well-charred film clearly visible.

A team of CORONA officers, ostensibly representing USAF, flew to Caracas to recover the remains. The capsule was lugged out by peasants to a point where the Venezuelan Defense Ministry could pick it up for flight to Caracas. There the CORONA officers bought the crumpled bucket from the Venezuelan government, and quietly dismissed the event as an unimportant NASA space experiment gone awry.

1. *(Continued)*

The story rated only a dozen lines in the *New York Times* of 5 August, but the local Venezuelan press had a field day. *Diario Catolico*, of San Cristobal, along with a lengthy report, published three pictures of the capsule showing the charred roll of film on the take-up spool. The *Daily Journal* handled the story in lighter vein with this parody of Longfellow:

I shot an arrow into the air.
It fell to earth I know not where.
Cape Kennedy signalled: "Where is it at you are?"
Responded the rocket: "La Fria, Tachira."

The CORONA technicians who examined the capsule after its arrival in the States concluded that the re-entry of the SRV was a result of normal orbit degeneration, with separation from the instrument fairing caused by re-entry forces. The thrust cone was sheared during separation but was retained by its harness long enough to act as a drogue chute, thus preventing the capsule from burning up during re-entry and stabilizing it for a hard, nose-down landing.

The Final Touches

The final major modification of the CORONA system got under way in the spring of 1965, when about a dozen and a half of the two-bucket J-1 systems had been flown. The J-1 was performing superbly, but it had little potential for within-system growth. The new CORONA improvement program was begun with a series of meetings among representatives of Lockheed, General Electric, Itek, and the various CORONA program offices to examine ways of bettering the performance of the panoramic and stellar/index cameras, and of providing a more versatile command system. These were the resulting design goals established for a new panoramic camera:

Improved photographic performance by removal of camera system oscillating members and reduction of vibration from other moving components.

Improvement of the velocity-over-height match to reduce image smear.

Improved photographic scale by accommodation of proper camera cycling rates at altitudes down to 80 n.m. (the minimum J-1 operating altitude was 100 n.m.).

Elimination of camera failures caused by film pulling out of the guide rails (an occasional problem with the J-1 system).

Improved exposure control through variable slit selection. (The J-1 system had a single exposure throughout the orbit resulting in poor performance at low sun angles.)

Capability of handling alternate film types and split film loads. An in-flight changeable filter and film change detector was added for this purpose.

Capability of handling ultra-thin base film (yielding a 50% increase in coverage with no increase in weight).

The panoramic camera that was developed to meet those design goals was known as the constant rotator. The predecessor C''' camera employed a com-

1. *(Continued)*

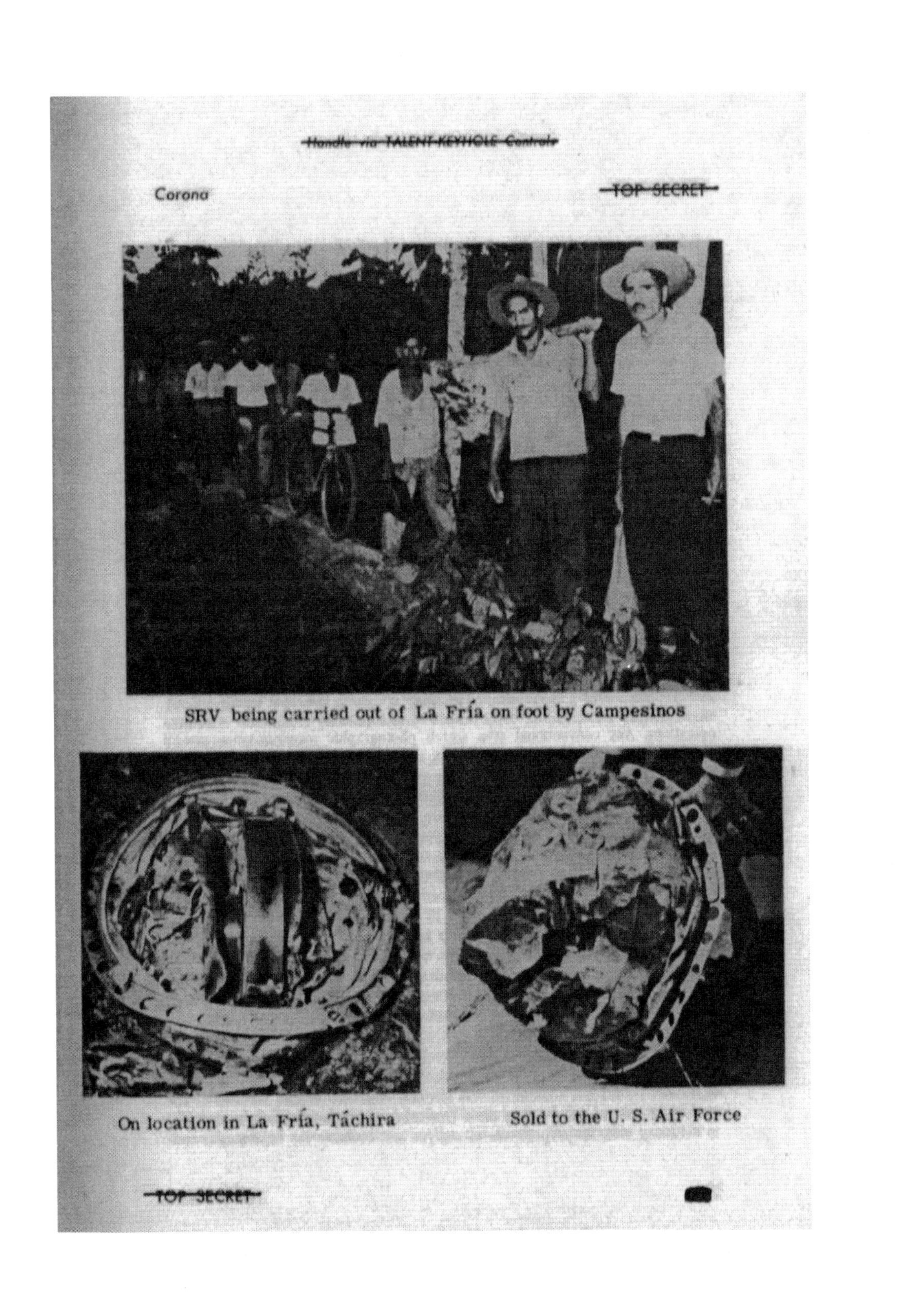

~~Handle via TALENT-KEYHOLE Controls~~

Corona ~~TOP SECRET~~

SRV being carried out of La Fría on foot by Campesinos

On location in La Fría, Táchira

Sold to the U. S. Air Force

~~TOP SECRET~~

bination of rotating lens cell and reciprocating camera members. In the constant rotator, the lens cell and the balance of the camera's optical system is mounted in a drum, and the entire drum assembly is continuously rotated, thus eliminating the reciprocating elements from the camera system. The film is exposed during a 70-degree angular segment of the drum's circular sweep. The capability of using ultra-thin base (UTB) film was one of the design goals, but the camera design was not to be constrained by requirements to accommodate the thinner film. UTB was successfully flown on several flights but ground test results showed a loss of reliability and attempts to use it in the contant rotator were eventully abandoned. In all other respects, however, the constant rotator was a resounding success. It yielded substantially better ground resolution in the photography. It also permitted versatility in operation far exceeding that available in the earlier cameras.

The stellar/index camera in use was a delicate instrument with a short (1.5") focal length and a history of erratic performance. The efforts at upgrading the performance of the stellar/index camera resulted in an instrument with a 3" focal length (like ARGON) and a dual-looking stellar element. The new camera had the jaw-breaking designation of Dual Improved Stellar Index Camera, commonly referred to by its acronym: DISIC.

The new payload system, which was designated the J-3, consisted of a pair of constant rotator panoramic cameras, a pair of horizon cameras, and a DISIC. The J-3 system naturally retained the stereo capability begun with the MURAL cameras and the two-bucket recovery concept of the J-1. Apart from the improved picture-taking capability of the hardware itself, the most significant advance of the J-3 was the flexibility it allowed in command and control of camera operations. Any conventional area search photographic reconnaissance system is film-limited. (When the film runs out, the mission is finished—assuming, of course, that other mission-limiting components of the system survive that long.) Consequently, the ultimate goal of all the CORONA improvement efforts was to pack the maximum of the best possible quality of photography of important intelligence targets into each roll of exposed film. The built-in flexibility of the J-3 system greatly increased the variety and degree of controls that could be applied to camera operations, thus substantially boosting the potential intelligence content of the photography.

The first J-3 system was launched on 15 September 1967, and it proved to be the one major modification with no bugs in it. In its nearly five years of operation, it yielded even better photographic intelligence and higher reliability than the remarkably successful predecessor J-1 system.

An early series of tests demonstrated the unusual flexibility of the J-3. It could not only accommodate a variety of film loads, including special camouflage-detection color and high-speed, high-resolution black and white; the camera also had two changeable filters and four changeable exposure slits on each camera.

These tests drew such interest throughout the intelligence community that a CORONA J-3 Ad Hoc Committee was formally convened by the Director of the National Reconnaissance Office on 4 December 1967, and formally constituted in February 1968. Its purpose was to analyze and evaluate the experiments conducted on these five test flights. Specific findings of the Committee included the recommendations that further testing of color films and techniques should be conducted, against specific intelligence requirements and that a special sub-committee of the Committee on Imagery Requirements and Exploitation (COMIREX) should be constituted to evaluate the utility of satellite color photography; and that a well-planned color collection program be worked out with the close cooperation of the system program offices, the Satellite Operations Center (SOC), the intelligence analysts, and the photo interpreters.

In Retrospect

Looking back on CORONA, it is not always easy to keep in mind that it was merely an assemblage of inanimate objects designed and put together to perform a mechanical task. The program began as a short-term interim system, suffered through adversity in its formative years, and then survived in glory throughout a decade. Those who were associated with the program or came to depend upon its product developed an affection for the beast that bordered on the personal. They suffered with it in failure and revelled in its successes.

The technological improvements engineered under CORONA advanced the system in eight years from a single panoramic camera system having a design goal of 20 to 25 feet ground resolution and an orbital life of one day, to a twin camera panoramic system producing stereo-photography at the same ground resolution; then to a dual recovery system with an improvement in ground resolution to approximately 7 to 10 feet, and doubling the film payload; and finally, to the J-3 system with a constant rotator camera, selectable exposure and filter controls, a planned orbital life of 18 to 20 days, and yielding nadir resolution of 5-7 feet.

The totality of CORONA's contributions to U.S. intelligence holdings on denied areas and to the U.S. space program in general is virtually unmeasurable. Its progress was marked by a series of notable firsts: the first to recover objects from orbit, the first to deliver intelligence information from a satellite, the first to produce stereoscopic satellite photography, the first to employ multiple re-entry vehicles, and the first satellite reconnaissance program to pass the 100-mission mark. By March 1964, CORONA had photographed 23 of the 25 Soviet ICBM complexes then in existence; three months later it had photographed all of them.

The value of CORONA to the U.S. intelligence effort is given dimension by this statement in a 1968 intelligence report: "No new ICBM complexes have been established in the USSR during the past year." So unequivocal a statement could be made only because of the confidence held by the analysts that if they were there, CORONA photography would have disclosed them.

CORONA coverage of the Middle East during the June 1967 war was of great value in estimating the relative military strengths of the opposing sides after the short combat period. Evidence of the extensive damage inflicted by the Israeli air attacks was produced by actual count of aircraft destroyed on the ground in Egypt, Syria, and Jordan. The claims of the Israelis might have been discounted as exaggerations but for this timely photographic proof.

In 1970, CORONA was called on to provide proof of Israeli-Egyptian claims with regard to cease-fire compliance or violation. CORONA Mission 1111, launched on 23 July 1970, successfully carried out the directions for this coverage, which brought the following praise from Dr. John McLucas, Under Secretary of the Air Force and Director, NRO, who said in a message to the Director of Special Projects, DD/S&T, on 25 August 1970:

> I extend my sincere thanks and a well done to you and your staff for your outstanding response to an urgent Intelligence Community requirement.
>
> The extension of . . . Mission 1111 to 19 days, without benefit of solar panels, and the change in the satellite orbit to permit photography of the Middle East on 10 August provided information which could not be obtained through any other means. This photography is being used as a baseline for determining compliance with the Suez cease-fire provisions.

CORONA's Decade of Glory is now history. The first, the longest, and the most successful of the nation's space recovery programs, CORONA explored and conquered the technological unknowns of space reconnaissance, lifted the curtain of secrecy that screened developments within the Soviet Union and Communist China, and opened the way for the even more sophisticated follow-on satellite reconnaissance systems. The 145th and final CORONA launch took place on 25 May 1972 with the final recovery on 31 May 1972. That was the 165th recovery in the CORONA program, more than the total of all of the other U.S. programs combined. CORONA provided photographic coverage of approximately 750.000,000 square nautical miles of the earth's surface. This dramatic achievement was surpassed only by intelligence derived from the photography.

In placing a value on the intelligence obtained by the U.S. through its photographic reconnaissance satellite programs between 1960 and 1970, a first consideration, on the positive side, would be that it had made it possible for the President in office to react more wisely to crucial international situations when armed with the knowledge provided by these programs. Conversely, it can be said that without the intelligence which this program furnished, we might have misguidedly been pressured into a World War III.

The intelligence collected by the reconnaissance programs makes a vital contribution to the National Intelligence Estimates upon which the defense of the U.S. and the strategic plans of the military services are based. Principal among those estimates are the ones which deal with the Soviet and Chinese Communist strategic weapons, space, and nuclear energy programs.

The intelligence from overhead reconnaissance counts heavily not only in planning our defense, but also in programming and budgeting for it. It helps to avoid the kind of floundering that occurred during the time of the projection of the "Missile Gap." Without the kind of intelligence which the CORONA program provided, the U.S. budget for the defense of our own territory, and for military assistance to our allies, would doubtless have been increased by billions.

The total cost for all CORONA activities of both the Air Force and the CIA over the 16-year period was [redacted]

The CORONA program was so efficiently managed that even the qualification models of each series were refurbished and flown. As a result, there was little hardware available at the termination of the program when it was suggested that a museum display should be set up to illustrate and to preserve this remarkable program. Using recovered hardware from the last flight, developmental models from the J-3 program, and photographic records from the memorable flights, a classified museum display was set up in Washington, D. C. In his speech dedicating the Museum, Mr. Richard Helms, the Director of Central Intelligence said:

> It was confidence in the ability of intelligence to monitor Soviet compliance with the commitments that enabled President Nixon to enter into the Strategic Arms Limitation Talks and to sign the Arms Limitation Treaty. Much, but by no means all, of the intelligence necessary to verify Soviet compliance with SALT will come from photoreconnaissance satellites. CORONA, the program which pioneered the way in satellite reconnaissance, deserves the place in history which we are preserving through this small Museum display.
>
> "A Decade of Glory," as the display is entitled, must for the present remain classified. We hope, however, that as the world grows to accept satellite reconnaissance, it can be transferred to the Smithsonian Institution. Then the American public can view this work, and then the men of CORONA, like the Wright Brothers, can be recognized for the role they played in the shaping of history.

Part II

The Committee on Overhead Reconnaissance

Part II: The Committee on Overhead Reconnaissance

Before 1958, the Director of Central Intelligence's management or coordination of what is now called the Intelligence Community had been unsteady, if not haphazard. In 1956 President Eisenhower formed his own President's Board of Consultants on Foreign Intelligence Activities (PBCFIA), which soon worried that the United States was insufficiently prepared to counter the Soviet missile threat. Out of this concern the Board suggested that the DCI should better coordinate US intelligence efforts for early warning, wartime operational planning, and intelligence on new Soviet weaponry. By the 1960 election year, the "Missile Gap" issue—the charge that the Soviets were about to take a commanding lead over the United States in ballistic missiles—had fostered even greater worries about Soviet intentions and capabilities.

In 1958, after consolidating two principal interdepartmental intelligence committees into a single United States Intelligence Board (USIB), President Eisenhower issued a new National Security Council Intelligence Directive that gave the Director of Central Intelligence (DCI) clear orders to coordinate the foreign intelligence effort of the United States. The DCI was to be responsible for all forms of intelligence collection, including communications, electronic, missile, and space intelligence. In early 1959, DCI Allen Dulles formed the Satellite Intelligence Requirements Committee (SIRC) to manage satellite programs independently of the older Ad Hoc Requirements Committee (ARC), which dealt with collection and exploitation for the U-2 program.

After the Soviets shot down a U-2 over Russia in May 1960, the DCI in August established the Committee on Overhead Reconnaissance (COMOR) to coordinate the development of intelligence requirements for reconnaissance missions over the Soviet Union and other denied areas. COMOR superseded both ARC and SIRC.

Initially, COMOR's responsibilities were limited, since U-2s could no longer fly over the Soviet Union. This dramatically changed with the success of DISCOVERER XIV, the first CORONA mission to bring back photographs of the Soviet Union. Most of this section's documents offer examples of how COMOR's first chairman, James Q. Reber, set out to coordinate the analysis of CORONA material and establish procedures for handling TALENT-KEYHOLE material. Perhaps the section's most interesting record is Document No. 4, COMOR's 18 August 1960 "List of Highest Priority Targets, USSR," which identified primary targets for the U-2 just as CORONA's KH-1 satellite arrived on the scene.

2. Director of Central Intelligence Directive, Committee on Overhead Reconnaissance (COMOR), 9 August 1960

SECRET

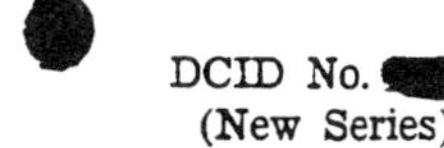

DCID No.
(New Series)

DIRECTOR OF CENTRAL INTELLIGENCE DIRECTIVE NO. 

COMMITTEE ON OVERHEAD RECONNAISSANCE (COMOR)

(Effective 9 August 1960)

Pursuant to the provisions of NSCID Nos. 1, 2, 3 and 6, and for the purpose of providing a focal point for information on, and for the coordinated development of foreign-intelligence [1] requirements for, overhead reconnaissance projects and activities of the Government over denied areas [2] (including foreign-intelligence requirements during research and development phases of such projects and activities) a Committee on Overhead Reconnaissance (COMOR) of the U.S. Intelligence Board is hereby established.

1. For the purposes of this directive the term "overhead reconnaissance" includes all reconnaissance for foreign-intelligence purposes by satellite, or by any vehicle over denied areas, whether by photographic, [redacted] or other means, but does not include reconnaissance and aerial surveillance in direct support of actively combatant forces.

2. The Committee shall coordinate the adaptation of priority foreign-intelligence objectives and requirements established by USIB, members of USIB, or other committees of USIB, to the capabilities of existing and potential overhead reconnaissance systems; and shall examine and make recommendations, as appropriate, on such related matters as dissemination and any special security controls required, but shall not undertake to provide operational guidance. ([redacted])

3. The Committee on Overhead Reconnaissance (COMOR) shall be composed of designated representatives of Intelligence Board departments and agencies. Representatives of other agencies may be invited by the Chairman to participate in selected discussions as agreed by the Committee.

4. The Chairman of the Committee shall be designated by the Director of Central Intelligence in consultation with and with the concurrence of the Intelligence Board.

ALLEN W. DULLES
Director of Central Intelligence

---

[1] As distinguished from operational "early warning" information and other operational-support intelligence.

[2] For purposes of this directive "denied areas" include all territory and territorial waters claimed by members of the Sino-Soviet Bloc, as well as such other areas of priority intelligence interest as may be determined by USIB.

SECRET

3. Col. James E. Mahon, JCS to COMOR, "Urgent Requirements for CORONA and ARGON," 18 August 1960

corona / copy Kept

~~TOP SECRET~~
~~HANDLE VIA TALENT CHANNELS ONLY~~

~~HANDLE VIA TALENT-KEYHOLE CHANNELS ONLY~~

18 August 1960

all but this copy Destroyed 13 SEP 1960

MEMORANDUM FOR: COMOR

FROM: Col. James E. Mahon, JCS

SUBJECT: Urgent Requirements for CORONA and ARGON

1. The successful recovery of Discoverer XIII necessitates a relook at scheduling of future launchings. It is understood that Discoverer XIV will incorporate components which have demonstrated success, i.e., the C camera as carried on Discoverer XI and the recovery system of Discoverer XIII and a continuance of the Agena A engine. Through this combination it is hoped that a systems capability can be demonstrated in total.

2. Accomplishment of the basic milestones of operating a camera in space and recovering a payload from orbit focuses attention on refinement and improvement of this collection system. These operations must also be considered in context with intelligence requirements. The requirements upon which Discoverer was based have become increasingly critical with the loss of time during the development phase, particularly when coupled with the loss of the U-2 capability during this interim. Wherever possible, efforts should now be expended to gain gross coverage of the USSR, on a selective priority basis, as expeditiously as possible. The C' camera and the second generation recovery device have been designed for improved performance and reliability. The Agena B engine has been developed to provide multiple day operations, allowing for programming reconnaissance coverage for all of the USSR as well as a greater altitude capability required by the geodetic satellite. The national requirements for reconnaissance and geodesy are both critical and it is difficult to assign relative priorities, i.e., reconnaissance

~~HANDLE VIA TALENT-KEYHOLE CHANNELS ONLY~~

~~HANDLE VIA TALENT CHANNELS ONLY~~
~~TOP SECRET~~

3. *(Continued)*

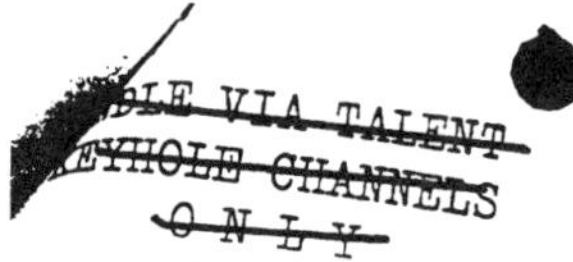

TOP SECRET
HANDLE VIA TALENT CHANNELS ONLY

is urgently needed to assess the threat of the USSR; and the geodetic locations must be acquired to ensure effectiveness of weapons systems, in being, or soon to be deployed, as well as to maintain an effective deterrent posture. An additional complicating factor is the limited May to October operational season for the geodetic satellite.

3. It is proposed that the COMOR recommend to the CORONA and ARGON operators that, if it is technically feasible, at the earliest possible date a CORONA shot with the C prime camera and Agena B engine be utilized to obtain reconnaissance of the whole of Russia with special reference to target areas as set forth in [redacted] 25 May 1960, "List of Highest Priority Targets," to be followed as soon as possible by the ARGON camera with Agena B engine to fulfill geodesy requirements. While it is recognized that the implementation of this recommendation would alter the schedule established for these programs the COMOR view is that the urgency of both national strategic targets (the objective of CORONA) and the geodesy requirements (ARGON) are of such urgency as to warrant the change in schedule if compatible with sound technical communications.

HANDLE VIA TALENT-
KEYHOLE CHANNELS
O N L Y

TOP SECRET
HANDLE VIA TALENT CHANNELS ONLY

3. *(Continued)*

TOP SECRET
HANDLE VIA TALENT CHANNELS ONLY
HANDLE VIA TALENT
KEYHOLE CHANNELS
ONLY
HANDLE VIA TALENT
KEYHOLE CHANNELS
ONLY
HANDLE VIA TALENT CHANNELS ONLY
TOP SECRET

4. James Q. Reber, "List of Highest Priority Targets, USSR," 18 August 1960

~~TOP SECRET~~

COMMITTEE ON OVERHEAD RECONNAISSANCE

18 August 1960

LIST OF HIGHEST PRIORITY TARGETS

USSR

1. The list of 32 highest priority targets for TALENT collection against the principal objective, USSR, is attached. As in previous lists, the priority interest centers on: (a) The ICBM, IRBM, sub-launched ballistic missiles; (b) The heavy bomber, and (c) Nuclear energy. However, the principal emphasis is the ICBM and the question of its deployment. At the moment, this objective transcends all others. In the main, it is expressed in this target list in terms of the search of sections of rail lines which are judged to be, among the total of USSR rails, the most likely related in some way to ICBM deployment and which are short enough in length to be considered as a terminal objective within operational capabilities. Major targets which are almost certainly to be covered if the given rails were searched are listed under each rail target. These specific geographic points vary in their importance some individually capable of sustaining the highest priority label by themselves. In any given case, the significance of all the individual targets subtended should assist in weighing the desirability for recommending missions as circumstances require.

2. The possibility of the association of long range air bases or air frame plants with missile activity has heightened our interest in the bomber question. It could well be that future coverage of long range air bases would cause us to include long range air bases covered three and four years ago prior to the serial production and deployment of the ICBM.

3. The limited role which nuclear energy plays in the list may be attributed to the past coverage which has been fairly extensive in terms of nuclear energy installations as well as the role of collateral in reducing our ignorance on critical nuclear energy questions. However, there are specific nuclear energy questions still unanswered. It may well be that the passage of time may cause us to wish to reexamine installations previously photographed which are critical in the determination production and stockpile of Soviet fissionable materials.

~~TOP SECRET~~

~~HANDLE VIA TALENT CONTROL SYSTEM ONLY~~

~~TOP SECRET~~

4. Consideration is also now being given the anti-ballistic missile missile problem, this being one of the reasons for recoverage of SARA SHAGAN [redacted] and of KAPUSTIN YAR [redacted].

5. In addition to the remarks above regarding rails, it should be noted that in the attached list there are several area targets, to wit: SARY SHAGAN (12); KLYUCHI Impact Area (23); the IULTIN/ANADYR Area (25)--representing interests which are not confined to single coordinates and where search is required to discover whether suspected developments exist.

6. The Soviet surface to air (SAM) threat has been kept very much in mind in the preparation of the target list because of the evidence of extensive deployment in the vicinity of critical industrial and military centers in the USSR. Complete information on SAM development is recognized as of very high interest to SAC. The collection against this target list should provide extensive information on this high priority requirement.

7. This list differs from previous lists of Highest Priority Targets in that most targets on those lists were supported by considerable firm evidence concerning their importance. Many targets on this list, however, are supported by relatively little firm evidence. They are included here because, on a basis of deductive reasoning, they appear to be the most likely of all known targets to bear upon missile deployment and other highest priority matters at this time. This means that the receipt by the Intelligence Community of a modest amount of firm evidence on a number of problems could cause us to add targets not on the list, or withdraw targets now carried.

8. This paper is for reference and is not intended, in its present form, to indicate an order to priority within itself. Such distinction would be the subject of specific recommendation by the COMOR when required.

James Q. Reber

JAMES Q. REBER
Chairman
Committee on Overhead Reconnaissance

[redacted] (cy 2)
[redacted] (cys 3, 4, 5)
[redacted] (cys 6, 7)
[redacted] (cy 8)
[redacted] (cy 9)
[redacted] (cy 10)

~~HANDLE VIA TALENT CONTROL SYSTEM ONLY~~

~~TOP SECRET~~

4. *(Continued)*

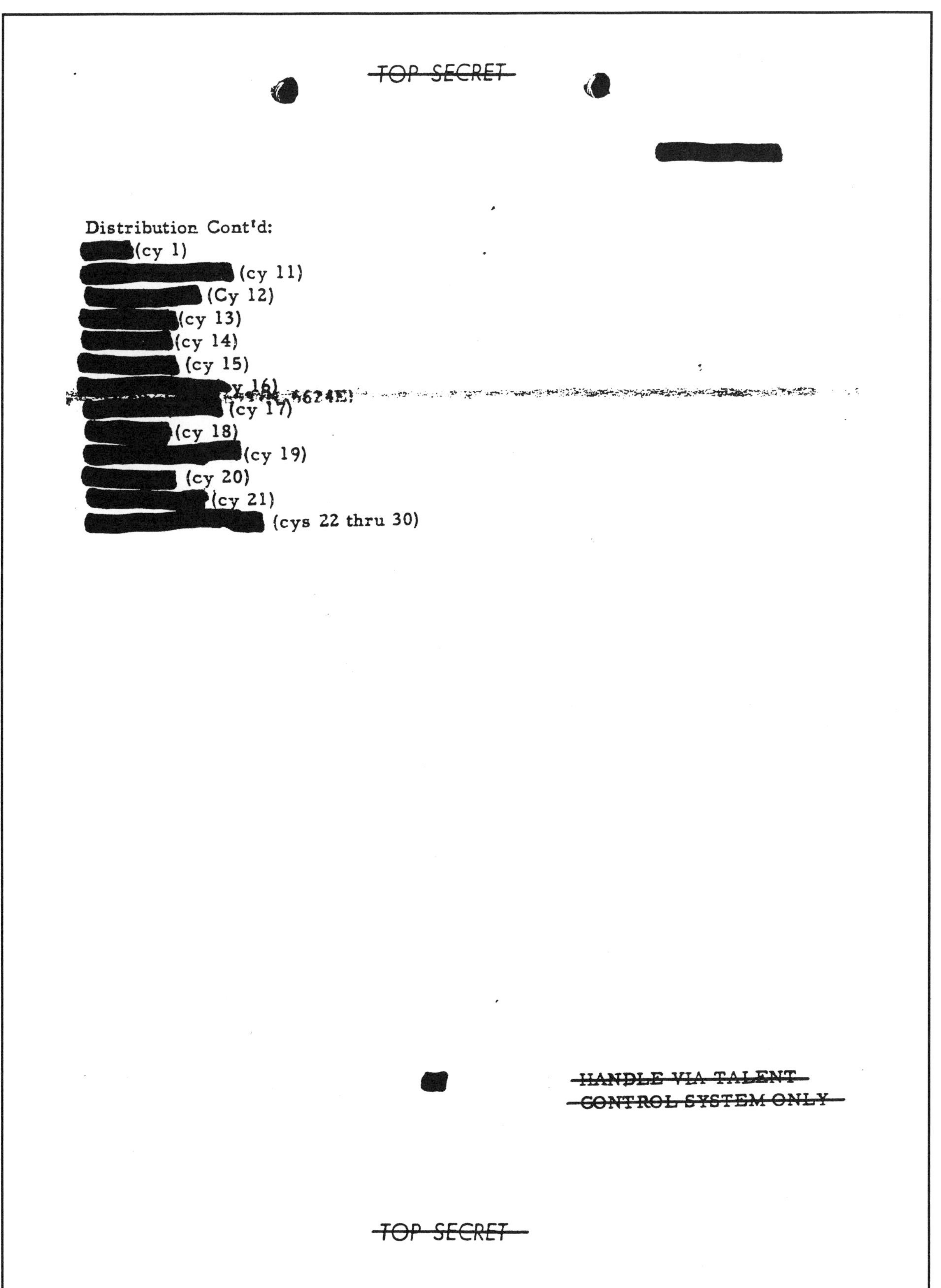

~~TOP SECRET~~

Distribution Cont'd:
[redacted] (cy 1)
[redacted] (cy 11)
[redacted] (Cy 12)
[redacted] (cy 13)
[redacted] (cy 14)
[redacted] (cy 15)
[redacted] y 16)
[redacted] (cy 17)
[redacted] (cy 18)
[redacted] (cy 19)
[redacted] (cy 20)
[redacted] (cy 21)
[redacted] (cys 22 thru 30)

~~HANDLE VIA TALENT~~
~~CONTROL SYSTEM ONLY~~

~~TOP SECRET~~

4. *(Continued)*

~~TOP SECRET~~

1. Kotlas (6116N-4635E) - Salekhard (6630N-6640E) Rail Line

   Ust Ukhta (6338N-5353E)
   Vorkuta (6730N-6403E)
   Polyarni Ural (6602N-6510E)
   Khal'mer-yu (6757N-6507E)
   Yeletskiy (6710N-6410E)
   Obskaya (6537N-6624E)
   Anderma (6940N-6145E)
   Kara (6915N-6457E)
   Muzhi (6523N-6645E)

   Category of interest: ICBM Deployment

2. Vologda (5913N-3953E) - Perm (5800N-5615E) Rail Line

   Kirov (5836N-4942E)
   Danilov (5812N-4010E)

   Category of interest: ICBM Deployment

3. Vologda (5913N-3953E) - Archangelsk (6434N-4032E) Rail Line

   Konosha (6058N-4009E)
   Severodvinsk (6434N-3950E)
   Plesetskaya (6243N-4017E)

   Category of interest: ICBM Deployment, Submarine Launch

4. Petrozavodsk (6149N-3420E) - Pechenga (6933N-3112E) Rail Line

   Belomorsk (6432N-3447E)
   Olenya (6809N-3315E)
   Murmansk (6858N-3305E)
   Kandalaksha (6709N-3226E)
   Sayda Guba (6915N-3315E)
   Kil'din (6920N-3410E)
   Severomorsk (6905N-3327E)
   Polyarnyy (6912N-3328E)

~~HANDLE VIA TALENT CONTROL SYSTEM ONLY~~

~~TOP SECRET~~

4. *(Continued)*

~~TOP SECRET~~

5. Trans-Siberian Rail Line Ufa (5443N-5558E) - Omsk (5500N-7324E)

Kurgan (5526N-6520E)
Chelyabinsk (5510N-6124E)
Zlatoust (5510N-5940E)

6. Trans-Siberian Rail Line Novosibirsk (5502N-8253E) - Irkutsk (5216N-1042E)

Angarsk (5235N-10354E)
Krasnoyarsk (Dodnovo) (5602N-9248E)
Belaya (5251N-10333E)

Categories of interest: ICBM Deployment, Long Range Aircraft Nuclear Energy

7. Chelyabinsk (5510N-6124E) - Ivdel (7042N-6028E) Rail Line

Kyshtym (5544N-6033E)
Sverdlovsk (5650N-6036E)
Nizhnaya Salda (5805N-6043E)
Nizhnaya Tura (5837N-5950E)

Categories of interest: ICBM Deployment, Missile Production, Nuclear Energy

8. Komsomolsk (6115N-13907E) - Vladivostoit (4308N-13150E) Rail Line

Khabarovsk (4839N-13506E)
Spaask Dal'niy (4437N-13248E)
Khorol (4425N-13204E)
Kremovo (4402N-13216E)

Categories of interest: IRBM Deployment, ICBM Deployment, Submarine Launch

9. Grodekovo (4425N-13123E) - Kraskino (4243N-13048E) Rail Line

Slavyan'ka (4929N-13045E)

Categories of interest: IRBM Deployment

10. Odessa (4628N-3043E) - Leningrad (5955N-3020E) Rail Line

Vinnitsa (4913N-2829E)
Zhitomir (5016N-2840E)
Mogilev (5355N-3021E)

~~TOP SECRET~~

~~HANDLE VIA TALENT CONTROL SYSTEM ONLY~~

4. *(Continued)*

~~TOP SECRET~~

Vitebsk (5512N-3013E)
Soltsy (5807N-3019E)

Categories of interest: IRBM Deployment, Longe Range Aircraft

11. Tallin (5926N-2444E) - Vyborg (6043N-2844E) Rail Line (Gulf of Finland)

Leningrad (5955N-3020E)
Kronshtadt (5959N-2947E)

Categories of interest: IRBM Deployment, Submarine Launch

12. Berezovka (5112N-4557E)

Category of interest: ICBM Deployment

13. Moscow Complex (5545N-3735E)

Shelkovo
Ramenskoye
Khimki
Fili

Category of interest: Long Range Aircraft, Missile Production, Missile Research and Development

14. Dnepropetrovsk (4828N-3500E)

Category of interest: Missile Production

15. Tyura Tam Rangehead (4555N-6318E)

Category of interest: Missile Research & Development

16. Gorkiy (5708N-4135E0

Category of interest: Long Range Aircraft

~~HANDLE VIA TALENT~~
~~TOP SECRET CONTROL SYSTEM ONLY~~

4. *(Continued)*

~~TOP SECRET~~

Category of interest: Long Range Aircraft

17. Mozhayak (5530N-3602E)

Category of interest: Nuclear Energy

18. Tiksi (7135N-1285E)

Categories of interest: ICBM Deployment, Nuclear Energy

19. Caspian Sea Test Range

Baku (4023N-4955E)
Fort Shevchenko (4430N-5016E)
Gurev (4707N-5115E)
Krasnovodsk (4000N-5300E)
Makhachkala (4258N-4730E)

Category of interest: Missile Research & Development

20. Priluki (5035N-3224E)

Category of interest: Long Range Aircraft

21. Black Sea Coastline

Sukhumi (4300N-4101E)
Kerch (4523N-3626E)
Novorossiysk (4444N-3748E)
Odessa (4628N-3044E)
Sevastapol (4437N-3332E)
Balakalave (4430N-3335E)
Ay-Petri (4435N-3412E)
Batumi (4139N-4139E)
Yalta (4430N-3410E)
Feodosiya (4502N-3523E)
Nikolayev (4658N-3200E)
Sudak (4458N-3502E)
Karangit (4502N-3558E)

~~HANDLE VIA TALENT CONTROL SYSTEM ONLY~~

~~TOP SECRET~~

4. *(Continued)*

~~TOP SECRET~~

Categories of interest: IRBM Deployment, Submarine Launch, Anti-ICBM Research and Development

22. Kluyuchi Impact Area

Uka (5749N-1620E)
Khutor (5309N-16205E)
~~Petropavlovsk (5300N-15840E)~~
Peschanny (5750N-16205E)

Categories of interest: Missile Research & Development, Submarine Launch

23. Baranovichi (5307N-2602E)

Category of interest: Long Range Aircraft

24. Anadyr Area

Ugolni Kopi (6430N-17738E)
Anadyr/Leninka (6445N-17910E)
Ugol'nyy (6225N-17910E)
Bukhta Ugolnaya (6258N-17917E)

Categories of interest: Missile Deployment, Nuclear Energy, Long Range Aircraft

25. Kapustin Yar (4835N-4545E) - Vladimirovka (4818N-4610E)
Rangehead Zone 9, Zone 10

Categories of interest: Missile Troop Training, Missile Research & Development

26. Mukachavo (4826N-2245E)
Uzhgorod (4838N-2217E)
Svalyava (4835N-2300E)
Lvov (4950N-2400E)
Stryy (4915N-2352E)
Delyatin (4828N-2438E)

~~HANDLE VIA TALENT~~
~~CONTROL SYSTEM ONLY~~

~~TOP SECRET~~

~~CONTROL SYSTEM ONLY~~

4. *(Continued)*

~~TOP SECRET~~

Categories of interest: IRBM Deployment, Long Range Aircraft

27. Kalingrad/Baltiysk (5443N-2030E) - Riga (5657N-2405E) Tallin (5926N-2444E) Rail Line

Leipaja (5632N-2100E)
Ventspile (5724N-2134E)
Dago Island (asta) (5855N-2240E)
Taurage (5515N-2218E)
Paplaka (5626N-2127E)
Klaypeda (5543N-2109E)

Categories of interest: ICBM Deployment, IRBM Deployment, Submarine Launch

28. Vinnitaa (4914N-2828E) - Kharkov (4958N-3615E) Rail Line

Borispol (5020N-3057E)
Mirispol (4958N-3057E)
Poltava (4936N-3434E)
Kiev (5037N-3032E)
Uzin/Chepalivka (4950N-3025E)

Categories of interest: IRBM Deployment, Long Range Aircraft

29. Malaya Sazanka/Ukraina (5114N-12804E)

Categories of interest: Long Range Aircraft Nuclear Energy

30. Sukhumi (4242E-4102E) - Dzhulfa (3854N-4538E)

Category of interest: IRBM Search

31. Ulyanovsk (5420N-4824E) - Saransk (5411N-4512E) - Murom (5536N-4202E) Rail Line

Arzamas (5523N-4305E)
Shatki (5511N-4408E)
Tashino (5452N-4349E)

~~HANDLE VIA TALENT CONTROL SYSTEM ONLY~~

~~TOP SECRET~~

4. *(Continued)*

~~TOP SECRET~~

Categories of interest: ICBM Deployment, Nuclear Energy

32. Sary Shagan (4610N-7355E)

1050 N. M. Impact Area (4617N-7201E)
950 N. M. Impact Area (4653N-6936E)
Sary Shagan Base Area (4610N-7335E)
~~Sary Shagan Test Area Installations~~
Range Staff Headquarters (4617N-7055E)
Vladimiravka Range Outstation (4654N-7047E)
Zone B (4550N-7230E)
Zone A (4617N-7300E)
Zone C (4530N-7250E)
Suspect Zone (4652N-7215E)
Suspect Area (4530N-7225E)

Categories of interest: ICBM Deployment, Anti-ICBM Research and Development, ABM Missile, Nuclear Energy.

~~HANDLE VIA TALENT CONTROL SYSTEM ONLY~~

~~TOP SECRET~~

~~CONTROL SYSTEM ONLY~~

5. James Q. Reber, Memorandum of Agreement, "Procedures for the Handling of T-KH Materials," 22 August 1960

~~HANDLE VIA TALENT KEYHOLE CHANNELS ONLY~~ ~~TOP SECRET~~

22 August 1960

MEMORANDUM OF AGREEMENT

SUBJECT: Procedures for the Handling of T-KH Materials

1. The following agreements were reached today by Army, Navy, Air Force, and PIC representatives.

2. It is recommended that all of the duplicate materials from T-KH on this shot be developed on a twenty-four hour basis. This is recommended not only because of the intelligence urgency at this particular time, but also because this material is new and photographic interpretation problems and procedures will require attention in every center where this material will be handled. It is recommended that after the reproduction of the materials for joint use in PIC subsequent materials should be reproduced in the following order of priority:

SAC	1 DP	1 DN
AFIC	1 DP	
Navy	1 DP	
ATIC	1 DP	1 DN
Army	1 DP	
AFIC	1 DN	
ACIC	1 DN	
Navy	1 DN	
AFIC	1 DP	
Army	1 DN	

~~TOP SECRET~~ ~~HANDLE VIA TALENT KEYHOLE CHANNELS ONLY~~

5. *(Continued)*

~~HANDLE VIA TALENT-KEYHOLE CHANNELS ONLY~~ ~~TOP SECRET~~ [redacted]

3. If it is necessary for budget reasons to cut back to eight hour production as soon as possible, we recommend that the cutback point for twenty-four hour processing be after the first eight listed duplicate materials.

4. The first three listed items will be picked up immediately upon completion. The SAC copies will be delivered to Westover for handling by the Air Force. The AFIC copy will be delivered to Washington. The next five copies listed will be picked up by Operations and delivered to Washington. The last four items will be picked up in the third pickup. PIC will be the recipient, will notify addressee agencies upon arrival of the material, and will be responsible for transshipment.

5. The memorandum for handling of TALENT photography in preliminary phases ([redacted]) was reviewed for applicability to the T-KH.

6. Only authorized personnel from each participating agency will be permitted in the area for OAK preparation. It was agreed that an ACIC representative would be allowed to view the material in order to provide advance planning to ATIC.

7. Concerning operations at PIC it was agreed that an OAK report would be issued daily. A negative report will be issued if no OAK report is available. It was further agreed that there would be no briefings, telephone calls, or written memoranda from PIC bearing upon the substance of the interpretation until the OAK for the day is produced. Inquiries into PIC on previous days' OAKs should be avoided. It was further agreed that in accordance with the procedure for handling of ODES PIC would be responsible for transmitting the OAK report through the Operations channel to SAC. PIC will cut the tape and handle transmission directly with Project Communications. ([redacted] to check with [redacted] on slug for transmission to SAC with necessary coordination with [redacted]). It is agreed at this time that there will not be transmission of OAK to Theater Commands because of the sensitivity of the material and the fact that secure arrangements have not yet been established for T-KH material overseas.

8. It is agreed that the Director/PIC in consultation with other representatives participating in the OAK will select enlargements for presentation to USIB. Briefing boards made with these materials

~~HANDLE VIA TALENT-KEYHOLE CHANNELS ONLY~~

~~TOP SECRET~~

5. *(Continued)*

~~HANDLE VIA TALENT-KEYHOLE CHANNELS ONLY~~ ~~TOP SECRET~~ [redacted]

will be disseminated to the Army, Navy, Air Force, and JCS [redacted] in view-graph form (reproduction of the briefing board) in the same fashion as was accomplished in the handling of [redacted] photography. The fact that there is greater coverage in the satellite photography has no effect upon the number of briefing boards to be prepared. This is left to the judgment of the Director/PIC in keeping with the time schedule of readout and problems of presentation. When the Director/PIC has materials available he will inform the DDI who is responsible for convening the USIB so that the presentation may be made.

9. It was agreed that each agency would examine its departmental views with regard to the need for establishing security arrangements to meet the needs of overseas commands. To be discussed at an early date.

James Q. Reber

JAMES Q. REBER
Chairman
Committee on Overhead Reconnaissance

~~HANDLE VIA TALENT-KEYHOLE CHANNELS ONLY~~

~~TOP SECRET~~

5. *(Continued)*

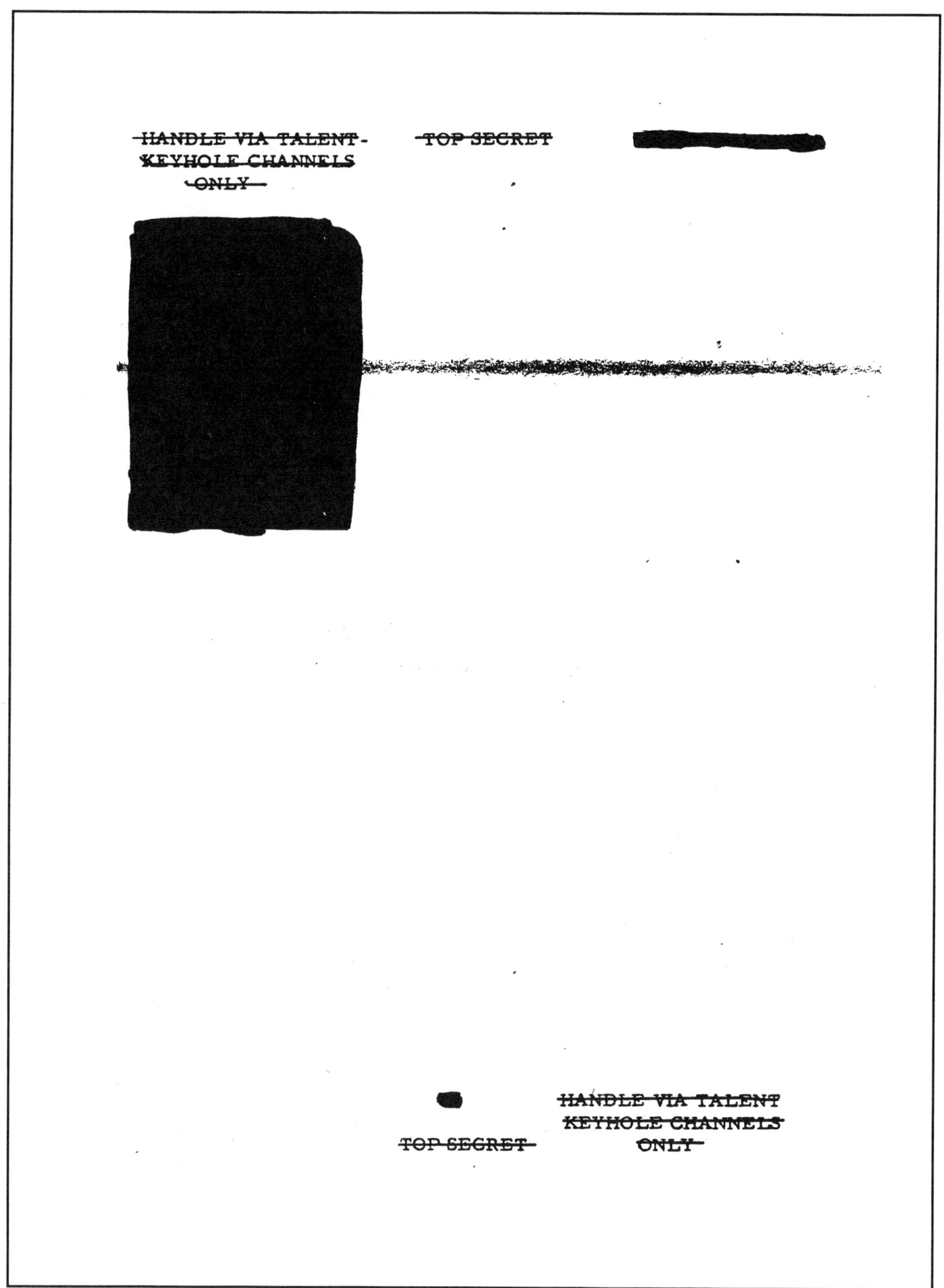
HANDLE VIA TALENT-KEYHOLE CHANNELS ONLY

TOP SECRET

TOP SECRET

HANDLE VIA TALENT KEYHOLE CHANNELS ONLY

6. James Q. Reber, Memorandum for Brig. Gen. Andrew J. Goodpaster, "Categories of Billets Planned for T-KH Certification," 26 August 1960

~~HANDLE VIA TALENT-KEYHOLE CHANNELS ONLY~~ ~~TOP SECRET~~

26 August 1960

MEMORANDUM FOR: Brig. Gen. Andrew J. Goodpaster

SUBJECT: Categories of Billets Planned for T-KH Certification

1. In January of 1959 USIB agencies participating in the TALENT Program established arrangements for the secure handling and control of satellite reconnaissance materials and information when available. Included in this planning were estimates based upon our experience in handling the larger photography of the U-2, the anticipated readout potential of the smaller scale satellite photography, the new problems involved in the handling of it from a photogrammetric point of view, and the nature of the intelligence anticipated. All estimates submitted to CIA and approved by CIA originated directly with the senior intelligence chief in each participating agency and included a justification and rationale.

2. For the purpose of clarification of the gross estimate-figures for the planned use of TALENT-KEYHOLE material (the photographic product of CORONA) each agency's estimate has been broken down into three categories as follows:

Category I   Senior Officials in the Participating Departments and Military Services

This category in each case lists the title of the office and in some cases the name of the person.

Category II   Substantive Intelligence Analysis and Estimators

This category includes substantive experts (not photo technicians) in the various agencies who must take the information prepared by photo

~~HANDLE VIA TALENT-KEYHOLE CHANNELS ONLY~~

~~TOP SECRET~~

6. *(Continued)*

~~HANDLE VIA TALENT-KEYHOLE CHANNELS ONLY~~ ~~TOP SECRET~~ [redacted]

interpretors, correlate it with other sources, and prepare reports and estimates for the senior intelligence officer or for his reporting to superiors, or contribution to the estimates produced by the United States Intelligence Board.

Category III Photo Interpretation

This category includes the technical photo interpretors, administrative, communication, and logistics support personnel handling the TALENT-KEYHOLE materials in the centers of the various agencies considered in this report.

James Q. Reber
JAMES Q. REBER
TALENT Control Officer, CIA

[redacted] ~~TOP SECRET~~ ~~HANDLE VIA TALENT-KEYHOLE CHANNELS ONLY~~

6. *(Continued)*

~~HANDLE VIA TALENT-KEYHOLE CHANNELS ONLY~~ ~~TOP SECRET~~ [redacted]

Central Intelligence Agency

I.	Senior Officials	22
	DCI	Mr. Allen W. Dulles
	DDCI	Gen. C. P. Cabell
	Inspector General	Mr. Lyman B. Kirkpatrick
	SA/DCI	Mr. John S. Earman
	DD/I	Mr. Robert Amory, Jr.
	A/DDI	Mr. William A. Tidwell
	DD/P	Mr. Richard M. Bissell
	A/DDP/A	Mr. C. Tracy Barnes
	ADD/S	Mr. H. Gates Lloyd
	Comptroller	Mr. Edward R. Saunders
	C/Budget/Compt.	Mr. Charles W. Mason
	General Counsel	Mr. Lawrence R. Houston
	Legis. Liaison/OGC	Mr. John S. Warner
	D/Communications	Gen. Harold M. McClelland
	D/Security	Col. Sheffield Edwards
	D/Personnel	Mr. Emmett D. Echols
	AD/OCI	Mr. Huntington Sheldon
	AD/ORR	Mr. Otto E. Guthe
	AD/OCR	Mr. Paul A. Borel
	AD/OSI	Dr. Herbert Scoville, Jr.
	AD/ONE	Dr. Sherman Kent
	D/PIC	Mr. Arthur C. Lundahl
II.	Substantive Intelligence Analysts and Estimators	100
III.	Photo Interpretation	164
	TOTAL	286

[redacted] ~~TOP SECRET~~ ~~HANDLE VIA TALENT-KEYHOLE CHANNELS ONLY~~

6. *(Continued)*

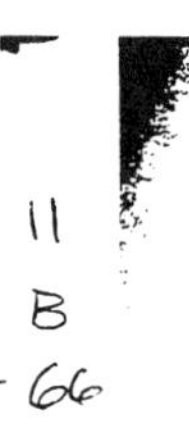

~~TOP SECRET~~

~~HANDLE VIA TALENT KEYHOLE CHANNEL ONLY~~

Office of the Secretary of Defense

I. Senior Officials 13 ~~12~~

Honorable Thomas S. Gates,
Secretary of Defense

Honorable James H. Douglas, Jr.,
Deputy Secretary of Defense

Honorable Herbert F. York,
Director of Research and Engineering

Gen. G. G. Erskine, Retired, Special Assistant
to the Secretary for Special Operations

Lt. Gen. Donald N. Yates, Deputy Director
Research and Engineering

Lt. Gen. William P. Ennis, Director Weapons
System Evaluation Group

Brig. Gen. Austin W. Betts, Director ARPA
(Advance Research Project Agency)

Col. Edwin F. Black, Military Assistant to the
Deputy Secretary of Defense

Brig. Gen. George S. Brown, Military Assistant
to the Secretary of Defense

Brig. Gen. Edward C. Lansdale,
Deputy to General Erskine

Capt. Means Johnston, Jr., Military Assistant
to the Secretary of Defense

Brig. Gen. William T. Seawell, Military Assistant
to the Deputy Secretary of Defense

Dr. Bruce H. Billings, Deputy Director Research + Engineering — Transferred from JC

~~HANDLE VIA TALENT KEYHOLE CHANNELS ONLY~~

II. Estimators (R&D and Security) 7 ~~8~~

~~TOP SECRET~~

TOTAL 20

6. *(Continued)*

~~TOP SECRET~~ ~~HANDLE VIA TALENT KEYHOLE CHANNELS ONLY~~

Joint Chiefs of Staff

I. Senior Officials 6 ~~7~~

General Nathan F. Twining, Chairman
Joint Chiefs of Staff

Lt. Gen. Earle G. Wheeler, U.S. Army,
Director, Joint Chiefs of Staff

Maj. Gen. James F. Whisenand,
Special Assistant to the Chairman

Maj. Gen. Robert A. Breitweiser, J-2

Rear Admiral William S. Post,
Deputy J-2

Brig. Gen. James C. Sherrill, Executive
to the Chairman

~~Dr. Bruce H. Billings, Deputy Director~~
~~Research and Engineering~~ add to OSD list

II. Estimators 10 ~~8~~

TOTAL 16 ~~15~~

~~TOP SECRET~~ ~~HANDLE VIA TALENT KEYHOLE CHANNELS ONLY~~

6. *(Continued)*

~~HANDLE VIA TALENT-KEYHOLE CHANNELS ONLY~~ ~~TOP SECRET~~

Strategic Air Command

II.	Substantive Intelligence Analysts and Estimators	73
III.	Photo Interpretation, Cartographic, Targets, Charts	127
	TOTAL	200

~~HANDLE VIA TALENT-KEYHOLE CHANNELS ONLY~~

~~TOP SECRET~~

6. *(Continued)*

~~TOP SECRET~~

United States Army

~~HANDLE VIA TALENT KEYHOLE CHANNELS ONLY~~

I.	Senior Service and Departmental	12
	Secretary of Army	
	Director of Research and Development, Army	
	Chief of Staff	
	Vice Chief of Staff	
	Secretary of General Staff	
	Comptroller of the Army	
	Director of Army Budget	
	Deputy Chief of Staff for Military Operations	
	Chief Research and Development	
	Deputy Chief Research and Development	
	Deputy Chief of Staff for Personnel	
	Deputy Chief of Staff for Logistics	
II.	Substantive Intelligence Analysts and Estimators	127
III.	Photo Interpretation	150
	TOTAL	289

~~TOP SECRET~~

~~HANDLE VIA TALENT KEYHOLE CHANNELS ONLY~~

6. *(Continued)*

~~TOP SECRET~~ ~~HANDLE VIA TALENT KEYHOLE CHANNELS ONLY~~

United States Navy

I.	Senior Service and Departmental	10
	Secretary of the Navy	
	Naval Aide to the Secretary of the Navy	
	The Under Secretary of the Navy	
	The Assistant Secretary of the Navy for R&D	
	The Chief of Naval Operations	
	The Aide to CNO	
	The Vice Chief of Naval Operations	
	The Aide to the Vice Chief of Naval Operations	
	The Deputy Chief of Naval Operations for Air	
	Deputy CNO for Development	
II.	Substantive Analysts and Estimators	47
III.	Photo Interpreters	43
	TOTAL	100

~~TOP SECRET~~ ~~HANDLE VIA TALENT KEYHOLE CHANNELS ONLY~~

6. *(Continued)*

~~TOP SECRET~~ ~~HANDLE VIA TALENT KEYHOLE CHANNELS ONLY~~

United States Air Force

I. Senior Service and Departmental		29
Secretary of the Air Force	1	
Under Secretary of the Air Force	1	
Chief of Staff of the Air Force	1	
Vice Chief of Staff	2	
Assistant Secretary for R&D	3	
Deputy Chief of Staff, Operations	9	
Deputy Chief of Staff, Developments	8	
Deputy Chief of Staff, Plans	4	
II. Substantive Intelligence Analysts and Estimators, Targets, Penetration, Aerial, Space (Technical)		~~250~~ 177
USAF	20	
ATIC	22	
ACIC	97	
AFIC	38	
III. Photo Interpretation, Cartographic and Air Targets, Charts		~~400~~ 273
USAF	51	
ATIC	28	
ACIC	132	
AFIC	62	
	TOTAL	479

~~TOP SECRET~~ ~~HANDLE VIA TALENT KEYHOLE CHANNELS ONLY~~

6. *(Continued)*

~~TOP SECRET~~ ~~HANDLE VIA TALENT KEYHOLE CHANNELS ONLY~~

National Security Agency

I. Senior Service and Departmental	3
Lt. General John A. Samford, USAF, Director NSA	
Louis W. Tordella, Deputy Director NSA	
Frank B. Rowlett, Special Assistant to Director NSA	
II. Substantive Intelligence Analysts and Estimators	7
TOTAL	10

~~TOP SECRET~~ ~~HANDLE VIA TALENT KEYHOLE CHANNELS ONLY~~

~~KEYHOLE CHANNELS ONLY~~

6. *(Continued)*

~~HANDLE VIA [redacted] ALENT KEYHOLE CHANNELS ONLY~~ ~~TOP SECRET~~

Department of State

I. Senior Service and Departmental

Secretary of State	1
Under Secretaries of State	2
Director of Intelligence and and Research	1
Deputy Director of Intelligence and Research	1
TOTAL	5

~~TOP SECRET~~ ~~HANDLE VIA TALENT KEYHOLE CHANNELS ONLY~~

~~KEYHOLE CHANNELS ONLY~~

6. *(Continued)*

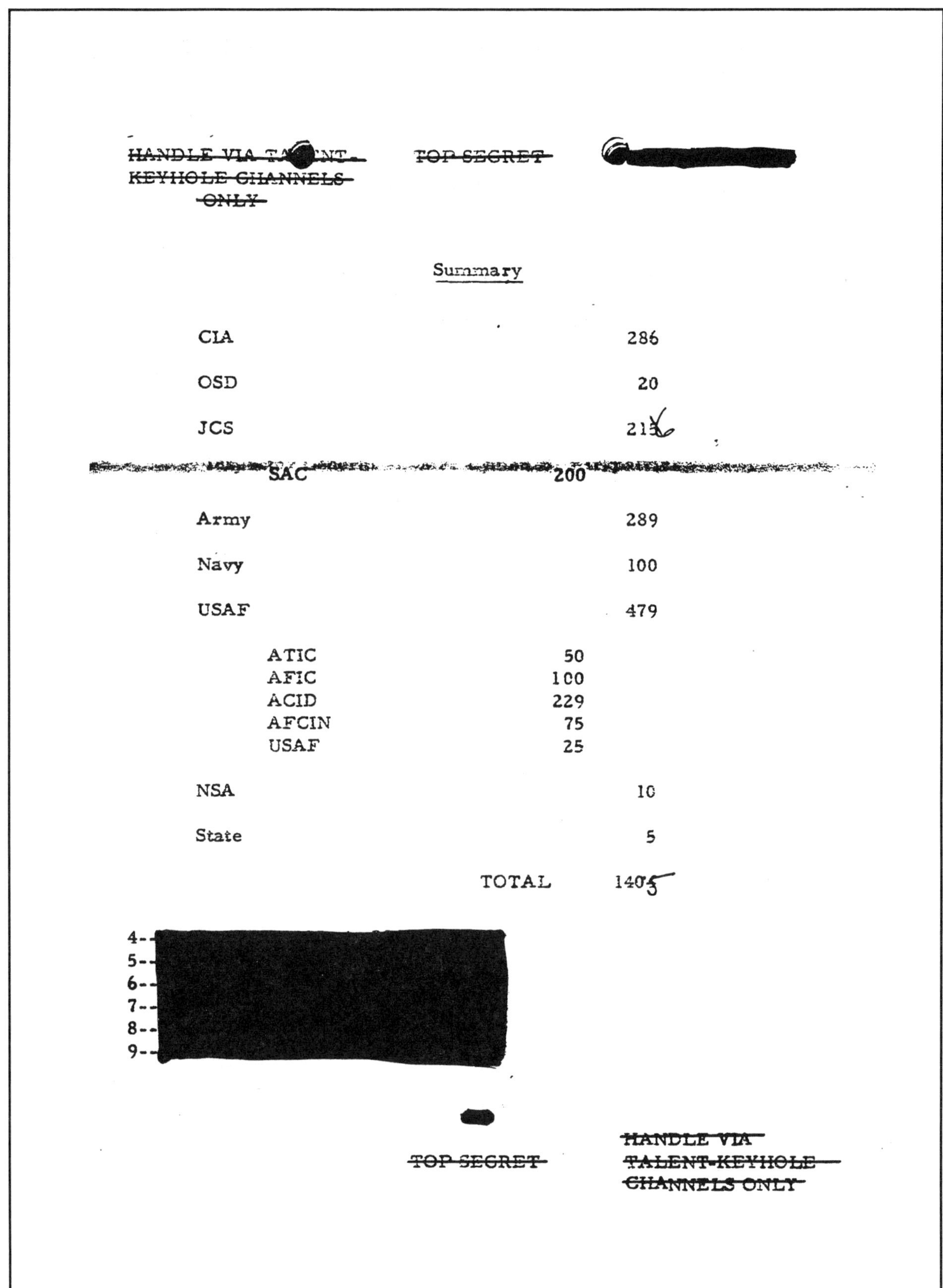

~~HANDLE VIA TALENT-KEYHOLE CHANNELS ONLY~~ ~~TOP SECRET~~

Summary

CIA		286
OSD		20
JCS		215
~~SAC~~	~~200~~	
Army		289
Navy		100
USAF		479
ATIC	50	
AFIC	100	
ACID	229	
AFCIN	75	
USAF	25	
NSA		10
State		5
	TOTAL	1405

4--
5--
6--
7--
8--
9--

~~TOP SECRET~~ ~~HANDLE VIA TALENT-KEYHOLE CHANNELS ONLY~~

7. Dwight D. Eisenhower, Memorandum for the Secretary of State, et al., 26 August 1960

~~TOP SECRET~~

~~HANDLE VIA TALENT-~~
~~KEYHOLE SYSTEM ONLY~~

THE WHITE HOUSE
Washington

August 26, 1960

MEMORANDUM FOR

The Secretary of State
The Secretary of Defense*
The Attorney General**
The Chairman, Atomic Energy Commission
The Director of Central Intelligence

I hereby direct that the products of satellite reconnaissance, and information of the fact of such reconnaissance revealed by the product, shall be given strict security handling under the provisions of a special security control system approved by me. I hereby approve the TALENT-KEYHOLE Security Control System for this purpose.

Within your agency, you shall be personally responsible for the selection of those personnel who will have access to the foregoing information and for determining the scope of that access. Access is to be on a "must know" basis related to major national security needs. A list of those selected shall be furnished to the Director of Central Intelligence, who will maintain and review the control roster. When they are indoctrinated, they shall be informed of my specific direction to them that the provisions of the special Security Control System I have approved be strictly complied with, including the prohibition upon them of imparting any information within this system to any person not specifically known to them to be on the list of those authorized to receive this material. The responsibility for the selection of personnel may be delegated only to the senior intelligence chief or chiefs within the agencies serving as members of the U. S. Intelligence Board.

The Director of Central Intelligence, in consultation with the U. S. Intelligence Board, will be responsible to me for determining all questions involved in the continued protection and control of the foregoing material and information, including the development of a common understanding as to the meaning of the term " 'must know' basis related to major national security needs, " and a broad consensus as to the numbers of personnel in each agency comprehended by this term.

signed Dwight D. Eisenhower

*For Department of Defense including OSD, JCS, Army, Navy, Air Force, and NSA
**For Director, FBI

~~HANDLE VIA TALENT-~~
~~KEYHOLE SYSTEM ONLY~~

~~TOP SECRET~~

8. James Q. Reber, Memorandum for US Intelligence Board Members, "TALENT-KEYHOLE Certification Plans," 27 August 1960

~~HANDLE VIA TALENT KEYHOLE CHANNELS ONLY~~ ~~TOP SECRET~~

27 August 1960

MEMORANDUM FOR: ~~General Graves B. Erskine, OSD~~
Major General John Willems, Army
Rear Admiral Laurence H. Frost, Navy
Major General James H. Walsh, US Air Force
Brigadier General Robert A. Breitweiser, JCS
Lt. General John A. Samford, NSA

SUBJECT: TALENT-KEYHOLE Certification Plans

1. In the course of last weeks discussions with General Goodpaster I was requested to submit a breakdown of the planned billets for the handling of TALENT-KEYHOLE material, the form of which breakdown will be readily evident from the attached paper. These figures and the names of positions where indicated are details in the custody of TALENT Control Officers in the representative organizations except the State Department. At the USIB meeting in the late afternoon of August 26 the members indicated they would like a copy of this paper. Accordingly, it is sent to you in pursuance of that request.

2. It is understood that by virtue of the President's Directive, the oral instructions of General Goodpaster, and the guideline indicated by the Director of Central Intelligence at the USIB meeting, it is now proper to proceed with the indoctrination of the billets as planned subject to the direction of the senior intelligence chief under the terms of the President's Directive, or in the case of the military services subject to further direction by the Secretary of Defense.

3. After the USIB meeting the necessary parties were informed in order that the duplicate film hitherto impounded would be released to the assigned recipients and the OAK report, the preliminary PI, would be disseminated as it became available through T-KH channels to T-KH-cleared people in the various agencies.

James Q. Reber
JAMES Q. REBER
TALENT Control Officer, CIA

Attachment:

~~TOP SECRET~~ ~~HANDLE VIA TALENT-KEYHOLE CHANNELS ONLY~~

8. *(Continued)*

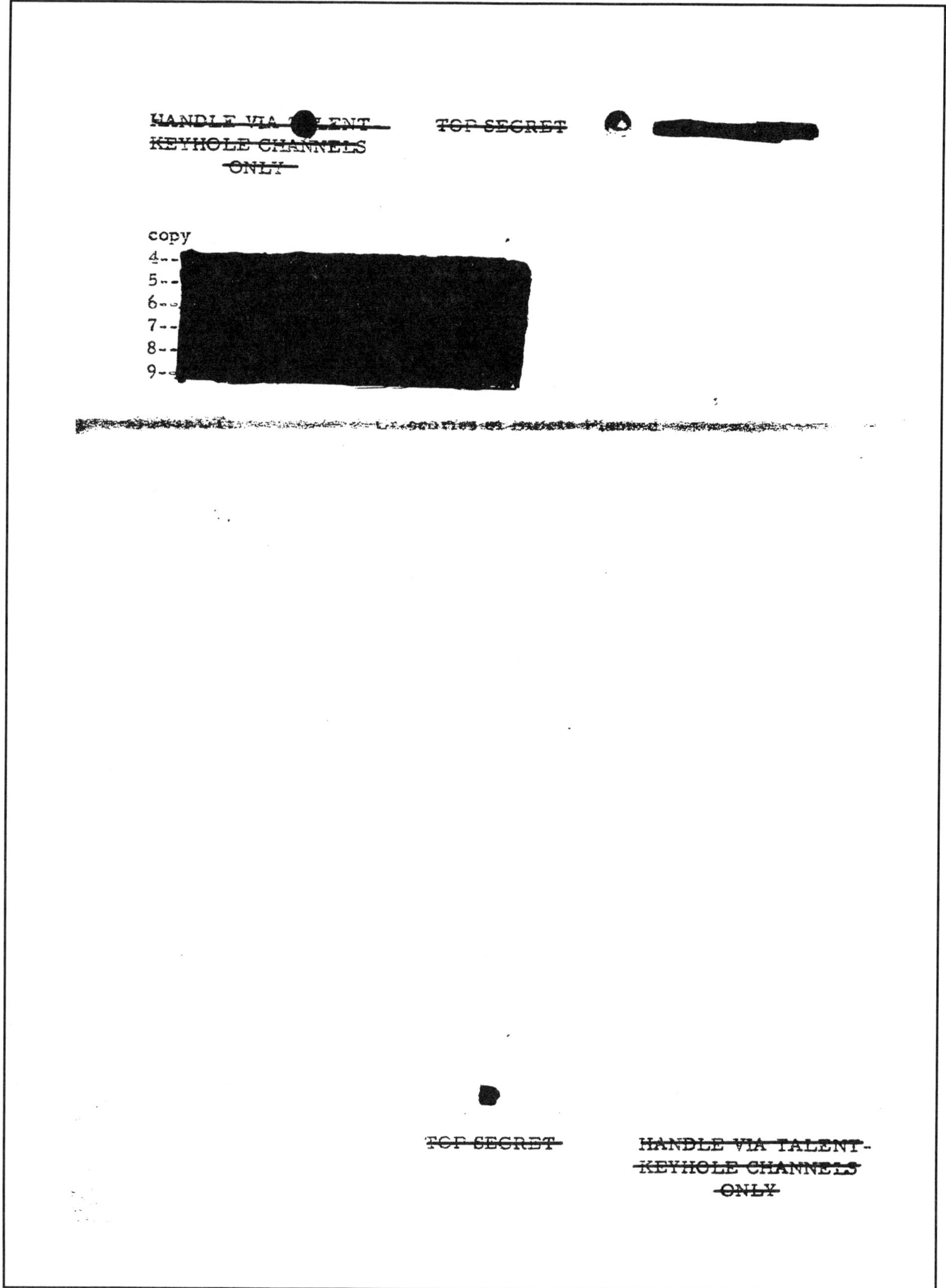
~~HANDLE VIA TALENT KEYHOLE CHANNELS ONLY~~ ~~TOP SECRET~~

copy
4--
5--
6--
7--
8--
9--

~~TOP SECRET~~ ~~HANDLE VIA TALENT-KEYHOLE CHANNELS ONLY~~

9. James Q. Reber, "Minutes of COMOR Meeting," 13 September 1960

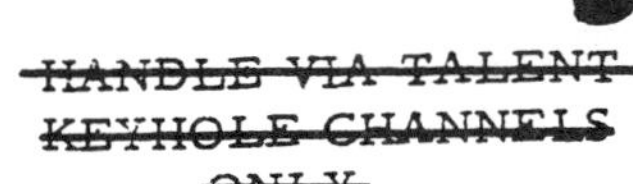

COMMITTEE ON OVERHEAD RECONNAISSANCE

Minutes of Meeting Held in Room 429
[redacted] Building, Central Intelligence Agency
at 1:00 p.m., 13 September 1960

PRESIDING

James Q. Reber
Chairman

MEMBERS PRESENT

CONSULTANTS PRESENT

Requirements for T-KH Duplicate Materials

1. The Committee took note of the experience on Mission 9009 calling attention to the fact that enlargement is required before exploitation takes place and that for this reason quality of reproduction of the duplicate materials is imperative. The Committee recommends that in the future processing of duplicate materials of T-KH photography the greatest emphasis should be placed upon quality and that insofar as quality reproduction takes a longer time such delays would have to be sustained by the consumer. It is recommended that operations make all duplicate positive materials from the original negative.

2. The COMOR requests the following schedule of reproduction be followed and that insofar as feasible the materials when accomplished be moved to their destinations in groups as indicated below:

~~TOP SECRET~~ ~~HANDLE VIA TALENT-KEYHOLE CHANNELS ONLY~~

9. *(Continued)*

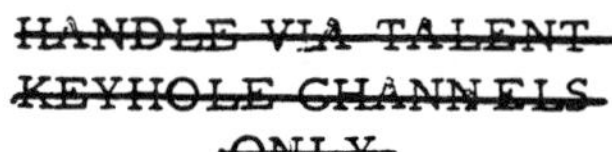

TOP SECRET

HANDLE VIA TALENT KEYHOLE CHANNELS ONLY

GROUP I	1 DN for PIC for joint interpretation 3 DPs for PIC for joint interpretation
GROUP II	1 DP & 1 DN for SAC 1 DP for AFIC
GROUP III	5 DPs for PIC for joint interpretation
GROUP IV	1 DP for Navy 1 DP for ATIC 1 DN for ATIC 1 DP for Army
GROUP V	1 DN for AFIC 1 DN for ACIC 1 DN for Navy 1 DP for ACIC 1 DN for Army
GROUP VI	1 DP for SAC 1 DP for ATIC 1 DP for ACIC 1 DP for Army

It is recommended that all of the foregoing duplicate materials be produced on a 24-hour basis except Group VI which may be on an 8-hour schedule.

Requirements for Future T-KH Collection

3. [redacted] of PIC/CIA presented a briefing showing the highest priority targets for the USSR in an overlay on the map of Russia along with another overlay showing the clustering of high and other lower targets and an overlay indicating the probable coverage in Mission 9009. Normally the last named chart will be distributed with the Mission coverage index produced by PIC/CIA. It was agreed that [redacted] (PIC) and [redacted] (AFCIN) would consult to get AFCIN assistance in the reproduction of these overlays for the benefit of the members in their requirements planning.

HANDLE VIA TALENT-KEYHOLE CHANNELS ONLY

TOP SECRET

9. *(Continued)*

~~TOP SECRET~~

~~HANDLE VIA TALENT-KEYHOLE CHANNELS ONLY~~

4. It was agreed that the principal target for planning multi-day orbit should be Polyarnyy Ural. If however this was confirmed as covered in the last single orbit series shortly to be delivered, then the principal highest priority target for planning purposes should be Ust Ukhta.

James Q. Reber

JAMES Q. REBER
Chairman
Committee on Overhead Reconnaissance

Copy 2--
3, 4, 5 --
6, 7 --
8--
9--
10--

3

~~HANDLE VIA TALENT-KEYHOLE CHANNELS ONLY~~

~~TOP SECRET~~

9. *(Continued)*

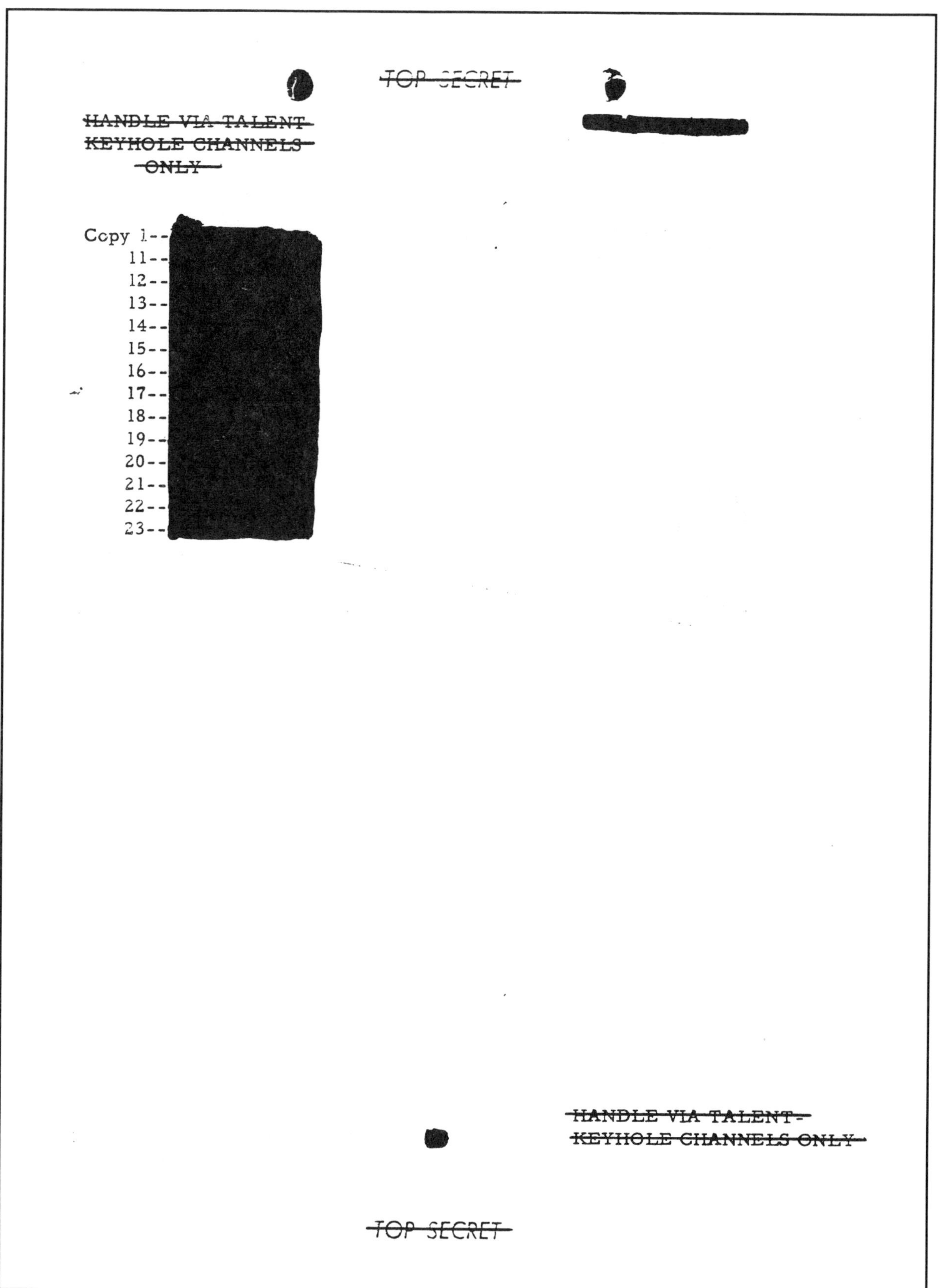

~~TOP SECRET~~

~~HANDLE VIA TALENT-KEYHOLE CHANNELS ONLY~~

Copy 1--
11--
12--
13--
14--
15--
16--
17--
18--
19--
20--
21--
22--
23--

~~HANDLE VIA TALENT-KEYHOLE CHANNELS ONLY~~

~~TOP SECRET~~

10. James Q. Reber, Memorandum for US Intelligence Board, "Proposed Expansion of Billets for the Exploitation and Use of TALENT-KEYHOLE Materials and Information," 14 September 1960

~~HANDLE VIA TALENT KEYHOLE CHANNELS ONLY~~ ~~TOP SECRET~~

14 September 1960

MEMORANDUM FOR: The United States Intelligence Board

SUBJECT : Proposed Expansion of Billets for the Exploitation and Use of TALENT-KEYHOLE Materials and Information

1. The OSD, JCS, Army, Navy, Air Force, NSA, and CIA members of the Committee on Overhead Reconnaissance have closely examined the problems incident to the exploitation of photography and the use of the information available from Mission 9009 satellite photography. They submit for approval of the USIB their conclusion as to the meaning of this photography for U.S. National Security purposes in the attached document. Along with this document are annexes devoted to the individual agencies showing the organization units (with their functions) which need to make use of the photography or information derived therefrom and the number of billets required in that use.

2. The billet structures planned for this purpose prior to receipt of Mission 9009 are judged by the agencies to be inadequate for exploitation and use of the material. The additional billets required by agencies are summarized in Annex A with a recapitulation by agency in subsequent annexes.

3. Recommendation: That the USIB approve the attached document with its annexes.

James Q. Reber

JAMES Q. REBER
Chairman
Committee on Overhead Reconnaissance

Attachment: As stated

Copy(s) 2
3, 4
5, 6
7
8

~~TOP SECRET~~ ~~HANDLE VIA TALENT KEYHOLE CHANNELS ONLY~~

10. *(Continued)*

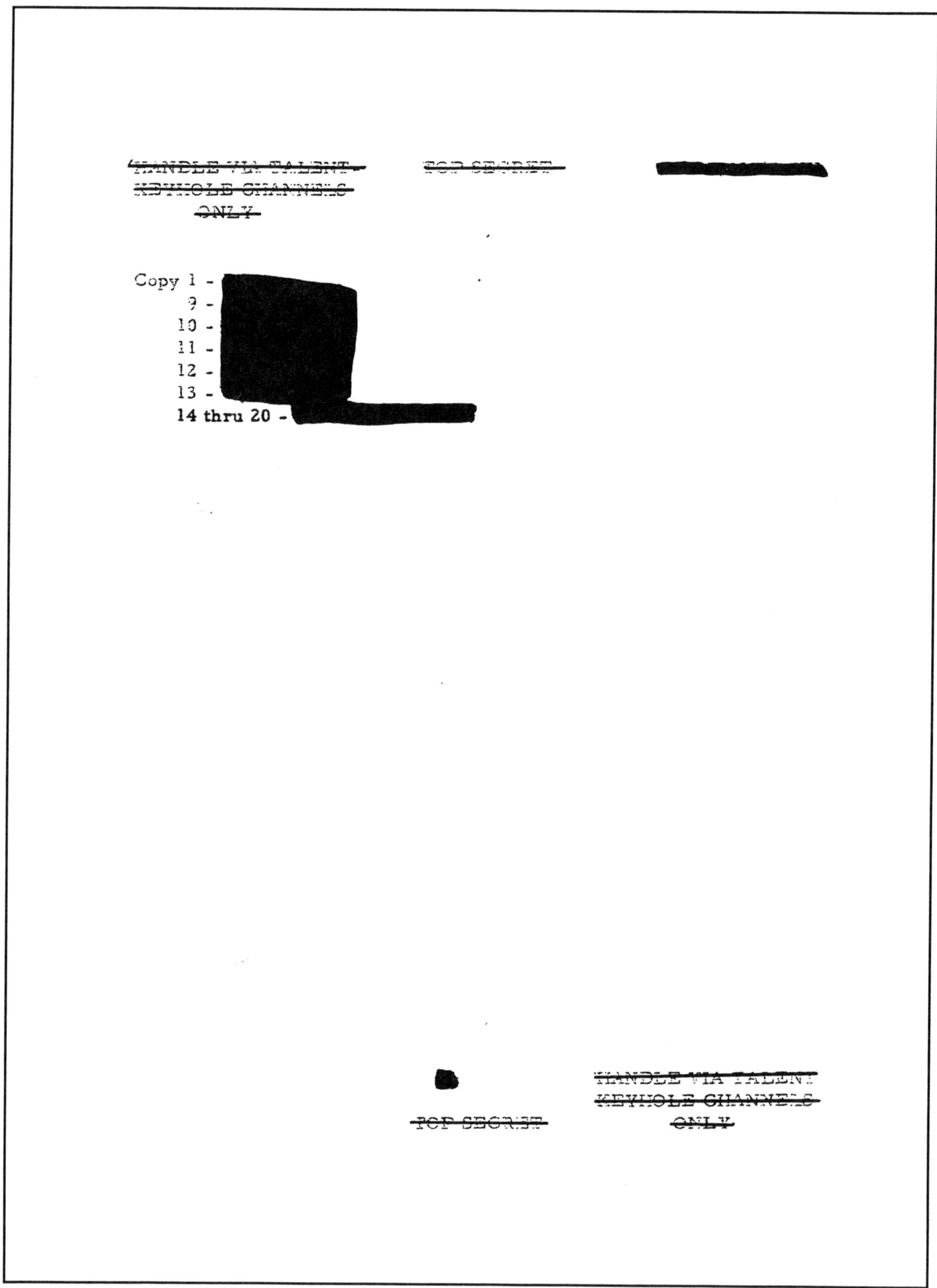

~~HANDLE VIA TALENT-KEYHOLE CHANNELS ONLY~~ ~~TOP SECRET~~

Copy 1 -
9 -
10 -
11 -
12 -
13 -
14 thru 20 -

~~TOP SECRET~~ ~~HANDLE VIA TALENT-KEYHOLE CHANNELS ONLY~~

10. *(Continued)*

~~HANDLE VIA TALENT-KEYHOLE CHANNELS ONLY~~ ~~TOP SECRET~~ Attachment to [redacted]

Proposed Expansion of Billets for the Exploitation and Use of TALENT-KEYHOLE Materials and Information

1. A Presidential Directive recently issued requires that satellite photography must be handled within the TALENT Security Control System in a separate compartment known as TALENT-KEYHOLE. It further requires that the United States Intelligence Board shall develop a broad consensus for determining those functions in the United States Government (and personnel within them) which must have access to satellite photography for National Security purposes.

2. The satellite photography from Mission 9009 is in hand and is currently being exploited and used within Washington Headquarters' intelligence agencies, and The Strategic Air Command, Aero-Space Technical Intelligence Center, Aeronautical Chart and Information Center, Army Map Service, and Navy Photographic Interpretation Center. The billet structure within these organizations was planned on an extremely limited basis in advance of the arrival of satellite photography and pending an evaluation of the exploitation potential.

3. For six years the U.S. intelligence agencies have had extensive experience with the larger scale photography from overflight held in the [redacted] and TALENT Systems. New equipment bearing upon the art of photographic interpretation has clearly expanded the quantity and quality of information derived from that photography. We have seen the extensive uses to which the material and the information derived therefrom can be put for strategic intelligence purposes, emergency war planning, intelligence purposes related to the responsibility of theater commanders, research and development requirements of the Department of Defense, and operational purposes of the military as well as intelligence operations.

4. The examination of satellite photography from Mission 9009 reveals that it can serve essentially the same purposes as the earlier TALENT photography. While it has definite limitations for technical intelligence purposes (as compared with TALENT), it serves those purposes

~~TOP SECRET~~ ~~HANDLE VIA TALENT-KEYHOLE CHANNELS ONLY~~

~~HANDLE VIA TALENT-KEYHOLE CHANNELS ONLY~~ ~~TOP SECRET~~ Attachment to [redacted]

through comparison with existing TALENT and through the use of collateral intelligence. Its vast geographic coverage clearly enhances our ability to search for guided missile sites of all sorts, and will permit the identification of installations with which we have become familiar under the TALENT Program. In addition to the uses for positive information on the USSR, it will materiallly assist in refining the targets for other collection programs and improving their potential.

5. The USIB has reviewed U.S. needs for National Security purposes in the light of the capabilities of the available satellite photography and with the purpose and injunction of the Presidential Directive uppermost in mind. The USIB considers the planned increase of TALENT-KEYHOLE billets set forth in the annexes hereto to be necessary and proper in terms of functional use of TALENT-KEYHOLE materials and information and the magnitude of the personnel forces to carry these functions at this time.

Annex A Summary of TALENT-KEYHOLE Billets for All Agencies [redacted]
B T-KH Billets [redacted]
C T-KH Billets [redacted]
D T-KH Billets [redacted]
E T-KH Billets [redacted]
F T-KH Billets [redacted]
G T-KH Billets [redacted]
H T-KH Billets [redacted]

~~TOP SECRET~~ ~~HANDLE VIA TALENT-KEYHOLE CHANNELS ONLY~~

11. James Q. Reber, Memorandum for US Intelligence Board, "Amendment to 'Proposed Expansion of Billets for the Exploitation and Use of TALENT-KEYHOLE Materials and Information,'" 14 October 1960

~~HANDLE VIA TALENT-KEYHOLE CHANNELS ONLY~~

14 October 1960

MEMORANDUM FOR: United States Intelligence Board

SUBJECT: Amendment to "Proposed Expansion of Billets for the Exploitation and Use of TALENT-KEYHOLE Materials and Information."

REFERENCE: [redacted]

1. On September 14, 1960 the Committee on Overhead Reconnaissance submitted to the United States Intelligence Board a memorandum with an attachment on Page 3 explaining the needs of the U.S. Government for the use of those materials along with Annexes A through H defining each participating agency's functional responsibilities and need for T-KH-certified personnel.

2. At that meeting General Erskine stated that he would not be able to approve those billets pertaining to the Department of Defense until he had secured approval of the Secretary of Defense. Also at that meeting the Acting Chairman, General Cabell, urged upon the members that further examination of the billet needs be undertaken with severest scrutiny from the point of view of functional need eliminating wherever possible those now included by virtue of position but who did not have a "must know" requirement for access to the material.

3. The Chairman of COMOR has received in one copy a memorandum from [redacted] the JCS/OSD Member of COMOR, which states that the Secretary of Defense has now approved revised billet schedules of the components of the Department of Defense. Time has not permitted a retyping and distribution of that document. However, each Department of Defense component has its own paper in this regard and the summary of all participating agencies' billet needs are set forth

~~HANDLE VIA TALENT-KEYHOLE CHANNELS ONLY~~

~~HANDLE VIA TALENT-KEYHOLE CHANNELS ONLY~~ ~~TOP SECRET~~

in Annex A to this document. The effect of the re-examination undertaken subsequent to the USIB meeting of September 20th is an overall reduction of between 20% and 25% from the billet needs set forth in the reference. In reviewing the attached paper it is suggested that special note be taken of paragraph 5.

5. Upon USIB approval of the attachment it will be submitted to the White House for comment in accordance with the request of General Goodpaster.

6. Recommendation: It is recommended that the United States Intelligence Board approve the attached paper.

James Q. Reber

JAMES Q. REBER
Chairman
Committee on Overhead Reconnaissance

Attachment A

Copy 2--
3, 4, 5---
6, 7---
8---
9---

~~TOP SECRET~~ ~~HANDLE VIA TALENT-KEYHOLE CHANNELS ONLY~~

11. *(Continued)*

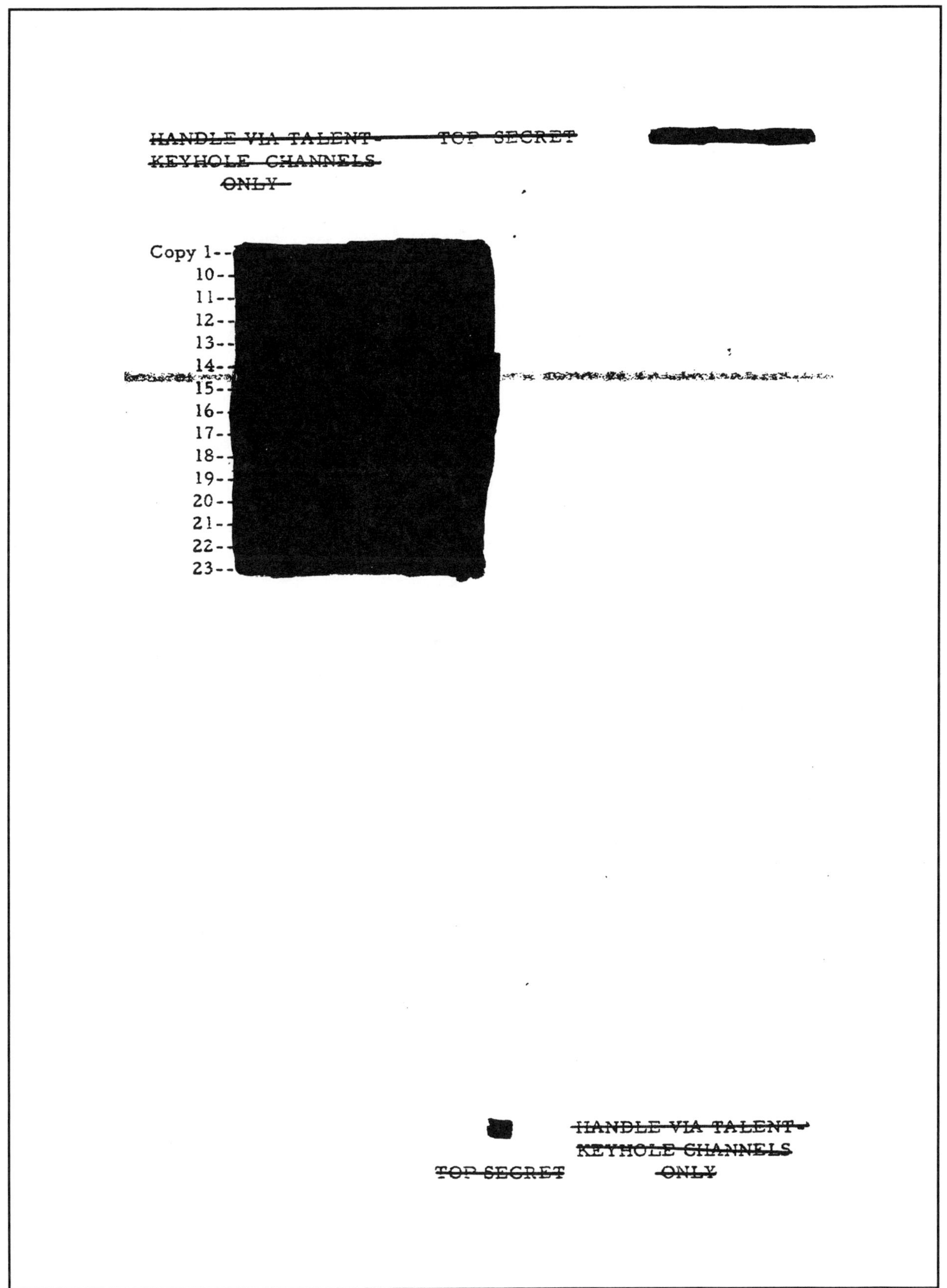
HANDLE VIA TALENT-KEYHOLE CHANNELS ONLY

TOP SECRET

Copy 1--
10--
11--
12--
13--
14--
15--
16--
17--
18--
19--
20--
21--
22--
23--

HANDLE VIA TALENT-KEYHOLE CHANNELS ONLY

TOP SECRET

11. *(Continued)*

~~HANDLE VIA TALENT-KEYHOLE CHANNELS ONLY~~ ~~TOP SECRET~~ Attachment to: [redacted]

Proposed Expansion of Billets for the Exploitation and Use of TALENT-KEYHOLE Materials and Information

1. A Presidential Directive recently issued requires that certain satellite photography must be handled within the TALENT Security Control System in a separate compartment known as TALENT-KEYHOLE. It further requires that the United States Intelligence Board shall develop a broad consensus for determining those functions in the United States Government (and personnel within them) which must have access to satellite photography for National Security purposes.

2. The satellite photography from Mission 9009 is in hand and is currently being exploited and used within Washington Headquarters' intelligence agencies, and The Strategic Air Command, Aero-Space Technical Intelligence Center, Aeronautical Chart and Information Center, Army Map Service, and Navy Photographic Interpretation Center. The billet structure within these organizations was planned on an extremely limited basis in advance of the arrival of satellite photography and pending an evaluation of the exploitation potential.

3. For six years the U.S. intelligence agencies have had extensive experience with the larger scale photography from overflight held in the [redacted] and TALENT Systems. New equipment bearing upon the art of photographic interpretation has clearly expanded the quantity and quality of information derived from that photography. We have seen the extensive uses to which the material and the information derived therefrom can be put for strategic intelligence purposes, emergency war planning, intelligence purposes related to the responsibility of theater commanders, research and development requirements of the Department of Defense, and operational purposes of the military as well as intelligence operations. add An important aspect of this material is its relevance to the formulation of foreign policy decisions.

4. The examination of satellite photography from Mission 9009 reveals that it can serve essentially the same purposes as the earlier TALENT photography. While it has definite limitations for technical intelligence purposes (as compared with TALENT), it serves those purposes

~~TOP SECRET~~ [redacted] ~~HANDLE VIA TALENT-KEYHOLE CHANNELS ONLY~~

~~HANDLE VIA TALENT KEYHOLE CHANNELS ONLY~~ ~~TOP SECRET~~ Attachment to

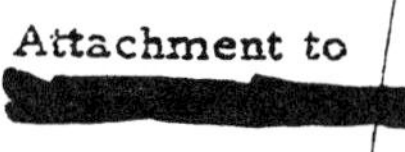

through comparison with existing TALENT and through the use of collateral intelligence. Its vast geographic coverage clearly enhances our ability to search for guided missile sites of all sorts, and will permit the identification of installations with which we have become familiar under the TALENT Program. In addition to the uses for positive information on the USSR, it will materially assist in refining the targets for other collection programs and improving their potential.

5. Insert The USIB has reviewed U.S. needs for National Security purposes in the light of the capabilities of the available satellite photography and with the purpose and injunction of the Presidential Directive uppermost in mind. The USIB considers the planned increase of TALENT-KEYHOLE billets set forth in the annexes hereto to be necessary and proper in terms of functional use of TALENT-KEYHOLE materials and information and the magnitude of the personnel forces to carry these functions at this time.

Annex A Summary of TALENT-KEYHOLE Billets for All Agencies.

~~The major exploiters of this material (the DOD & CIA) conducted an extensive review of their requirement for TKH Billets. Based upon these actions the~~

~~TOP SECRET~~ ~~HANDLE VIA TALENT-KEYHOLE CHANNELS ONLY~~

12. Allen W. Dulles, Memorandum for Brig. Gen. Andrew J. Goodpaster, "Proposed Additional Billets for the TALENT-KEYHOLE Security System," 19 October 1960

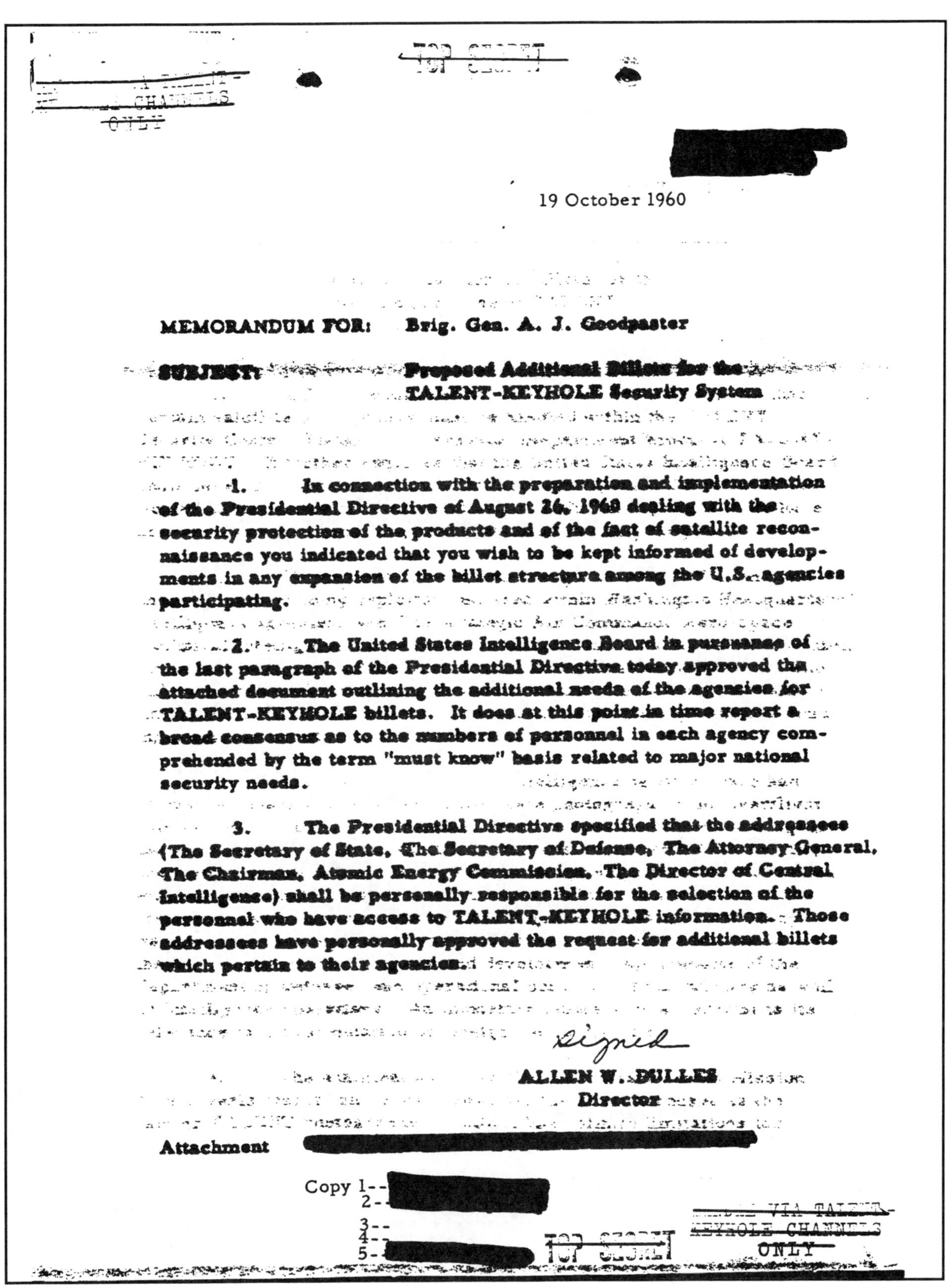

~~TOP SECRET~~

~~HANDLE VIA TALENT-KEYHOLE CHANNELS ONLY~~

19 October 1960

MEMORANDUM FOR: Brig. Gen. A. J. Goodpaster

SUBJECT: Proposed Additional Billets for the TALENT-KEYHOLE Security System

1. In connection with the preparation and implementation of the Presidential Directive of August 26, 1960 dealing with the security protection of the products and of the fact of satellite reconnaissance you indicated that you wish to be kept informed of developments in any expansion of the billet structure among the U.S. agencies participating.

2. The United States Intelligence Board in pursuance of the last paragraph of the Presidential Directive today approved the attached document outlining the additional needs of the agencies for TALENT-KEYHOLE billets. It does at this point in time report a broad consensus as to the numbers of personnel in each agency comprehended by the term "must know" basis related to major national security needs.

3. The Presidential Directive specified that the addressees (The Secretary of State, The Secretary of Defense, The Attorney General, The Chairman, Atomic Energy Commission, The Director of Central Intelligence) shall be personally responsible for the selection of the personnel who have access to TALENT-KEYHOLE information. Those addressees have personally approved the request for additional billets which pertain to their agencies.

Signed

ALLEN W. DULLES
Director

Attachment

Copy 1--
2-
3--
4--
5-

~~TOP SECRET~~

~~HANDLE VIA TALENT-KEYHOLE CHANNELS ONLY~~

TOP SECRET

Attachment to: [redacted]

Proposed Expansion of Billets for the Exploitation and Use of TALENT-KEYHOLE Materials and Information

1. A Presidential Directive recently issued requires that certain satellite photography must be handled within the TALENT Security Control System in a separate compartment known as TALENT-KEYHOLE. It further requires that the United States Intelligence Board shall develop a broad consensus for determining those functions in the United States Government (and personnel within them) which must have access to satellite photography for National Security purposes.

2. The satellite photography from Mission 9009 is in hand and is currently being exploited and used within Washington Headquarters' intelligence agencies, and The Strategic Air Command, Aero-Space Technical Intelligence Center, Aeronautical Chart and Information Center, Army Map Service, and Navy Photographic Interpretation Center. The billet structure within these organizations was planned on an extremely limited basis in advance of the arrival of satellite photography and pending an evaluation of the exploitation potential.

3. For six years the U.S. intelligence agencies have had extensive experience with the larger scale photography from overflight held in the [redacted] and TALENT Systems. New equipment bearing upon the art of photographic interpretation has clearly expanded the quantity and quality of information derived from that photography. We have seen the extensive uses to which the material and the information derived therefrom can be put for strategic intelligence purposes, emergency war planning, intelligence purposes related to the responsibility of theater commanders, research and development requirements of the Department of Defense, and operational purposes of the military as well as intelligence operations. An important aspect of this material is its relevance to the formulation of foreign policy decisions.

4. The examination of satellite photography from Mission 9009 reveals that it can serve essentially the same purposes as the earlier TALENT photography. While it has definite limitations for

TOP SECRET

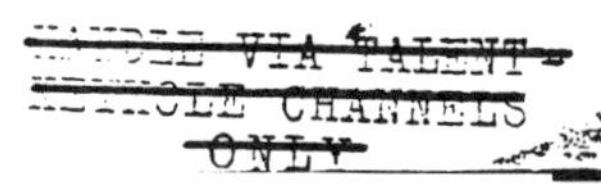

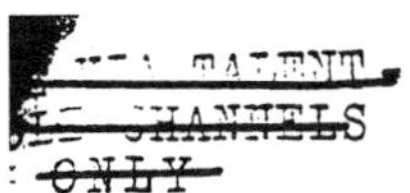

~~TOP SECRET~~

Attachment to:

technical intelligence purposes (as compared with TALENT), it serves those purposes through comparison with existing TALENT and through the use of collateral intelligence. Its vast geographic coverage clearly enhances our ability to search for guided missile sites of all sorts, and will permit the identification of installations with which we have become familiar under the TALENT Program. ~~In addition to the uses for positive information on the USSR~~, it will materially assist in refining the targets for other collection programs and improving their potential.

5. The exploiters of TALENT-KEYHOLE material conducted an extensive review of their requirement for T-KH billets. Based upon these actions, the USIB has reviewed U.S. needs for National Security purposes in the light of the capability of the available satellite photography and with the purpose and injunction of the Presidential Directive uppermost in mind. The USIB considers the planned increase of TALENT-KEYHOLE billets set forth in the annex hereto to be necessary and proper in terms of functional use of TALENT-KEYHOLE materials and information and the magnitude of the personnel forces to carry these functions at this time.

Annex A
Summary of TALENT-KEYHOLE
Billets for All Agencies.

~~TOP SECRET~~

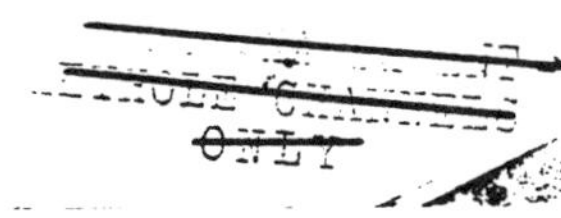

~~HANDLE VIA TALENT-KEYHOLE CHANNELS ONLY~~

~~TOP SECRET~~

ANNEX A
Attachment to: [redacted]

SUMMARY OF
UNITED STATES INTELLIGENCE BOARD
TALENT-TALENT KEYHOLE
BILLETS

AGENCY	TALENT Clearances	Present T-KH Billets	Additional T-KH Billets Needed	Total T-KH Billets Required
OSD	70	20	0	20
JCS	184**	17	111**	127**
Army	903	289	242	531
Navy	793	100	121	221
Air Force	2,712	679	890	1,569
NSA	316	10	31	41
CIA	1,115	305	113	418
State	70	5	0	5
AEC	19	3	0	3
FBI	7	3	0	3
TOTAL	6,189	1,431	1,508	2,938

**Includes 93 billets for newly established Joint Strategic Target Planning (JSTP)

~~TOP SECRET~~

~~HANDLE VIA TALENT-KEYHOLE CHANNELS~~

Part III

NPIC Products and Other Reports

Part III: NPIC Products and Other Reports

Modern day imagery analysis dates back to the development of aerial reconnaissance during World Wars I and II. The CIA's ability to process and interpret photographs advanced rapidly in its new Photo-Intelligence Division (PID) with the advent of the secret high-altitude U-2 aircraft in the 1950s. By 1958, with some additional Army and Navy photographic support, CIA expanded PID into the Photographic Intelligence Center (PIC). In January 1961, DCI Dulles, consolidated all US photographic interpretation into a single community organization, the National Photographic Interpretation Center (NPIC). NPIC proved invaluable during the Cuban Missile Crisis of 1962, when it provided key intelligence for the decisions of President Kennedy and his advisers.

Because PIC had the unique experience of processing and interpreting U-2 photographs, it was given a similar role after CORONA began to produce imagery in 1960. An example of CIA's pre-CORONA reporting is Document No. 13, "Visual-TALENT Coverage of the USSR in Relation to Soviet ICBM Deployment January 1959–June 1960," which the Office of Research and Reports produced in conjunction with PIC. This report succinctly summarizes how much CIA knew about the USSR from U-2 photography on the eve of CORONA's first successful mission.

With the advent of CORONA, CIA's reporting requirements surged to keep up with the growing amount of satellite imagery. Document No. 14 is the first Joint Mission Coverage Index of Eastern European targets identified by Mission 9009 in August 1960. Photographic Intelligence Reports (PIRs) described specific targets located during CORONA missions that merited further in-depth analysis. The Committee on Overhead Reconnaissance designated specific targets of interest for satellite reconnaissance. The PIRs integrated imagery from CORONA missions with earlier U-2 material and occasionally also with captured German World War II aerial reconnaissance photographs. In December 1961, for example, Document No. 16, "Uranium Ore Concentration Plant, Steiu, Rumania," drew on imagery from Mission 9009 and other CORONA overflights as well as from 1944 German Luftwaffe photography.

Several PIRs presented here demonstrate how CIA looked all over the world for varying types of targets and how the PIC, and later NPIC, skillfully brought together analysts, cartographers, artists, and modelmakers to produce succinct and accurate analysis.

A Photographic Evaluation Report (PER) was a "technical publication expressing the photo quality results of a mission of photography." NPIC primarily used PERs to enhance camera resolution for future missions, as Document No. 24, an April 1965 PER, illustrates. A Photographic Interpretation Report, or "OAK report," was a "first-phase photographic interpretation report presenting the results of the initial analysis of a new satellite photographic mission." Although OAK reports concentrated on highest priority COMOR targets, they could also cover other sites. This volume includes excerpts of three OAK reports from one KH-4A mission, which cover the Soviet Union, the Middle East, and Southeast Asia in mid-1967—a crucial period just after the Arab-Israeli war and while US combat operations were expanding in Vietnam. NPIC produced several other documents for each CORONA mission, such as Mission Control Plots (MCP) and Orbit Ephemeris Data. Although limitations of space make it impossible to include examples of these lengthy technical documents in this volume, they will be reviewed for declassification along with the rest of the CORONA material from which this collection has been compiled.

NPIC products, primarily basic working documents, were eventually incorporated into the national strategic analysis that formed the basis for National Intelligence Estimates (NIEs). This analysis drew heavily on the steady output from CORONA throughout the 1960s and into the early 1970s.

This section includes the September 1961 National Intelligence Estimate (NIE) 11-8/1-61, Document No. 15, which provided key supplemental information to CIA's earlier estimates of Soviet ICBM strength. This estimate, based primarily on CORONA imagery, offered US analysts and policymakers conclusive evidence about the strength and capabilities of the Soviet long-range ballistic missiles. The NIE answered many questions about the Soviet's strategic forces and put to rest the "Missile Gap" debate within the intelligence community. CIA previously released this estimate to the public with significant omissions for security reasons. Due to the overall downgrading of CORONA material, the Agency is now able to offer historians and other interested readers more information from this important NIE.

In addition, CORONA satellites provided increasingly important intelligence about Chinese nuclear developments in the 1960s. In late August 1964, Special National Intelligence Estimate 13-4-64, Document No. 23, provided clear evidence that the Chinese would soon obtain nuclear status. Indeed, the Chinese detonated their first nuclear device in October 1964, two months after the special estimate.

CORONA quickly proved its great value and played a major role in the intelligence revolution. The records in this part—the PIRs, PERs, OAK Reports, and NIEs—all derived their wealth of information from satellite imagery.

13. Office of Research and Reports, "Visual-TALENT Coverage of the USSR in Relation to Soviet ICBM Deployment, January 1959–June 1960," 11 July 1960

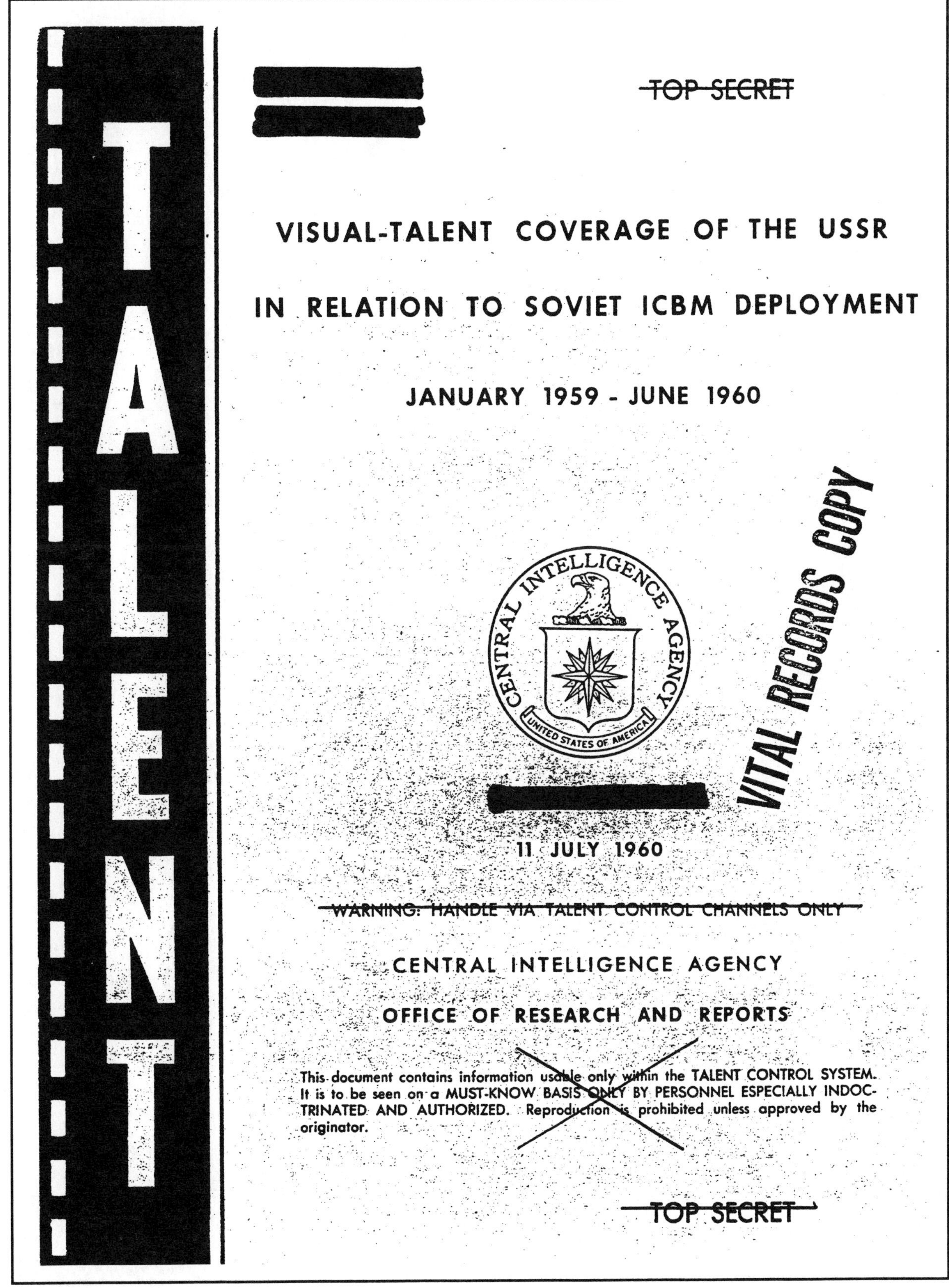

TALENT

~~TOP SECRET~~

VISUAL-TALENT COVERAGE OF THE USSR

IN RELATION TO SOVIET ICBM DEPLOYMENT

JANUARY 1959 - JUNE 1960

CENTRAL INTELLIGENCE AGENCY
UNITED STATES OF AMERICA

VITAL RECORDS COPY

11 JULY 1960

~~WARNING: HANDLE VIA TALENT CONTROL CHANNELS ONLY~~

CENTRAL INTELLIGENCE AGENCY

OFFICE OF RESEARCH AND REPORTS

~~This document contains information usable only within the TALENT CONTROL SYSTEM. It is to be seen on a MUST-KNOW BASIS ONLY BY PERSONNEL ESPECIALLY INDOCTRINATED AND AUTHORIZED. Reproduction is prohibited unless approved by the originator.~~

~~TOP SECRET~~

13. *(Continued)*

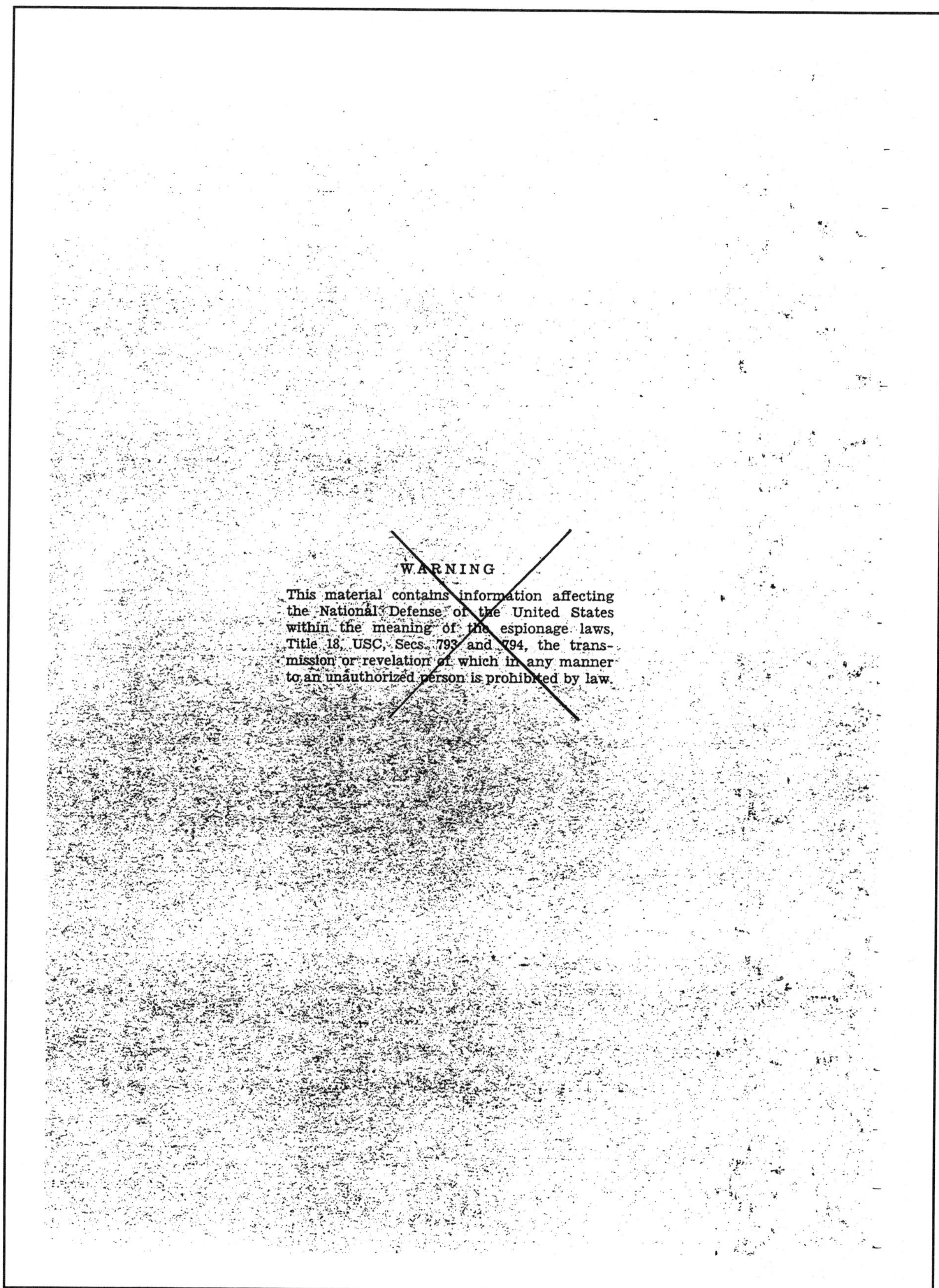
WARNING

This material contains information affecting the National Defense of the United States within the meaning of the espionage laws, Title 18, USC, Secs. 793 and 794, the transmission or revelation of which in any manner to an unauthorized person is prohibited by law.

13. *(Continued)*

~~TOP SECRET~~ [redacted]

VISUAL-TALENT COVERAGE OF THE USSR

IN RELATION TO SOVIET ICBM DEPLOYMENT

JANUARY 1959 - JUNE 1960

11 JULY 1960

~~TOP SECRET~~ [redacted]

13. *(Continued)*

~~TOP SECRET~~ [redacted]

ACKNOWLEDGMENT

The Office of Research and Reports wishes to acknowledge the assistance provided by the CIA Photographic Intelligence Center in publishing this report.

~~TOP SECRET~~ [redacted]

~~TOP SECRET~~ [redacted]

FOREWORD

This report provides an estimate of the portions of the USSR that have been [redacted] covered by [redacted] intelligence sources during the period January 1959 to June 1960. The extent of this coverage is compared with the total area of the USSR, with the total area considered suitable for long-range missile deployment, and with the area of those established priority regions that are believed to be most suitable for missile deployment. Similar comparisons based on railroad route mileage are also presented.

~~TOP SECRET~~ [redacted]

~~TOP SECRET~~ [redacted]

VISUAL-TALENT COVERAGE OF THE USSR IN RELATION TO SOVIET ICBM DEPLOYMENT

JANUARY 1959 - JUNE 1960

I. Summary

During the period January 1959 through June 1960, about 7.5 percent of the total land area of the USSR is covered by useable TALENT photography. Since about 45 percent of the terrain of the USSR is unsuitable for long-range ballistic missile deployment (especially for deployment of the first few units), a more meaningful statistic is the coverage of Soviet land area suitable* for such deployment. About 13.6 percent of the suitable area has been covered by useable TALENT photography.

Eight areas* (about 24 percent of the land area of the USSR) have been designated for priority search for deployed long-range ballistic missiles by the intelligence community. About 3.6 percent of the total of these areas is covered by useable TALENT photography.

The intelligence community has concluded that the Soviet ICBM system depends very heavily on railroad transportation; therefore, the portion of the Soviet railroad network covered during this period is probably the most meaningful statistic. Useable TALENT coverage of the total rail route mileage amounts to about 11.5 percent, or about 8.5 percent of such mileage in the priority areas. Over 35 percent of the rail route mileage in priority area 2 and more than 10 percent of priority areas 1 and 3 have been covered with useable TALENT. There has been no useable TALENT coverage of the other priority areas.

In addition to this highly reliable TALENT coverage, certain portions of the USSR have been subject to observation by other [redacted] intelligence sources. Although some 4.5 percent of the total land area of the USSR was observable** to these sources during the period, less than one percent of the area is estimated to have been observed**. The estimated observed coverage of suitable areas by these sources is about 1.5 percent, and such coverage of the priority areas is about 2 percent. Roughly 35 percent of the rail route mileage in the total land area, suitable areas, and priority areas of the USSR was traveled during this period, and it is estimated that useable observations were made along about 7 percent of the rail routes in these areas.

* [redacted]

** Definitions and method of calculation are presented below.

~~TOP SECRET~~ [redacted]

~~TOP SECRET~~ [redacted]

In conclusion, it is estimated that more than 85 percent of the suitable area, 95 percent of the priority areas, and 85 percent of the rail route mileage in priority areas have not been observed or covered by useable TALENT during the period. In view of the large areas still uncovered and the limited number of ICBMs that are likely to be deployed so early in the Soviet program, it is not surprising that none of these sites has been positively identified.

II. Suitable and Priority Areas

The total area of the USSR suitable for ICBM deployment is estimated to be 4,764,000 sq. miles. The area considered unsuitable for ICBM deployment is 45 percent of the total area of the USSR (8,647,000 sq. miles) and includes areas of continuous permafrost, high mountains, marshes, swamps, open bodies of water, towns and cities. Because of difficulties of construction and logistics, it is unlikely that any of the earlier long-range missiles would be deployed in such areas; these areas might be used only for some of the very last missiles deployed, if at all. [redacted]

Within the USSR, eight areas have been recognized by the intelligence community as being of priority interest in the search for long-range missile launching sites. [redacted]

The intelligence community has concluded that the Soviet ICBM system depends very heavily on railroad transportation. If the launching facilities are fixed, the railroad network is the primary means of logistic support; or if mobile, these facilities are rail mobile. For this reason, the coverage of the Soviet railroad network, primarily in the priority areas, is probably the most meaningful of the various measures presented in this paper.

III. TALENT Coverage

The total area of the USSR covered by useable TALENT photography since January 1959 has been calculated as 650,000 square miles. In making this calculation, linear photo mileage obtained by the four most recent TALENT missions was multiplied by 55 miles considered to be the width of effective coverage. The resulting figure of gross square mile coverage was then reduced to compensate for varying degrees of cloud cover (heavy clouded areas were assumed to have yielded only 25% coverage and scattered cloud areas 75% coverage).

Table 1 presents data on the portions of the total area, suitable area and priority areas covered by useable TALENT during the period.

~~TOP SECRET~~ [redacted]

13. *(Continued)*

~~TOP SECRET~~ [redacted]

Table 1

Areas of the USSR Covered by Useable TALENT Photography
January 1959-June 1960

Area	Total Land Area (Square Miles)	Estimated TALENT Coverage (Square Miles)	Estimated TALENT Coverage (Percent)
Total USSR	8,647,000	650,000	7.5
Suitable for Deployment	4,764,000	650,000	13.6
Priority Areas			
Total	2,081,000	75,150	3.6
Area 1	467,800	10,750	2.3
Area 2	315,600	60,270	19.1
Area 3	170,700	4,130	2.4
Areas 4-8	1,126,900	0	0

Table 2 presents data on the portions of the rail route mileage covered by useable TALENT photography.

Table 2

Rail Route Mileage of the USSR Covered by Useable TALENT Photography
January 1959-June 1960

Area	Total Rail Route (Miles)	Estimated TALENT Coverage (Miles)	Estimated TALENT Coverage (Percent)
Total USSR	75,900	8,750	11.5
Suitable for Deployment	75,400	8,750	11.6
Priority Areas			
Total	46,000	3,910	8.5
Area 1	6,200	620	10.0
Area 2	8,300	2,950	35.6
Area 3	3,000	340	11.4
Areas 4-8	28,500	0	0

~~TOP SECRET~~ [redacted]

~~TOP SECRET~~ [redacted]

IV. Non-TALENT Coverage

A. Observable Area

Large areas of the USSR have been subject to observation [redacted] during the period January 1959 to June 1960. The maximum area that could have been observed if ideal conditions prevailed -- that is, if there were absolutely no obstructions or limitations to vision along the routes traveled -- was calculated by multiplying the total route miles traveled by the width of the maximum observation belt. The observable belt for air travel is estimated to be 10 miles and for rail, water, and highway travel to be 5 miles. Table 3 presents data on the maximum observable areas for total USSR land area, suitable area, and priority areas.

Table 3

Maximum Observable Areas in the USSR
January 1959-June 1960

Area	Total Area (Sq. Mi.)	Maximum Observable Area (Sq. Mi.)	Maximum Observable Area (Percent)
Total USSR	8,647,000	387,500	4.5
Suitable for Deployment	4,764,000	368,000	7.7
Priority Areas			
Total	2,081,000	220,800	10.6
Area 1	467,800	12,000	2.6
Area 2	315,600	33,000	10.5
Area 3	170,700	13,900	8.2
Area 4	195,600	58,200	29.8
Area 5	290,700	52,100	17.9
Area 6	469,800	39,900	8.5
Area 7	108,000	8,400	7.7
Area 8	62,800	3,300	5.3

B. Estimated Observed Area

The total area actually observed [redacted] is estimated to be far less than the maximum observable area based on route miles traveled, for conditions for observations are frequently far from ideal. Limiting factors considered in calculating the actual extent of the area observed are as follows:

~~TOP SECRET~~ [redacted]

13. *(Continued)*

TOP SECRET

1. Visibility restrictions, including terrain, vegetation, rain, snow, fog, time of day (light or darkness), and man-made obstacles of various types.

2. Limitation of vision to one side of the vehicle (nullified somewhat if the route is frequently traveled; applies least to auto travel).

3. Limitation to air observation by altitude, cloud cover, and seat location.

4. Speed of travel (particularly by train), which limits the time span for recognition of features, thus reducing the width of the area that can be effectively observed.

5. Harassment by security personnel, which is particularly likely at points where sensitive installations might be observed.

In view of the above limitations, the area observed by travelers was calculated by multiplying the maximum observable area by an estimated percentage of effectiveness of observation. The fact that many routes were traveled a number of times is taken into consideration in determining the percentage of effectiveness. The percentages used to estimate the portion of observable area actually observed are as follows:

Type of Travel	Effective Observation (Percent)
Air	15
Rail	20
Water	10
Highway	35

The estimated observed coverage for each type of area under consideration is presented in Table 4.

~~TOP SECRET~~ [redacted]

Table 4

Estimated Observed Area in the USSR
January 1959-June 1960

Area	Square Miles	Percent of Area
Total USSR	71,900	0.8
Suitable for Deployment	68,700	1.4
Priority Areas		
Total	41,800*	2.0
Area 1	2,300	0.5
Area 2	5,800	1.8
Area 3	2,600	1.5
Area 4	11,900	6.1
Area 5	10,500	3.6
Area 6	6,500	1.4
Area 7	1,500	1.4
Area 8	600	1.0

C. Railroad Route Mileage Traveled

Table 5 presents data on the Soviet railroad route mileage traveled by [redacted] observers during the period. The mileage traveled is also reduced for observational difficulties; to arrive at an estimate of useable traveler observations, the factor of 20 percent was used (see paragraph IV B above).

* Numbers have been rounded; total is based on unrounded data.

~~TOP SECRET~~ [redacted]

13. *(Continued)*

~~TOP SECRET~~

Table 5

Railroad Route Mileage of the USSR Traveled by [redacted] Observers
Adjusted for Effective Coverage
January 1959-June 1960

Area	Total Miles	Traveled		Useable Percent
		Miles	Percent	
USSR	75,900	25,700	34	7
Suitable for Deployment	75,400	25,100	33	7
Priority Areas				
Total	46,000	16,940	37	7
Area 1	6,200	1,850	30	6
Area 2	8,300	2,700	33	7
Area 3	3,000	790	26	5
Areas 4-5	22,000	8,370	38	8
Area 6	4,000	2,020	50	10
Area 7	1,250	620	50	10
Area 8	1,250	590	47	9

V. Total Visual-TALENT Coverage

In order to get an appreciation of total useable visual and TALENT coverage of the various areas of the USSR during the period, a range of values is estimated; the lower end of the range reflects the useable TALENT coverage and the upper end includes the useable visual coverage with an allowance for possible duplication. These estimates are presented in Table 6.

~~TOP SECRET~~

13. *(Continued)*

~~TOP SECRET~~ [redacted]

Table 6

Useable Visual-TALENT Coverage of the USSR
January 1959-June 1960

Area	Land Area (Percent)	Rail Route Mileage (Percent)
USSR	7-8	12-18
Suitable for Deployment	14-15	12-18
Priority Areas		
Total	4-6	9-15
Area 1	2-3	10-15
Area 2	19-21	36-42
Area 3	2-4	11-16
Areas 4-5	0-5	0-8
Area 6	0-1	0-10
Area 7	0-1	0-10
Area 8	0-1	0-9

~~TOP SECRET~~ [redacted]

14. CIA/PIC, Joint Mission Coverage Index, "Mission 9009, 18 August 1960," September 1960 (Excerpt)

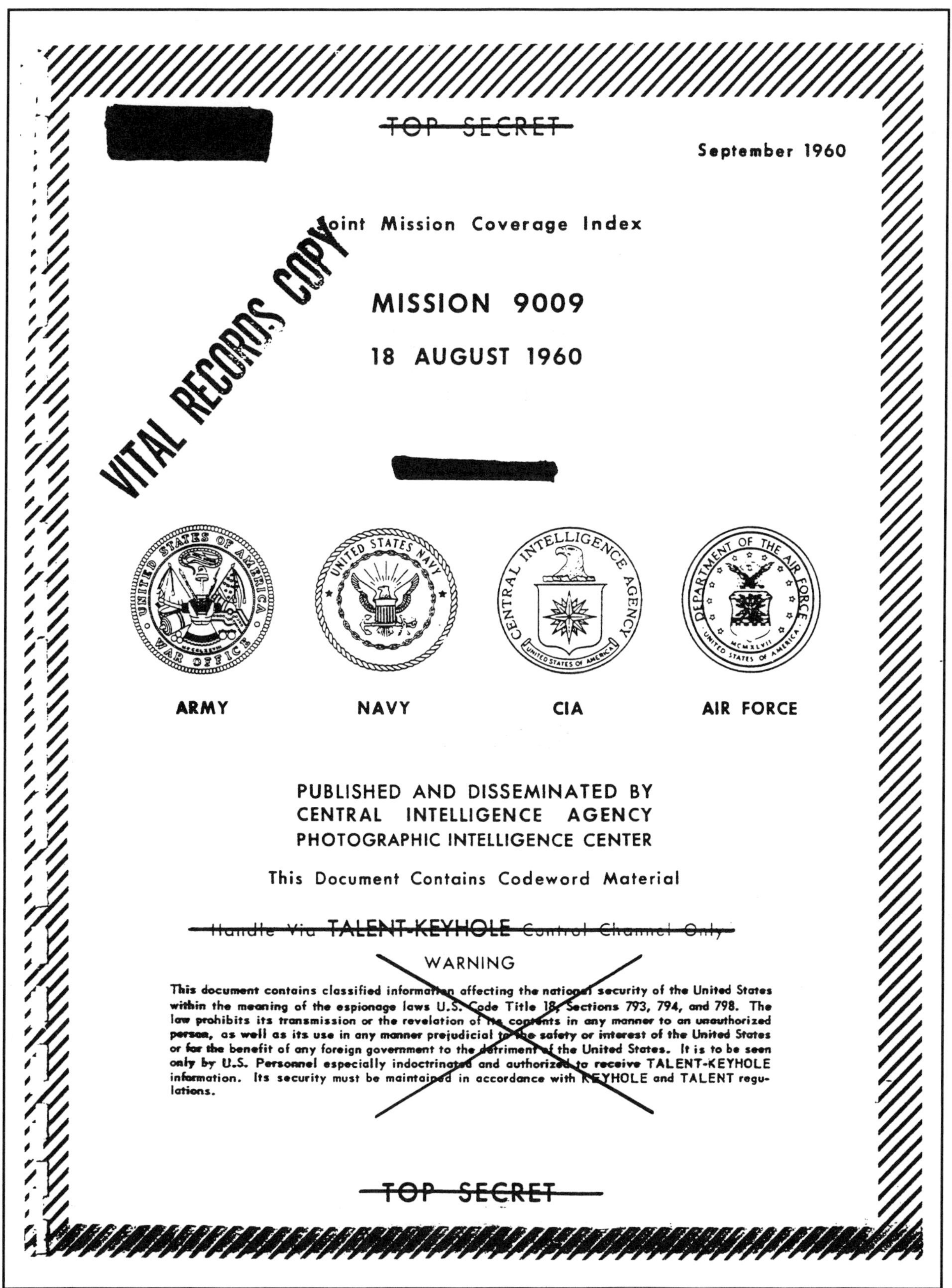

TOP SECRET

September 1960

Joint Mission Coverage Index

VITAL RECORDS COPY

MISSION 9009

18 AUGUST 1960

ARMY NAVY CIA AIR FORCE

PUBLISHED AND DISSEMINATED BY
CENTRAL INTELLIGENCE AGENCY
PHOTOGRAPHIC INTELLIGENCE CENTER

This Document Contains Codeword Material

Handle Via TALENT-KEYHOLE Control Channel Only

WARNING

This document contains classified information affecting the national security of the United States within the meaning of the espionage laws U.S. Code Title 18, Sections 793, 794, and 798. The law prohibits its transmission or the revelation of its contents in any manner to an unauthorized person, as well as its use in any manner prejudicial to the safety or interest of the United States or for the benefit of any foreign government to the detriment of the United States. It is to be seen only by U.S. Personnel especially indoctrinated and authorized to receive TALENT-KEYHOLE information. Its security must be maintained in accordance with KEYHOLE and TALENT regulations.

TOP SECRET

14. *(Continued)*

~~TOP SECRET RUFF~~

MISSION 9009

18 AUGUST 1960

SEPTEMBER 1960

~~TOP SECRET RUFF~~

14. *(Continued)*

~~TOP SECRET RUFF~~

PREFACE

This Joint Mission Coverage Index (JMCI) furnishes a listing of intelligence targets covered by Mission 9009. All priority items of intelligence significance reported in the six installments of the OAK 9009 immediate report have been included in this index. Detailed descriptions appearing in the OAK Report are not repeated.

Items are arranged by (1) country, (2) WAC area within the country, (3) subject, and (4) coordinates (grouped by degree square from north to south within the subject grouping).

For an explanation of the codes used in presenting information in this report see the appendix.

~~TOP SECRET RUFF~~

14. *(Continued)*

~~TOP SECRET RUFF~~

TABLE OF CONTENTS

~~TOP SECRET RUFF~~

14. *(Continued)*

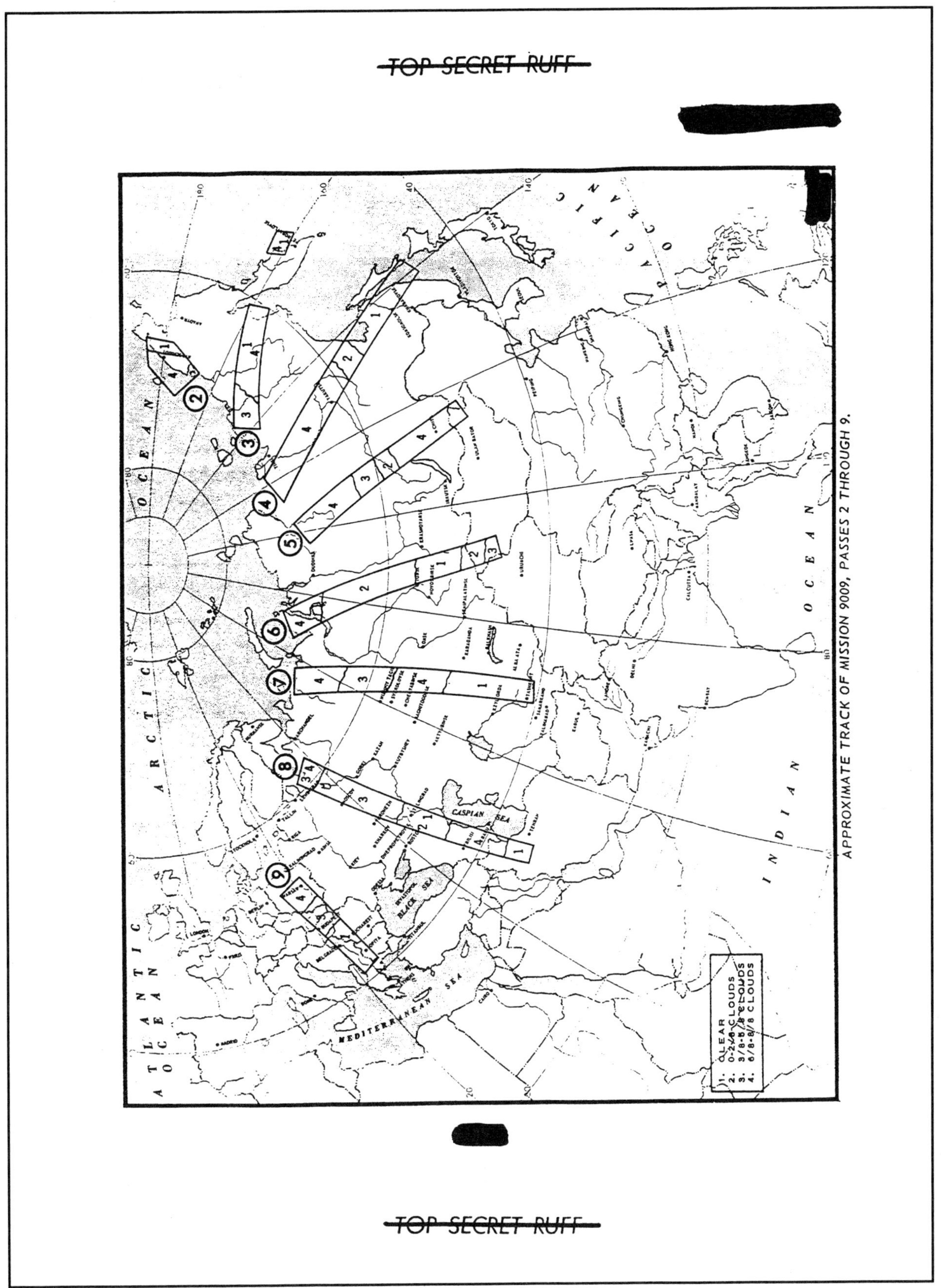

APPROXIMATE TRACK OF MISSION 9009, PASSES 2 THROUGH 9.

14. *(Continued)*

SUMMARY

Mission 9009 was accomplished on 18 August 1960. It consists of eight north-south passes over the USSR and includes portions of China, the Satellites and Yugoslavia (see accompanying coverage map).

Approximately 25 percent of the coverage is cloud free, with light-scattered to heavy clouds covering the remainder of the photography. The PI quality of the unobscured coverage ranges from good to very good.

The scale of the photography is estimated to range from 1:300,000 to 1:450,000. Average ground resolution is in the order of 20 to 30 feet on a side.

Major items of intelligence significance covered by Mission 9009 include the Kapustin Yar Missile Test Range (KYMTR), the western portion of the presumed 1,050 nm impact area of the KYMTR, 20 newly identified hexadic SA-2 surface-to-air missile sites and six possible SA-2 sites under construction, the Sarova Nuclear Weapons Research and Development Center, several new airfields, and numerous urban complexes.

~~TOP SECRET RUFF~~

14. *(Continued)*

~~TOP SECRET RUFF~~

Ctry	Installation	PIC Target No WAC	Target	Coordinates	Sbj
	-USSR-				
UR	MYS SHMIDTA A/F 2/53-62 X10Y2(59) C	65	5-A	6853N 17924W	01
UR	CHOKURDAKH A/F PROB OPERATIONAL, HARD SURFACED 3/7 X27Y2 H	67	2-A	7039N 14752E	01
UR	DUDEVO A/F 3/13-14 X45Y1(14) C	67	3	6913N 14712E	01
UR	U/I INSTALLATION 8 NM SW OF KADZHEROM ADJACENT TO KOTLAS-VORKUTA RR 7/25-26 X67Y2(25) SC	93	8	6438N 05542E	13
UR	U/I CONSTRUCTION ACTIVITY ROAD CONSTR AND OTHER ACTIVITY LOCATED AT POLUNOCHNOYE 7/49-54 X53Y3(51) SC	100	1	6052N 06025E	13
UR	NEW RR SPUR CONST NUMEROUS SPURS, THREE GROUPS OF BLDGS 15 NM SW KONOSHA 8/32-33 X27Y4(32) C	102	3-C	6048N 04000E	11
UR	KARGOPOL STORAGE AREA 1 NM N OF KARGOPOL 8/27-28 X27Y4(27) SC	102	26	6130N 03855E	12
UR	NYANDOMA 8/28-29 X14Y2(28) SC	102	11	6140N 04013E	12
UR	KONOSHA NEW RR CONSTRUCTION & STORAGE AREAS NO A/F NOTED 8/32 X17Y2 C	102	3-B	6058N 04015E	12
UR	U/I INSTALLATION NEW ROADS AND OTHER CONSTR ACTIVITY 8/23-26 X3Y3(25) SC	102	25	6220N 04105E	13
UR	NYANDOMA MINING AREA 4 AREAS. GROUND SCARING, NEW ROAD AND RR, LOCATED 3 NM SW NYANDOMA 8/28-29 X15Y4(28) SC	102	11-A	6139N 04015E	13

~~TOP SECRET RUFF~~

14. *(Continued)*

~~TOP SECRET RUFF~~

SUBJECT INDEX

-AIRFIELDS-

-USSR-

A/F AT INSTRUMENTATION SITE 5
A/F AT INSTRUMENTATION SITE 6
ABAKAN
ABGANEROVO
ACHINSK
ALEKSEYEVSKOYE
ARZAMAS
ATBASAR
AYAN
BYAUDE
CHIMKENT
CHIMKENT SE
CHIRCHIK
CHOKURDAKH
CHOP
CHUCHKOVO
CHUMIKAN
DE-KASTRI S
DE-KASTRI SEAPLANE BASE
DOBRYNSKOYE
DUDEVO
DZHAMBUL W
DZHEZKAZGAN
GIZHIA SEAPLANE STATION
GORKIY/FEDYAKOVO
GORKIY/SORMOVO
GORKIY/STRIGINO
IKETSUKI
IVANOVO N
KAMYSHIN
KAMYSHIN NE
KAPUSTIN YAR
KHUMMI
KOMSOMOLSK S
KOMSOMOLSK E
KOMSOMOLSK
KORSAKOV (OTOMARI)
KOSTROMA
KOSTROMSKOYE
KOTELNIKOVSKIY
KYZYLAGACH
LUKHOVITSY

~~TOP SECRET RUFF~~

14. *(Continued)*

~~TOP SECRET RUFF~~

-CZECHOSLOVAKIA-
OIL STGE FACIL U/C W OF MICHALOVCE

-RUMANIA-
POL STORAGE AREA S OF DRAGASANI

-YUGOSLAVIA-
SMEDEROVO PETROLEUM STORAGE
SMEDEROVO W PETROLEUM STORAGE

-MILITARY INSTALLATIONS-

-USSR-
ADADYM AMMO DEPOT
ARYS (PROB) CW STORAGE
CHIRCHIK MILITARY INSTALLATION
CHUCHKOVO AMMO STGE INSTALLATION
DZERZHINSK AMMO DEPOT W
DZHAMBUL MIL AREA
EXPLOSIVES STORAGE SSE OF KOVROV
EXPLOSIVES STORAGE S OF SHUYA
EXPLOSIVES STORAGE W OF MUROM
FROLISHCHEVA PUSTYA AMMO STORAGE
GOROKHOVYYSKIY LAGER CW TNG CENTER
KAMYSHIN OFFICERS TANK TRNG SCHOOL
KOMSOMOLSK AMMO STORAGE
MELENKI AMMO STORAGE
MIL PROCESSING, STORAGE & HANDLING N OF ARZAMAS
MILITARY AMMO STORAGE DEPOT SE OF ZHUKOVKA
MOZDOK AMMUNITION DEPOT
NAVOLOKI EXPLOSIVES STORAGE
NEREKHTA AMMO DEPOT, BURMAKINO
PENZE ARMY BKS AND TRAINING AREA
POSS ARMORED TRNG AREA MUKACHEVO
POSSIBLE MIL BKS SSE OF RYAZAN
PROBABLE EXPLOSIVE STGE AREA U/C WNW OF KASIMOV
SHUYA EXPLOSIVES STORAGE
SOVETSKAYA GAVAN AMMO STORAGE
SOVETSKAYA GAVAN AMMO STORAGE AREA
SOVETSKAYA GAVAN SUPPLY DEPOT, VANINO 5
STALINGRAD AMMO STOR AREA
STORAGE AREA ESE OF NIKOLAYEVSK-NA-AMURE

~~TOP SECRET RUFF~~

14. *(Continued)*

~~TOP SECRET RUFF~~

APPENDIX

Explanation of Codes Used in the JMCI

Individual items are, in general, arranged according to the following scheme.

1. Installation Index (First Line)

a. Country: The country is designated by the two-letter code used in the [redacted].

b. Installation: The name will be given, if known. If not, the installation will be titled according to an associated geographic name or according to obvious use, such as storage area, instrumentation station, etc. The [redacted] when known, will be given.

c. PIC Target Number: PIC Target numbers are comprised of two elements: (1) the WAC number for the area in which the installation lies, and (2) a numerical designation (occasionally followed by capital letters) for the specific target within that WAC area. For example, 246-6 designates target number 6 in WAC 246.

d. Coordinates: Coordinates are given to the nearest minute for the approximate center of the installation.

e. Subject: Thirteen categories are used; they are as follows:

1. Airfields
2. Atomic Energy
3. Electronics and Telecommunications
4. Industry
5. Liquid Fuels
6. Military Installations
7. Missiles
8. Naval Installations
9. Ports and Harbors
10. Storage Facilities, General

~~TOP SECRET~~

14. *(Continued)*

~~TOP SECRET RUFF~~

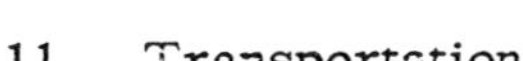

11. Transportation
12. Urban Areas
13. Miscellaneous

2. Significant Information

A very brief statement of significant information in connection with the installation will appear in the second and subsequent lines.

3. Photo Reference (Last Line)

This line is best explained by using an example:

3/729-31 x42Y3(730) HC

3 designates the pass number.

729-31 shows the frame numbers.

x42Y3(730) gives the Universal Reference Grid coordinates of the installation on frame 730.

HC - This designation indicates cloud conditions as they exist over the installation. The code used is as follows:

C Clear
SC Scattered Clouds
HC Heavy Clouds
O Overcast
H Haze (includes smoke, blowing snow and dust)
CS Cloud Shadow (cloud shadows cast on ground reducing interpretability)

15. National Intelligence Estimate, 11-8/1-61, "Strength and Deployment of Soviet Long Range Ballistic Missile Forces," 21 September 1961

~~TOP SECRET~~

NIE 11-8/1-61
21 September 1961

NATIONAL INTELLIGENCE ESTIMATE
NUMBER 11 - 8/1 - 61

STRENGTH AND DEPLOYMENT OF SOVIET LONG RANGE BALLISTIC MISSILE FORCES

(SUPPLEMENTS NIE 11 - 8 - 61)

This Document Contains Multiple Codeword Material

~~Handle Via~~ 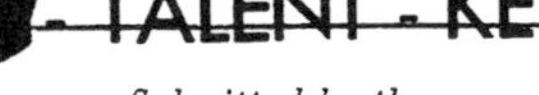 ~~- TALENT - KEYHOLE Controls Jointly~~

Submitted by the
DIRECTOR OF CENTRAL INTELLIGENCE

The following intelligence organizations participated in the preparation of this estimate: The Central Intelligence Agency and the intelligence organizations of the Departments of State, the Army, the Navy, the Air Force, The Joint Staff, the National Security Agency, and the Atomic Energy Commission.

Concurred in by the
UNITED STATES INTELLIGENCE BOARD

on 21 September 1961. Concurring were the Director of Intelligence and Research, Department of State; the Assistant Chief of Staff for Intelligence, Department of the Army; the Assistant Chief of Naval Operations for Intelligence, Department of the Navy; the Assistant Chief of Staff, Intelligence, USAF; the Director for Intelligence, The Joint Staff; the Atomic Energy Commission Representative to the USIB; the Assistant to the Secretary of Defense, Special Operations; and the Director of the National Security Agency. The Assistant Director, Federal Bureau of Investigation, abstained, the subject being outside of his jurisdiction.

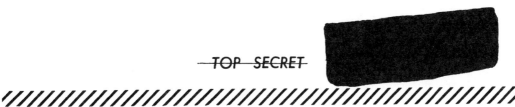

~~TOP SECRET~~

15. *(Continued)*

The title of this estimate when used separately from the text should be classified: ~~SECRET~~

WARNING

This document contains classified information affecting the national security of the United States within the meaning of the espionage laws U.S. Code Title 18, Sections 793, 794, and 798. The law prohibits its transmission or the revelation of its contents in any manner to an unauthorized person, as well as its use in any manner prejudicial to the safety or interest of the United States or for the benefit of any foreign government to the detriment of the United States. It is to be seen only by U.S. personnel especially indoctrinated and authorized to receive [redacted] and TALENT-KEYHOLE information. Its security must be maintained in accordance with [redacted] and KEYHOLE and TALENT regulations. No action is to be taken on any [redacted] which may be contained herein, regardless of the advantages to be gained, unless such action is first approved by the Director of Central Intelligence.

DISTRIBUTION:

White House
National Security Council
Department of State
Department of Defense
Atomic Energy Commission
Federal Bureau of Investigation

15. *(Continued)*

NATIONAL INTELLIGENCE ESTIMATE

NIE 11-8/1-61

STRENGTH AND DEPLOYMENT OF

SOVIET LONG RANGE BALLISTIC MISSILE FORCES*

(SUPPLEMENTS NIE 11 - 8 - 61)

THE PROBLEM

To estimate current Soviet operational strength in ICBM's and other ground-launched ballistic missiles with ranges of 700 n.m. or more, to identify present areas and methods of deployment, and to estimate the probable trends in strength and deployment over the next few years.

---

* NIE 11-8/1-61 revises and updates the estimates on this subject which were made in NIE 11-8-61: "Soviet Capabilities for Long Range Attack", TOP SECRET, 7 June 1961. It supplements the summary of evidence and analysis on the Soviet ICBM program appended to NIE 11-8-61 in Annexes C and D (TOP SECRET CODEWORD). The new estimate is issued at CODEWORD classification so that the reader can fully appreciate the quantity and quality of information on which it is based.

A brief summary of this estimate, at non-CODEWORD classification, will be included in the forthcoming NIE 11-4-61: "Main Trends in Soviet Capabilities and Policies, 1961-1966", now scheduled for completion in December 1961. In that estimate, the treatment of ground launched missiles will be incorporated into a summary of the entire Soviet long-range attack capability, including bombers, air-to-surface missiles, and submarine-launched missiles. For our current estimates on these latter elements of the long range striking force, see NIE 11-4-61, Annex A: "Soviet Military Forces and Capabilities", 24 August 1961, TOP SECRET, paragraphs 16-23.

15. *(Continued)*

CONCLUSIONS

1. New information, providing a much firmer base for estimates on Soviet long range ballistic missiles, has caused a sharp downward revision in our estimate of present Soviet ICBM strength but strongly supports our estimate of medium range missile strength.

2. We now estimate that the present Soviet ICBM strength is in the range of 10 - 25 launchers from which missiles can be fired against the US, and that this force level will not increase markedly during the months immediately ahead. 1/ We also estimate that the USSR now has about 250-300 operational launchers equipped with 700 and 1,100 n.m. ballistic missiles. The bulk of these MRBM launchers are in western USSR, within range of NATO targets in Europe; others are in southern USSR and in the Soviet Far East. ICBM and MRBM launchers probably have sufficient missiles to provide a reload capability and to fire additional missiles after a period of some hours, assuming that the launching facilities are not damaged by accident or attack.

3. The low present and near-term ICBM force level probably results chiefly from a Soviet decision to deploy only a small force of the cumbersome, first generation ICBMs, and to press the development of a smaller, second generation system. Under emergency conditions the existing force could be supplemented somewhat during the first half of 1962, but Soviet ICBM strength will probably not increase substantially until the new missile is ready for operational use, probably sometime in the latter half of 1962. After this point, we anticipate that the number of operational launchers will begin to increase significantly. On this basis, we estimate that the force level in mid-1963 will approximate 75-125 operational ICBM launchers. 2/

1/ The Assistant Chief of Staff, Intelligence, USAF, does not concur in this sentence. See his footnote following the Conclusions.

2/ The Assistant Chief of Staff, Intelligence, USAF, does not concur in paragraph 3. See his footnote following the Conclusions.

15. *(Continued)*

4. In addition to 700 and 1,100 n.m. missiles now available, the USSR will probably have a 2,000 n.m. system ready for operational use late this year or early next year. The USSR's combined strength in these missile categories will probably reach 350-450 operational launchers in the 1962-1963 period, and then level off.

5. Soviet professions of greatly enhanced striking power thus derive primarily from a massive capability to attack European and other peripheral targets. Although Soviet propaganda has assiduously cultivated an image of great ICBM strength, the bulk of the USSR's present capability to attack the US is in bombers and submarine-launched missiles rather than in a large ICBM force. While the present ICBM force poses a grave threat to a number of US urban areas, it represents only a limited threat to US-based nuclear striking forces. 3/

3/ The Assistant Chief of Staff, Intelligence, USAF, does not concur in paragraph 3 and the last sentence of paragraph 5. See his footnote following the Conclusions.

Position on ICBM force levels of the Assistant Chief of Staff, Intelligence, USAF:

1. The Assistant Chief of Staff, Intelligence, USAF believes that the Soviets had about 50 operational ICBM launchers in mid-1961 and that they will have about 100 in mid-1962 and about 250 in mid-1963. In his view, the early availability and high performance record of the first generation ICBM indicates the probability that, by mid-1961, substantial numbers of these missiles had been deployed on operational launchers. Four considerations weigh heavily in this judgment:

a. The continuance of [redacted] firings of the first generation ICBM;

b. The feasibility of adapting the type "C" pad - now identified as being deployed in the field - for use with the first generation system;

c. 

d. The USSR's current aggressive foreign policy indicates a substantial ICBM capability.

2. In view of the time that has passed since the first generation system became suitable for operational deployment, now over 18 months, the Assistant Chief of Staff, Intelligence, USAF believes that about 50 operational launchers in mid-1961 is likely, even though the Soviets may have elected to await development of second generation missiles before undertaking large-scale deployment.

3. The Assistant Chief of Staff, Intelligence, USAF believes that the force now deployed constitutes a serious threat to US-based nuclear striking forces.

4. As to the future, the Assistant Chief of Staff, Intelligence, USAF believes that the Soviets will continue to deploy first generation missiles, as an interim measure until the second generation missiles become available. He believes that the Soviets would prefer this approach to acceptance of an inordinate delay in the growth of their ICBM capabilities. Once the second generation system has become operational, which could be in early 1962, he believes that deployment will be accelerated, with first generation missiles being withdrawn from operational complexes and replaced by the new missiles. It is evident from their test program that the Soviets feel obliged to increase the tempo of their efforts. The Assistant Chief of Staff, Intelligence, USAF believes that this sense of urgency, plus the gains realizable from experience will result, in the next year or two, in a launcher deployment program more accelerated than that indicated in the text.

15. *(Continued)*

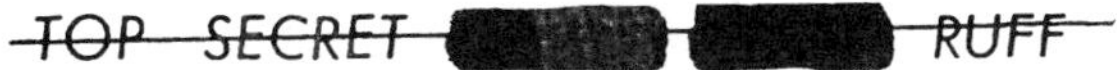

DISCUSSION

6. The requirement to revise our estimates on Soviet long range ballistic missile forces stems from significant recent evidence of three principal types. First, read-out of electronic data on the 1961 activities at the Soviet ICBM and space vehicle test range has provided information on the new types of ballistic vehicles now being developed and on the pace and progress of the development programs. Second, photographic coverage of large regions of the USSR has provided the first positive identification of long range ballistic missile deployment complexes, has given excellent guidance as to Soviet deployment methods, and has permitted detailed search of large areas of the USSR, including many previously suspected to contain missile complexes. Finally, reliable clandestine reports have provided useful evidence on the general status and organization of long range missile forces. Therefore, although significant gaps continue to exist and some of the available information is still open to alternate interpretations, the present estimate stands on firmer ground than any previous estimate on this critical subject.

ICBM Development

7. The test-firing program from the Tyuratam ICBM and space launching rangehead has been much more intensive in 1961, and has at the same time suffered many more failures, than in any other period in its four year history. Thirty-nine launching operations were undertaken between January and 17 September 1961. 4/ Of these, 13 involved either first generation ICBMs or space vehicles using essentially the same booster. All but one of these 13 were generally successful. The other 26 operations involved new vehicles not previously observed in range activities. Of these, only about half resulted in generally successful

4/ A more recent launching operation on 19 September 1961, which resulted in a failure, cannot as yet be categorized as to type of vehicle.

firings which reached the vicinity of the instrumented impact areas. Of the last seven operations involving new vehicles, however, six have been generally successful. (See Figure 1.)

8. One of the new vehicles (called Category B by US intelligence) is probably a second generation ICBM; the other (Category C) may be a competitive ICBM design or a special vehicle to test ICBM and space components. Both are tandem staged, that is, the upper stage is ignited at altitude as in the case of Titan, rather than at launch as in the case of Atlas and the first generation Soviet ICBM. Our data are sufficient to show that both of the new vehicles are liquid propelled, but not to establish whether the propellants are storable or non-storable. Some aspects of [redacted] performance of the upper stage of the Category B vehicle are similar to those of the 2,000 n.m. missile, which was tested intensively at Kapustin Yar for some months preceding the Category B operations at Tyuratam. The vehicles fired to a distance of 6,500 n.m. into the Pacific on 13 and 17 September 1961 were probably Category B vehicles. Some relationship seems to exist between the upper stages of the Category C vehicle and Venus probes. Despite this apparent relationship with space vehicles, it was a Category C firing which immediately preceded Khrushchev's remark to McCloy last July, that a "new ICBM" had been launched successfully. No further details are known about the configuration, propulsion, guidance, range, or payload of the new vehicles. 5/

9. The 1961 tests confirm our previous estimate that the Soviets would develop a new ICBM system, and we continue to believe that a major requirement for such a system is a missile which can be more readily handled and deployed than their original ICBM. This belief is supported by a reliable clandestine source who learned, in 1960 or early 1961, that the Soviet leadership desired an ICBM using higher-energy fuel which

5/ We have taken note of Soviet statements concerning a 100 megaton weapon. We do not believe that present Soviet capabilities include a missile warhead with 100 megaton yield or a ballistic vehicle capable of delivering such a warhead to intercontinental ranges. We will examine this matter in fuller detail in an early estimate.

15. *(Continued)*

~~TOP SECRET RUFF~~

would require less bulk. In order to be flight tested in early 1961, design work on a new missile was certainly underway in 1958. Nuclear tests appropriate to the development of lighter warheads were conducted in 1957 and 1958; the current nuclear testing program may serve further to prove the warhead design.

10. Although the flight-test failures in the first half of 1961 probably set back the Soviet schedule for development of second generation missiles, it is clear from the test range activities that the R&D program has been pursued with great vigor. The recent successes with the Category B vehicle, and the probable firing of such vehicles to 6,500 n.m. after only about 8 months of testing to Kamchatka, suggest that the initial difficulties with this system may now have been largely overcome. Moreover, it is probable that one or both the new vehicles have borrowed components or at least design techniques from proven systems, thereby aiding the R&D program. We believe that the program will continue to be pursued with vigor, and that a smaller, second generation ICBM will have been proven satisfactory for initial operational deployment in the latter half of 1962.

11. Thus we believe that the first generation system will be the only Soviet ICBM system in operational use for the months immediately ahead and probably for about the next year. Despite its inordinate bulk and the other disadvantages inherent in a non-storable liquid fueled system, the first generation system is capable of delivering a high yield nuclear warhead with good accuracy and reliability against targets anywhere in the US. (For a summary of its estimated operational characteristics, see Figure 2.) Test range launchings of first generation missiles (now called Category A) continued from January through July.

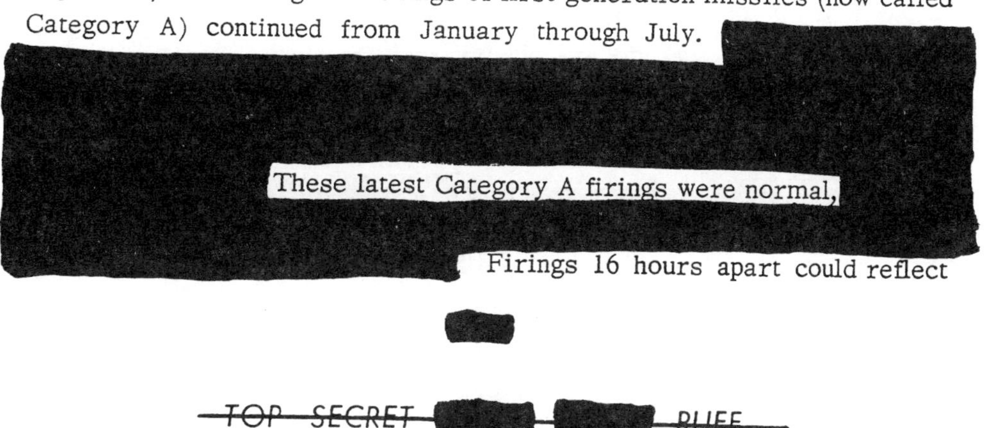

These latest Category A firings were normal,

Firings 16 hours apart could reflect

~~TOP SECRET RUFF~~

15. *(Continued)*

the training of operational crews for launching second salvos, but it cannot be determined whether these firings were from a single pad. Accuracy could not be determined, but reliability continued high. 6/

Utilization of Launching Pads

12. Soviet ICBM capabilities at present depend in part, and in the near future will depend in considerable measure, upon whether or not the deployment complexes now being discovered through KEYHOLE photography can be used to fire first generation missiles, or whether they cannot become fully operational until a second generation missile becomes available. The first generation missile is obviously compatible with massive, fully rail-served launchers similar to those at Tyuratam Areas A and B. But the launchers at confirmed field complexes, whose construction began only in late 1959 or thereafter, resemble the simplified pair of pads at Tyuratam Area C, where missiles are transported to the pad by road and some of the support equipment is mounted on vans. (For artists' conceptions of the launchers at Tyuratam and a layout of the rangehead, see Figures 3-5.)

13. From our examination of the 1961 test firing program, the physical dimensions of various items at Areas A and C, and the requirements for handling and firing the first generation missile, we conclude that the simplified Area C was designed for a new and smaller missile now being test fired. Although it is technically feasible for the Soviets to adapt the rail-based first generation missile to road served launchers of the type at Area C, it would be necessary to redesign much of the check-out, handling, erecting, and fueling equipment. This redesigned equipment would differ from both that at Area A and that designed for use with the

6/ To date we have no firm evidence to indicate that the Soviets have experimentally investigated the decoy problem in ICBM flights to Kamchatka. We believe that the Soviets can and will provide decoy protection, should they deem it necessary.

new missile. Such action might have been taken as an interim measure if a long delay in the advent of the second generation system had been anticipated well in advance.

ICBM Deployment

14. Through KEYHOLE photography over the past three months, we have positively identified three ICBM complexes under construction. Two are near Yur'ya and Yoshkar-Ola, in a region several hundred miles northeast of Moscow, and the third is near Verkhnyaya Salda in the Urals. The paired, road-served pads at these complexes closely resemble those at Tyuratam Area C. Near Kostroma, in the same general region but closer to Moscow, the photography revealed a new clearing suitable for a pair of pads, and we believe this is possibly a fourth complex similar to the others. Portions of the installation at Plesetsk, farther to the northwest, were covered again in mid-1961, but the new photography was too limited either to confirm or rule out this location as an ICBM deployment complex. (The locations of presently known and suspected areas of ICBM deployment activities are shown in Figure 9.)

15. The new evidence confirms that the present Soviet deployment concept involves large, fixed complexes, with multiple pads and extensive support facilities. The identified deployment complexes are served by rail spurs which provide their major logistic support. The complexes are highly vulnerable to attack. For example, although the Yur'ya complex is quite large, the entire installation is soft and each pair of pads is separated from its neighbor by only 3-4 n.m. [redacted] concealment from ground observation has been achieved by locating the installations in remote, densely wooded areas. For active defense against aircraft, SA-2 surface-to-air missile sites are being installed near the complexes.

16. At Yur'ya, the confirmed complex whose construction appears most advanced, eight launchers in four pairs were observed in various

stages of construction in mid-1961 (see Figure 6). Considerations of logistics and control, together with evidence from the MRBM program and other factors, lead us to believe that eight is the typical number of launchers for this type of complex. 7/ Each pair of launchers has checkout and ready buildings which are probably capable of housing a missile for each pad; however, the extent of the support facilities strongly suggests that additional missiles are to be held there to provide a reload or standby capability. The designed salvo capability of the complex is apparently to be eight missiles. There would be at least 5 minutes delay between groups of four missiles if the system is radio-inertial (as is the first generation ICBM) and if one set of guidance facilities is provided for each pair of launchers. A second salvo might be attempted after some hours, assuming the launching facilities were not damaged by accident or attack. Although we have no direct evidence on this matter, we believe it might be feasible to prepare a second salvo in 8-12 hours.

17. On the basis of evidence dating back to 1957 and other more recent information, we have estimated that Plesetsk is an ICBM complex with rail-served launchers designed to employ the first generation ICBM. The installation at Plesetsk (see Figure 7) is even larger than the Yur'ya complex. Although the presence of ICBM launchers has not been confirmed, there are SAM sites, several very large support areas, and numerous buildings, including what appears to be housing for some 5,000 to 15,000 persons. The photographic and other evidence is inadequate to establish the number of launchers which may be at Plesetsk. We believe that the number may be as few as two, but four or more is also possible. An ICBM complex involving this much equipment, investment, and personnel would probably have a reload of at least one missile per pad. Based on

7/ The Assistant Chief of Staff, Intelligence, USAF, believes that this typical number may be larger than eight. He agrees, however, that if guidance facilities are provided for each pair of launchers, the sequence of launching would be as described in the text.

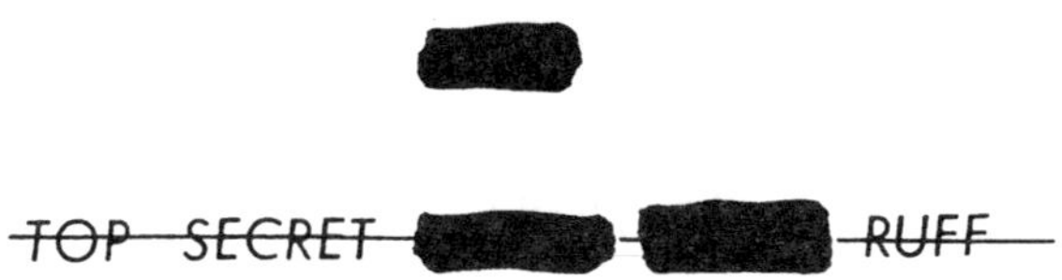

Tyuratam experience, we estimate the time to prepare a second salvo at about 16 hours. 8/

18. The new evidence gives a better measure of the timing of some ICBM deployment activities. Based on its size, the extent of its facilities, and its present state of construction, the Yur'ya complex must have been started in the autumn of 1959, concurrent with or very shortly after the start of construction at Tyuratam launch Area C. Yur'ya is probably one of the earliest complexes of its type. Construction and installation of equipment will probably be completed some time early in 1962. The similar complex at Yoshkar-Ola is many months behind Yur'ya; the evidence is less conclusive with respect to Kostroma and Verkhnyaya Salda, but what can be seen is apparently in the early stages of construction. From the evidence, therefore, we have reasonably firm indications that at least two years were used for the construction of even the simpler ICBM complexes, although this may be reduced to about 18 months as experience is gained.

Adequacy of Recent Intelligence Coverage

19. Through KEYHOLE operations since mid-1960, our coverage of suspected deployment areas in the USSR has been substantially augmented. This photography has been studied in detail by photo-interpreters with knowledge of US and Soviet missile programs. The search has been aided by photography of Soviet missile test range installations, which are now known to bear a close resemblance to deployment sites in the field. On the basis of this activity, combined with other information and analysis, we now estimate that we have good intelligence coverage of approximately 50 percent of the total railroad route mileage in the USSR. This coverage is not uniform, however; certain portions of the railroad route mileage

8/ The Assistant Chief of Naval Operations (Intelligence), Department of the Navy, believes that evidence of ICBM deployment at Plesetsk is indeterminate but that, in the aggregate, it points against such deployment.

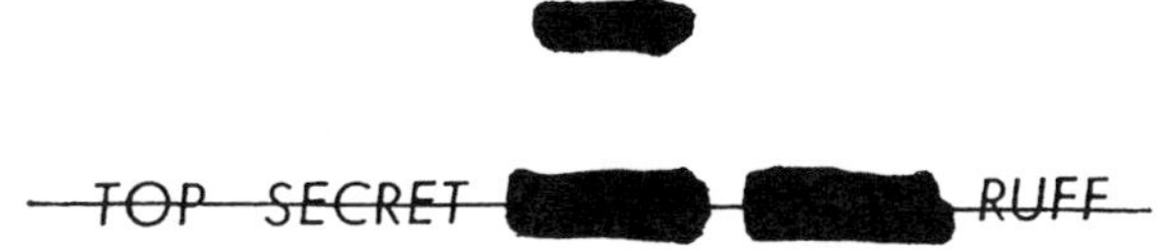

15. *(Continued)*

~~TOP SECRET~~ ~~RUFF~~

in suspected deployment areas, including some in the northwest, have not yet been well covered by KEYHOLE photography. Moreover, positive identifications of ICBM deployment have actually been made in areas where our photographic and other coverage is only fair. We therefore believe that we have useful intelligence coverage for more than 50 percent of those portions of the USSR within which ICBM deployment is most likely. 9/

20. Of the five confirmed or possible ICBM complexes located in KEYHOLE photography, Yur'ya, Plesetsk, and Verkhnyaya Salda were previously suspected on the basis of other information, We previously had not suspected Yoshkar-Ola or Kostroma. The discovery of these latter two complexes demonstrates our ability to detect and recognize ICBM complexes of present types in KEYHOLE photography of unsuspected areas.

21. The KEYHOLE search has shown that many previously suspected areas did not contain ICBM complexes as of the summer of 1961. Four areas not covered by recent, good quality KEYHOLE photography remain under active consideration as suspected locations of ICBM deployment activity (see Figure 9). Past experience indicates that some or all of the areas now under active consideration may prove to be negative, and conversely, that deployment activity may now be under way in other unsuspected areas. It is extremely unlikely, however, that undiscovered ICBM complexes exist in areas on which there is recent KEYHOLE photography of good quality.

9/ Annex D of NIE 11-8-61, which dealt with intelligence coverage relating to ICBM deployment, will be revised following completion of a detailed survey of photographic coverage.

Probable ICBM Force Levels 10/

22. We believe that our coverage of both test range activities and potential deployment areas is adequate to support the judgment that at present there are only a few ICBM complexes operational or under construction. While there are differences within the intelligence community as to the progress of the Soviet program to date and the precise composition of the current force, we estimate that the present Soviet ICBM capability is in the range of 10-25 launchers from which missiles can be fired against the US. The low side of this range allows for the possibility that the Soviets could now fire only a token ICBM salvo from a few launchers, located at the Tyuratam rangehead and an operational complex, perhaps Plesetsk. The high side, however, takes into account the limitations of our coverage and allows for the existence of a few other complexes equipped with first generation missiles, now operational but undetected.

23. The Soviet system is probably designed to have a refire capability from each launcher. The USSR may therefore be able to fire a second salvo some hours after the first, assuming that the launching facilities are not damaged by accident or attack.

24. The reasons for the small current capability are important to an estimate of the future Soviet buildup. The first generation system, designed at an early stage of Soviet nuclear and missile technology, proved to be powerful and reliable but was probably too cumbersome to be deployed on a large scale. One or more first generation sites may have been started but cancelled. The urgent development of at least one second generation system probably began in about 1958, and an intensive firing program is now underway concurrent with the construction of simplified deployment complexes.

10/ The Assistant Chief of Staff, Intelligence, USAF, does not concur in the estimate of ICBM force levels. For his position, see his footnote following the Conclusions.

15. *(Continued)*

~~TOP SECRET~~ [redacted] ~~RUFF~~

We therefore believe that in about 1958 the Soviet leaders decided to deploy only a small force of first generation ICBMs while pressing toward second generation systems.

25. The net effect of this Soviet decision, together with whatever slippage is occurring in the development of second generation systems, has been to produce a low plateau of ICBM strength. Under emergency conditions the existing force could be supplemented during the first half of 1962 by putting some second generation ICBMs on launcher at one or two completed complexes before the weapon system has been thoroughly tested. However, the Soviets could not have very much confidence in the reliability, accuracy and effectiveness of such a force. In any event, operational ICBM strength will probably not increase substantially until the new missile has been proved satisfactory for operational use, probably some time in the latter half of 1962. Alternatively, the possibility cannot be excluded that second generation ICBMs could be proved satisfactory for operational use somewhat earlier in 1962, possibly as soon as the first simplified complex is completed. After this point, we anticipate that the number of operational launchers will begin to increase significantly.

26. We continue to believe, for the many reasons adduced in NIE 11-8-61, that the Soviet leaders have desired a force of several hundred operational ICBM launchers, to be acquired as soon as practicable over the next few years. In addition to the complexes known to be under construction, it is probably that work is under way on other undiscovered complexes and that the construction of still others is scheduled to begin soon. Taking account of this probability, together with our present intelligence coverage and our information on site activation lead-time, we estimate that the force level in mid-1963 will approximate 75-125 operational ICBM launchers. The high side of this range allows for eight complexes of eight launchers each under construction at the present time, with four more scheduled to begin by the end of the year; it would

15. *(Continued)*

require site activation time to decrease to about 18 months by the end of the year; it builds from a present force level of about 25 operational launchers. The low side of the mid-1963 range would be achieved if six complexes were now under construction, two more were begun by the end of the year, and the present force level were only about 10 launchers.

27. As noted in NIE 11-8-61, Soviet force goals for the period to 1966 will be increasingly affected by developments in US and Soviet military technology, including the multiplication of hardened US missile sites, the possible advent of more advanced Soviet missiles which can better be protected, and by developments in both antimissile defenses and space weapons. The international political situation will also affect Soviet force goals, and there is a good chance that the Soviet leaders themselves have not yet come to a definite decision. We have not been able as yet to review, in the light of the new evidence, these and other considerations pertaining to the probable future pace of the Soviet ICBM program. Therefore we are unable to project a numerical estimate beyond mid-1963. Considering the problems involved in site activation, however, we believe that a rate of 100 or possibly even 150 launchers per year beginning in about 1963 would be feasible. To accomplish such a schedule, the USSR would have to lay on a major program of site construction within the next year, which we believe would be detected through continuing KEYHOLE operations and other means of intelligence collection.

Medium and Intermediate Range Ballistic Missiles

28. Recent KEYHOLE photography confirms the large-scale deployment of 700 and 1,100 n.m. ballistic missiles in western USSR. Through this photography, approximately 50 fixed sites with a total of about 200 pads suitable for launching these MRBMs have been firmly identified in a wide

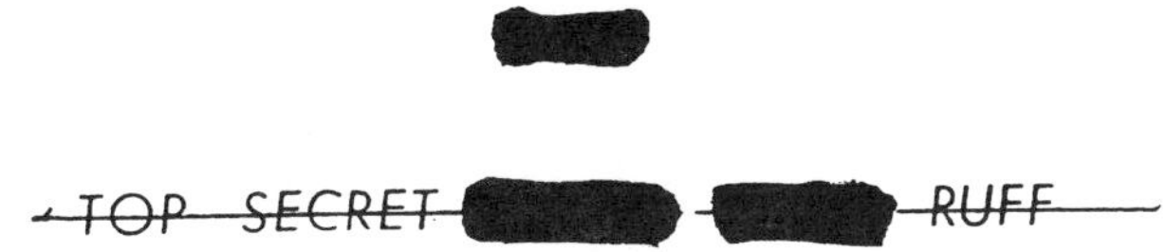

15. *(Continued)*

belt stretching from the Baltic to the southern Ukraine. Since photography establishes that the sites are paired, we are virtually certain that there are about 10 additional sites hidden by scattered clouds. Taking account of indicators pointing to still other locations not yet photographed, we estimate with high confidence that in the western belt alone there are now about 75 sites with a total of about 300 launch pads, completed or under construction. (For known and estimated site locations in this area, see Figure 9.)

29. The new information does not establish whether individual sites are fully operational, nor does it reveal which type of missile each is to employ. At the time of photography (obtained during a 3-month period in the summer of 1961) approximately three-quarters of the identified sites appeared to be complete or nearly so, some were under construction, and the evidence on others is ambiguous. Construction has probably been completed at some sites since the time of photography; the installation of support equipment and missiles could probably be accomplished relatively quickly thereafter, perhaps in a period of some weeks. Three basic site configurations have been observed, all of them bearing a strong resemblance to launch areas at the Kapustin Yar rangehead (see Figure 8). Any of the three types could employ either 700 or 1,100 n.m. missiles, whose size and truck-mounted support equipment are virtually identical. The sites could not employ ICBMs, but one type might be intended for the 2,000 n.m. IRBM which has been under development at Kapustin Yar.

30. On the basis of the new evidence and a wealth of other material on development, production, training and deployment, we estimate that in the western belt alone the USSR now has about 200-250 operational launchers equipped with 700 and 1,100 n.m. ballistic missiles, together with the necessary supporting equipment and trained personnel. From these launchers, missiles could be directed against NATO targets from Norway to Turkey. On less firm but consistent evidence, about 50 additional

15. *(Continued)*

~~TOP SECRET~~ [redacted] [redacted] ~~RUFF~~

launchers are believed to be operational in other areas: in the Transcaucasus and Turkestan, from which they could attack Middle Eastern targets from Suez to Pakistan; and in the southern portion of the Soviet Far East within range of Japan, Korea, and Okinawa. Very recent KEYHOLE photography confirms the presence of some sites in Turkestan and in the Soviet Far East, north of Vladivostok.

31. On this basis, we estimate that the USSR now has a total of about 250-300 operational launchers equipped with medium range ballistic missiles, the bulk of them within range of NATO targets in Europe. This is essentially the same numerical estimate as given in NIE 11-8-61, but it is now made with greater assurance.

32. Contrary to our previous view that MRBMs were deployed in mobile units, we now know that even though their support equipment is truck-mounted, most if not all MRBM units employ fixed sites. Like the ICBM complexes, these are soft, screened from ground observation by their placement in wooded areas, and protected against air attack by surface-to-air missile sites in the vicinity. The systems are probably designed so that all ready missiles at a site can be salvoed within a few minutes of each other. Two additional missiles are probably available for each launcher; a second salvo could probably be launched about 4-6 hours after the first. There is some evidence that after one or two salvos the units are to move from their fixed sites to reserve positions. Their mobility could thus be used for their immediate protection, or they could move to new launch points to support field forces in subsequent phases of a war.

33. The Soviet planners apparently see a larger total requirement for MRBMs and IRBMs than we had supposed. While the rate of deployment activity in the western belt is probably tapering off after a vigorous three-year program, some sites of all three basic types are still under construction. There will therefore be at least some increase in force levels in the coming months. The magnitude of the buildup thereafter will depend largely on the degree to which the 2,000 n.m. system is deployed,

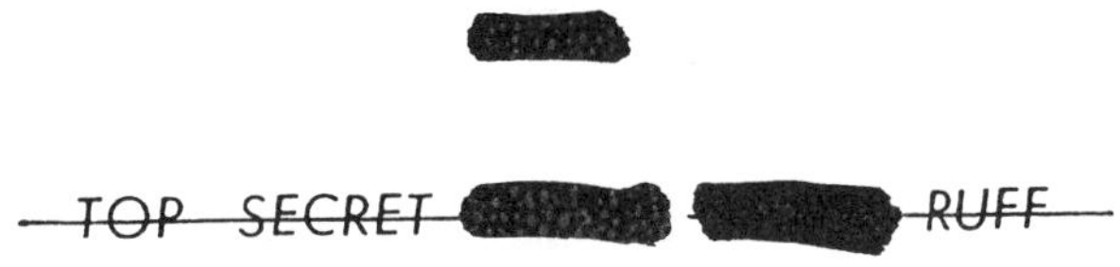

~~TOP SECRET~~ [redacted] [redacted] ~~RUFF~~

15. *(Continued)*

TOP SECRET RUFF

and whether or not it will supplement or replace medium range missiles.

34. With the advent of the 2,000 n.m. IRBM, probably in late 1961 or early 1962, the Soviets will acquire new ballistic missile capabilities against such areas as Spain, North Africa, and Taiwan. To this extent at least, they probably wish to supplement their present strength. They may also wish to deploy IRBMs or MRBMs to more northerly areas within range of targets in Greenland and Alaska. Moreover, evidence from clandestine sources indicates that the Soviet field forces are exerting pressure to acquire missiles of these ranges. In general, however, we believe that the future MRBM/IRBM program will emphasize changes in the mix among the existing systems, and later the introduction of second generation systems, rather than sheer numerical expansion. Taking these factors into account, we estimate that the USSR will achieve 350-450 operational MRBM and IRBM launchers sometime in the 1962-1963 period, and that the force level will be relatively stable thereafter.

15. *(Continued)*

~~TOP SECRET [redacted] [redacted] RUFF~~

FIGURES

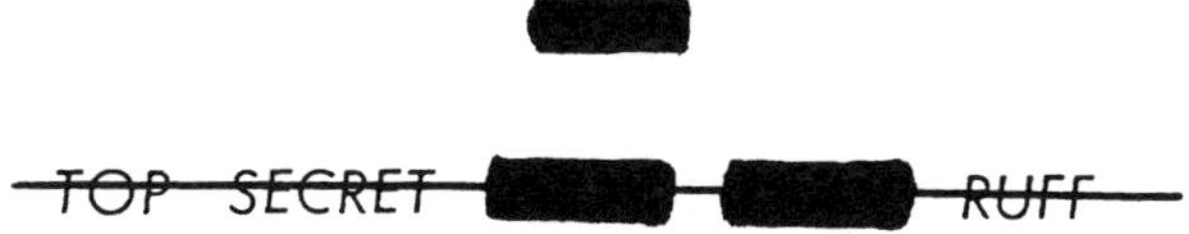

15. *(Continued)*

FIGURE 2

ESTIMATED CURRENT PERFORMANCE CHARACTERISTICS
SOVIET LONG RANGE BALLISTIC MISSILES

	SS-3	SS-4	SS-5[1]	SS-6[2]	Second Generation ICBM[1]
Max. Operational Range (nm)	700	1100	2000	5000 / 7000	at least 6500
Guidance	Radio/ Inertial	Radio/ Inertial	Radio/ Inertial	Radio/ Inertial	NA
Accuracy	1 nm	1½ nm	1½ nm or better	2 nm	NA
Configuration	Single Stage	Single	Single	Partial or Parallel	Tandem
Propellants	NonStor. Liquid	NonStor. Liquid	Liquid	NonStor. Liquid	Liquid
Gross Takeoff Weight (lbs)	60,000	75,000	NA	450,000-500,000	Prob less than SS-6
Warhead Weight (lbs)	3000	3000	3000-5000	6000-10000 / 6000	NA
Ready Missile Rate	85%	85%	75%	70-85%[3]	NA
Reliability, on Launcher	90%	95%	80%	85-90%[4]	NA
Reliability, in Flight	80%	80%	75%	70-85%[4]	NA
Reaction Time[5] - Condition I	1-3 hrs	1-3 hrs	1-3 hrs	1-3 hrs	NA
Reaction Time - Condition II	15-30 min	15-30 min	15-30 min	15-30 min	NA
Reaction Time - Condition III	5-10 min	5-10 min	5-10 min	5-10 min	NA
Refire Capability[6]	4-6 hrs	4-6 hrs	6-8 hrs	about 16 hrs	8-12 hrs

[1] Not yet operational.

[2] For this missile the range and warhead weight figures are for heavy nosecone (top figure) and lighter nosecone (bottom figure).

[3] The lower limit of this range approximates the percentage which might be maintained ready in continuous peacetime operations for an indefinite period. The upper limit might be achieved if the Soviets prepared their force for an attack at a specific time designated well in advance, i.e., maximum readiness.

[4] The upper limit would be more likely to be achieved if the Soviets had provided time for peaking their forces on launcher prior to an attack at a specific time.

[5] Condition I: Crews on routine standby, electrical equipment cold, missiles not fueled.
Condition II: Crews on alert, electrical equipment warmed up, missiles not fueled.
Condition III: Crews on alert, electrical equipment warmed up, missiles fueled and topped. This condition probably can not be maintained for more than an hour or so.

[6] From same pad, and dependent upon condition of alert.

15. *(Continued)*

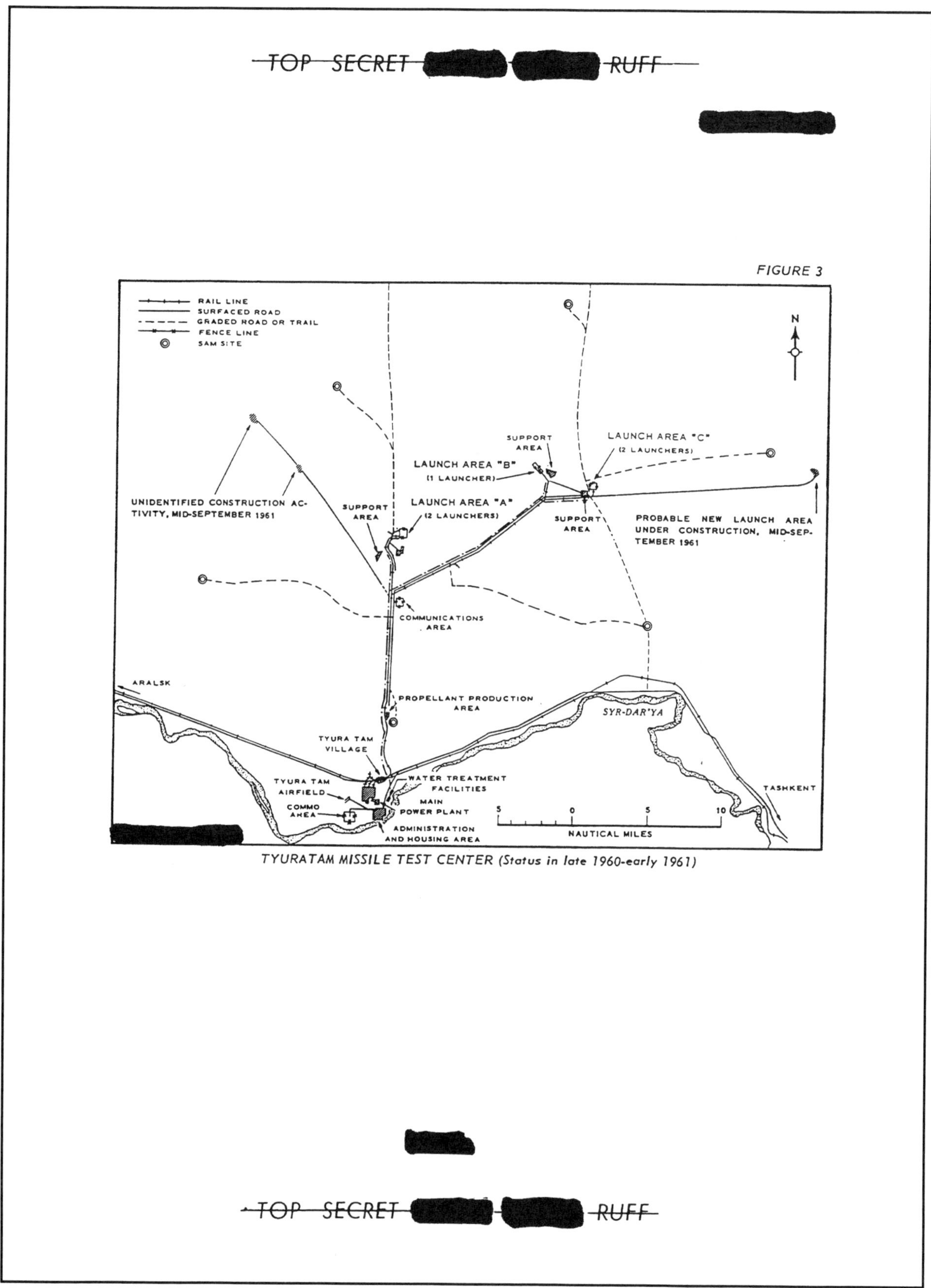

TYURATAM MISSILE TEST CENTER (Status in late 1960-early 1961)

15. *(Continued)*

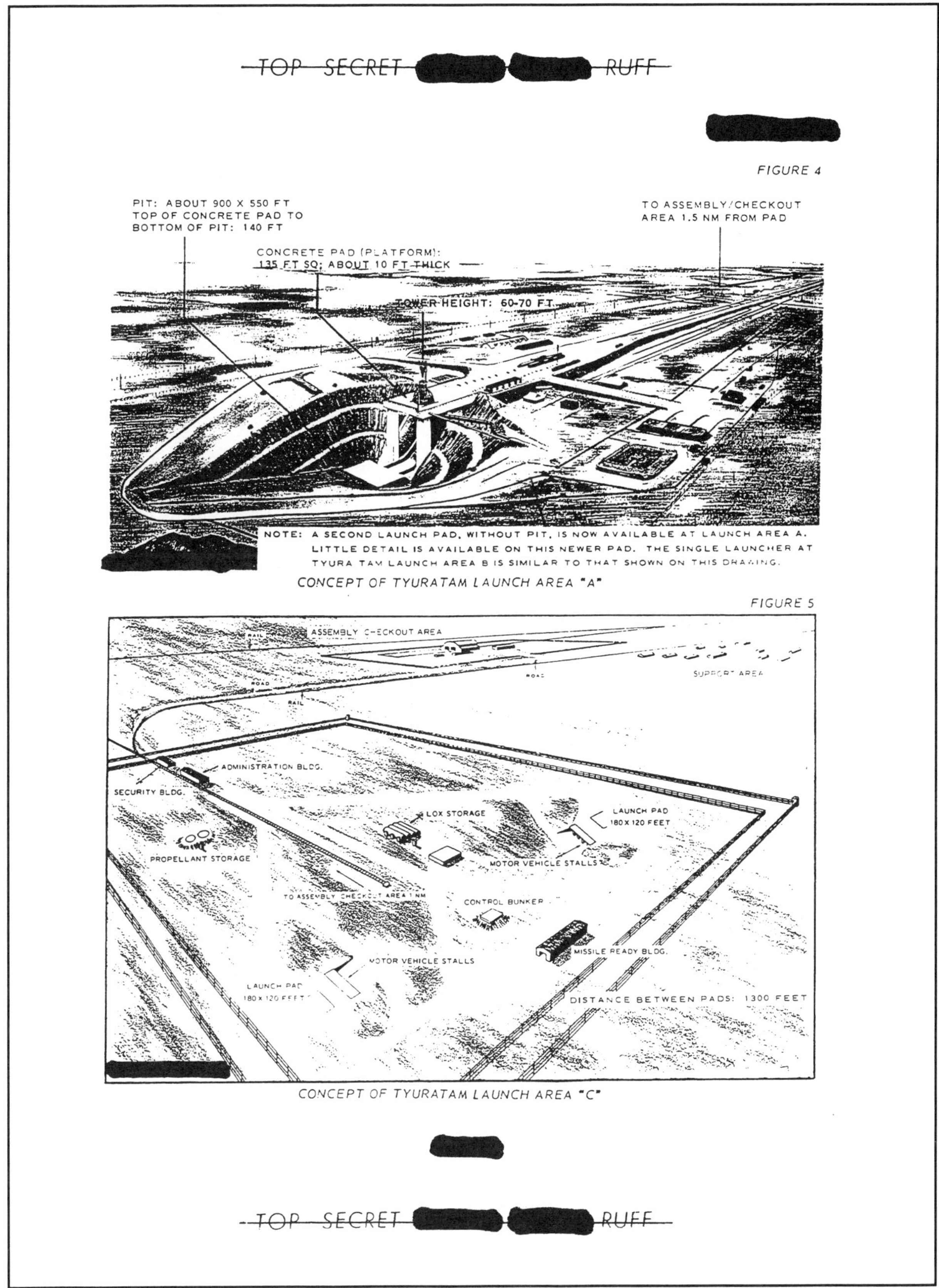

CONCEPT OF TYURATAM LAUNCH AREA "A"

CONCEPT OF TYURATAM LAUNCH AREA "C"

15. *(Continued)*

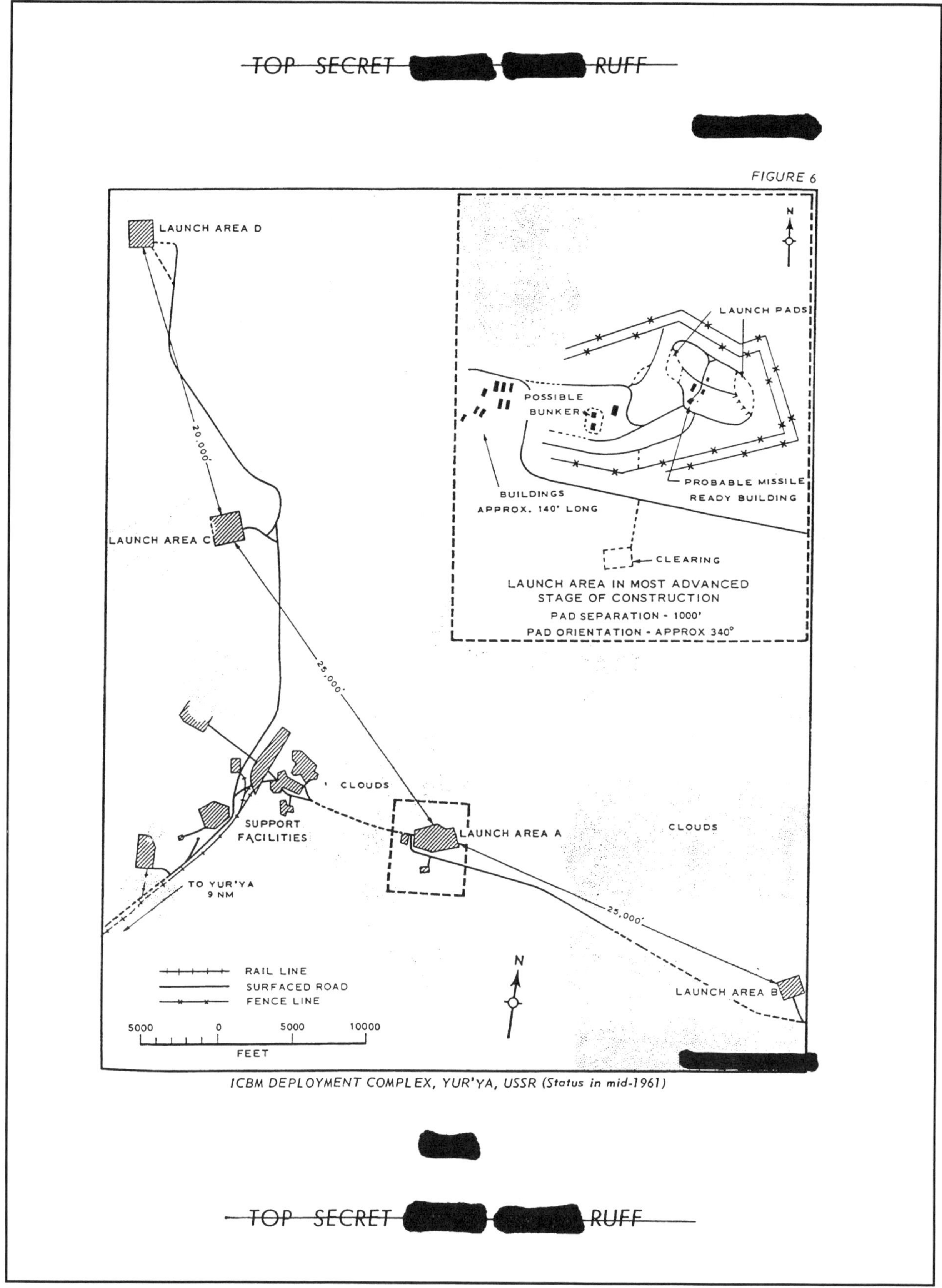

ICBM DEPLOYMENT COMPLEX, YUR'YA, USSR (Status in mid-1961)

15. *(Continued)*

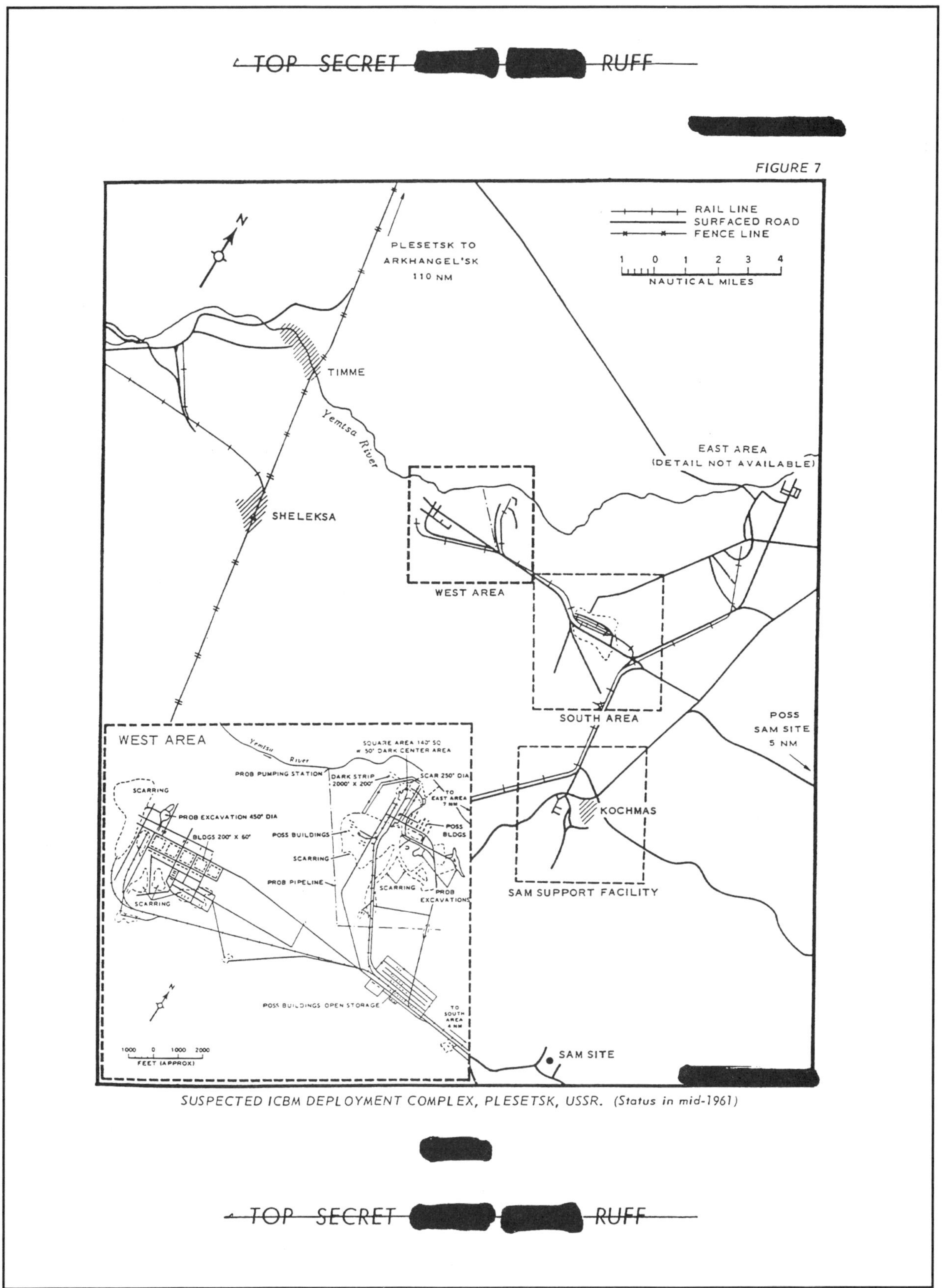

SUSPECTED ICBM DEPLOYMENT COMPLEX, PLESETSK, USSR. *(Status in mid-1961)*

15. *(Continued)*

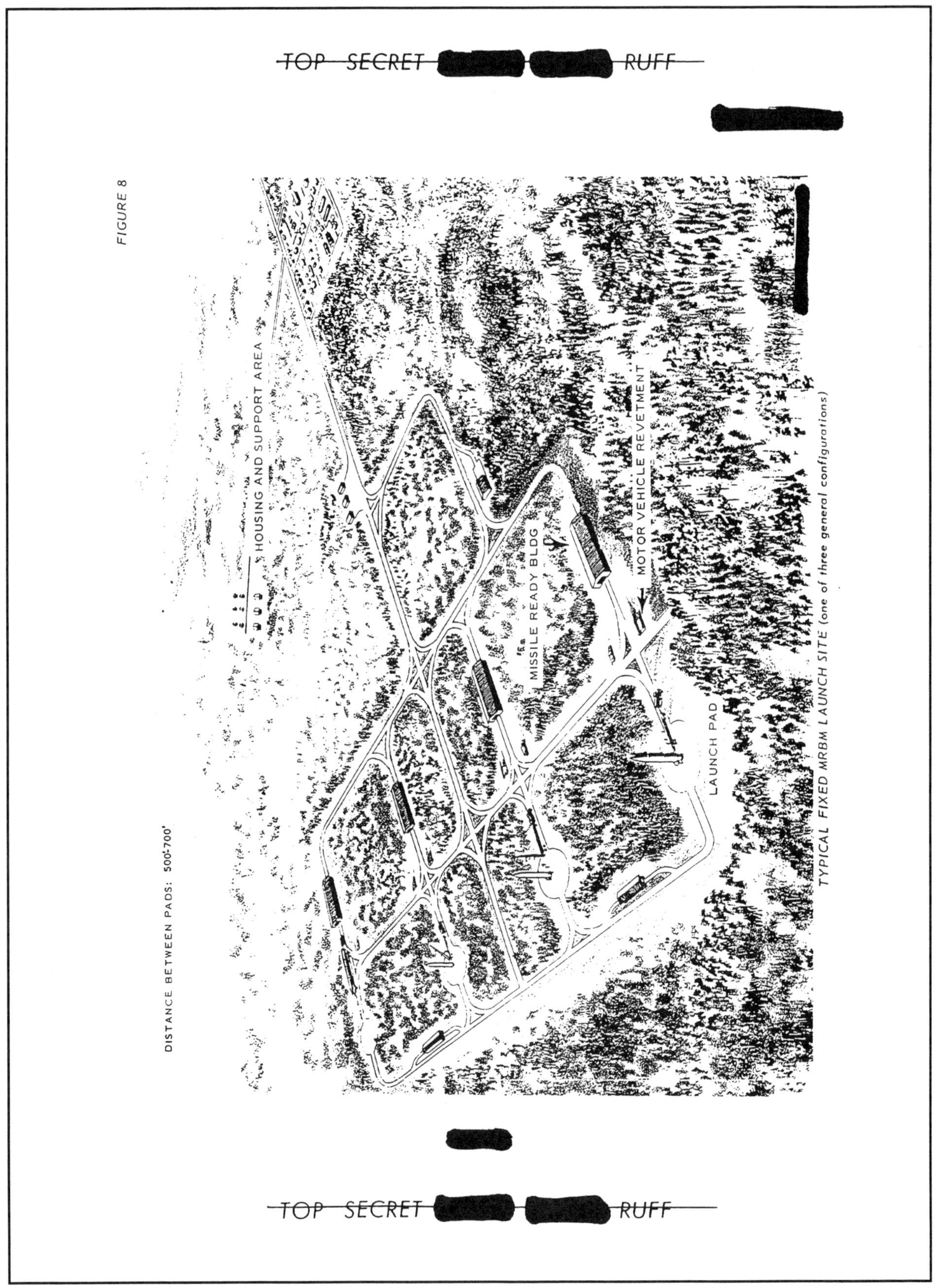

TYPICAL FIXED MRBM LAUNCH SITE (one of three general configurations)

15. *(Continued)*

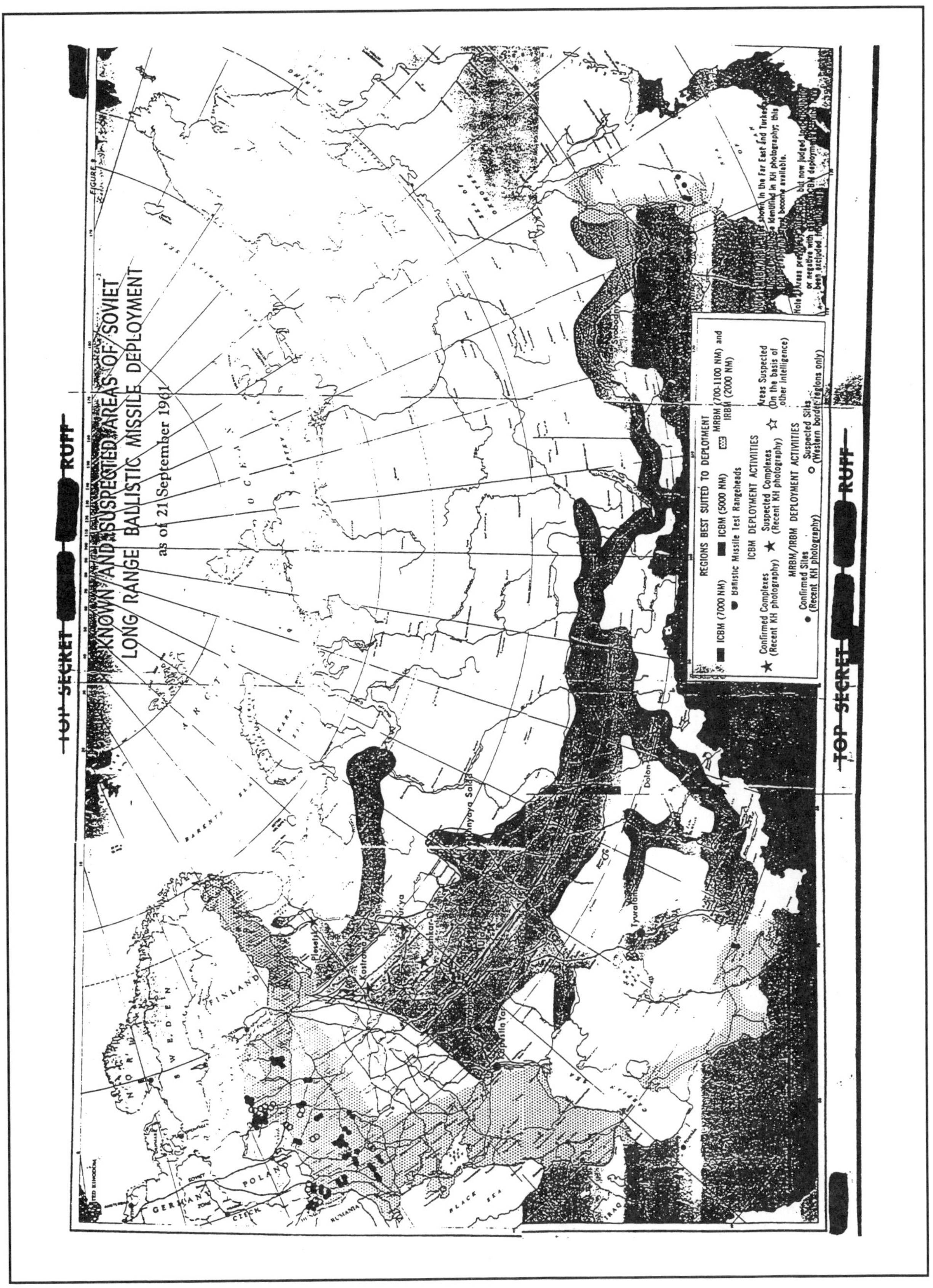

16. CIA/NPIC, Photographic Intelligence Report, "Uranium Ore Concentration Plant, Steiu, Rumania," December 1961

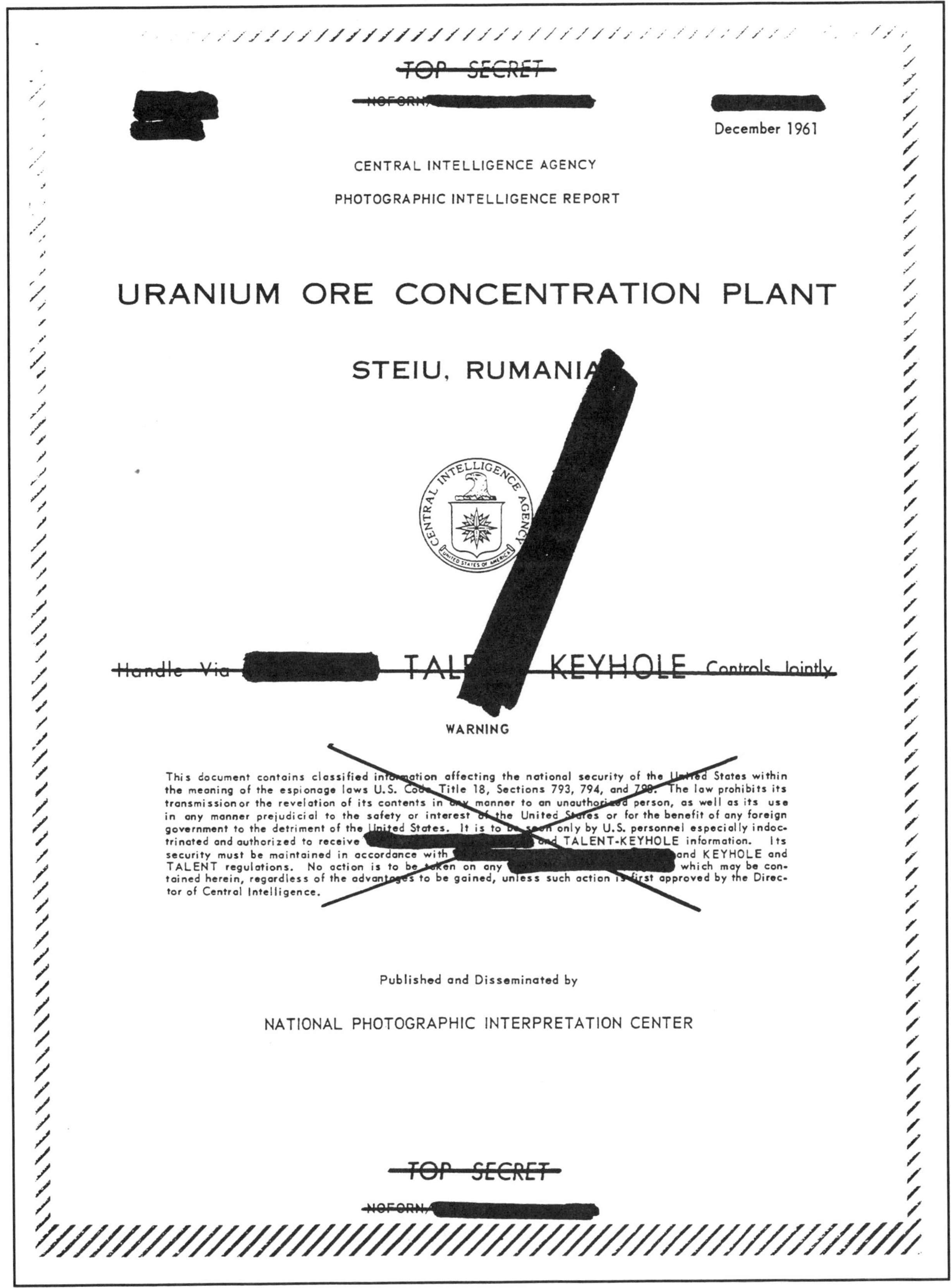

TOP SECRET

NOFORN/

December 1961

CENTRAL INTELLIGENCE AGENCY

PHOTOGRAPHIC INTELLIGENCE REPORT

URANIUM ORE CONCENTRATION PLANT

STEIU, RUMANIA

Handle Via TAL KEYHOLE Controls Jointly

WARNING

This document contains classified information affecting the national security of the United States within the meaning of the espionage laws U.S. Code Title 18, Sections 793, 794, and 798. The law prohibits its transmission or the revelation of its contents in any manner to an unauthorized person, as well as its use in any manner prejudicial to the safety or interest of the United States or for the benefit of any foreign government to the detriment of the United States. It is to be seen only by U.S. personnel especially indoctrinated and authorized to receive and TALENT-KEYHOLE information. Its security must be maintained in accordance with and KEYHOLE and TALENT regulations. No action is to be taken on any which may be contained herein, regardless of the advantages to be gained, unless such action is first approved by the Director of Central Intelligence.

Published and Disseminated by

NATIONAL PHOTOGRAPHIC INTERPRETATION CENTER

TOP SECRET

NOFORN/

16. *(Continued)*

PHOTOGRAPHIC INTELLIGENCE REPORT

URANIUM ORE CONCENTRATION PLANT

STEIU, RUMANIA

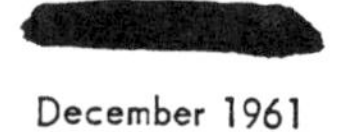

December 1961

Published and Disseminated by

NATIONAL PHOTOGRAPHIC INTERPRETATION CENTER

16. *(Continued)*

INTRODUCTION

The Steiu uranium ore concentration plant appears on fair KEYHOLE photography of 19 August 1960, 8 July 1961, and 1 September 1961 at 46-31N 22-28E, on the east-central edge of the newly constructed town of Steiu. Steiu is located in a pocketed valley along the Crisul Negru between the Bihorului and Codrului mountains in the northwest part of Rumania, approximately 35 nautical miles (nm) southeast of Oradea and 49 nm west-southwest of Cluj (Figure 1). The Steiu area is served by a good road and a single-track rail line running from Oradea and terminating 3 nm south at Vascau. Only road transportation is available from three mining areas in the vicinity to the plant. Strict security provisions are said to be in effect in the area.

The concentration process at the plant probably involves crushing of the ore, followed by ion-exchange of the solutions from the residue, and finally precipitation of uranium oxide. The uranium oxide is probably then shipped by rail to the Soviet Union, via a transshipment point at Halmeu, Rumania, approximately 87 nm north on the USSR-Rumanian border. An adjacent thermal power plant furnishes power. Possible servicing and repair facilities for the plant and mining areas are adjacent to the plant. A possible research institute is located on the southeast edge of the town. 1/ Several storage areas are located throughout the built-up area. 2/, 3/ west of the plant may be associated with it.

Annual production of the Steiu plant cannot be computed by estimating the volume of material in the possible tailings area and recovery ponds, due to the scale of the satellite photography. The small-scale photography can confirm only the general layout of the plant and provide a clue to the possible functions of the buildings at the plant. Building measurements given in this report are only approximate and their relative degree of error must be assumed to be quite large. Heights cannot be determined at all.

16. *(Continued)*

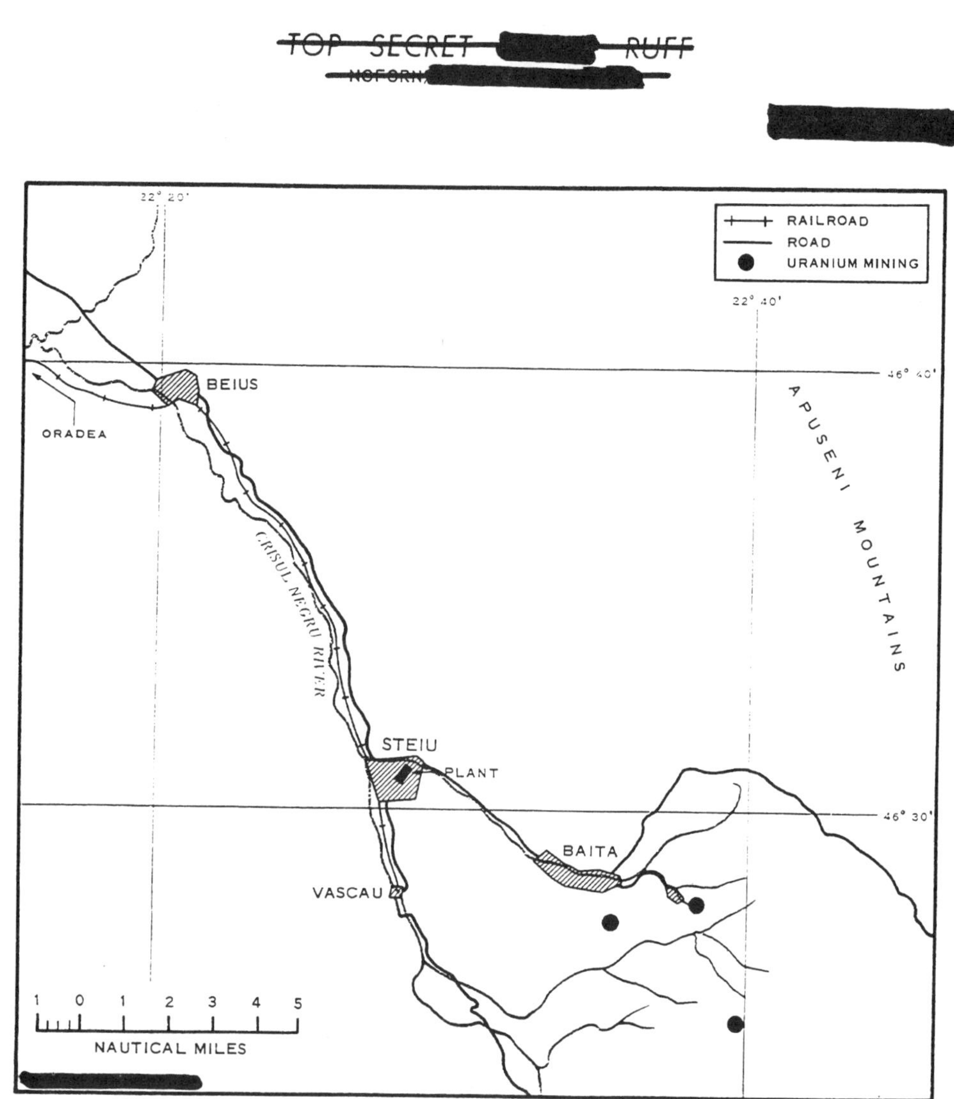

FIGURE 1. LOCATION OF STEIU URANIUM PLANT AND MINING AREAS IN VICINITY.

MINING AREAS

The uranium ore mining activity associated with the Steiu facility is 8 to 10 nm southeast of the plant in the Apuseni mountains at 46-28N 22-36E, 46-28N 22-38E, and 46-23N 22-40E (Figure 1). Ores are transported to the plant by road.

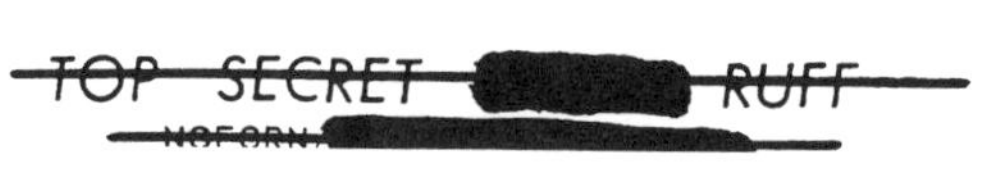

16. *(Continued)*

~~TOP SECRET [redacted] RUFF~~

~~NOFORN/~~ [redacted]

[redacted] 4/ Partial ore concentration would probably then have begun in 1955-56, with full-scale production (mining and concentration) probably being reached in 1957. These are open-pit mining operations, with two of them in the early stages of development. One mine was developed between the August 1960 and July 1961 KEYHOLE coverages. The deposits in the Apuseni mountains consist of siliceous siltstone, coated with flakes of metatorbermite. [redacted] 5/, 6/ Reserves of ore probably are adequate for a 10-year operation.

Recovery at the Steiu plant is probably on the order of 90 percent of the uranium present in the ore. If the mill was completed in 1957, it can be assumed that the production process would be comparable to the present US practice of ion exchange for the recovery of uranium oxide.

ORE CONCENTRATION PLANT

The ore concentration plant (Figure 2) occupies an area of approximately 170 acres. It contains four probable main processing buildings, a possible crusher building, associated buildings, and a possible Dorr-type thickener. Figure 2 represents a concept of the plant layout and structure based on the KEYHOLE photography and on the layout of other known plants of the same type. Approximate dimensions of the buildings are contained in the key to annotations accompanying this illustration.

The main structures visible in the plant area are a possible ore-receiving building (item 1), a possible ore classification, crusher, and grinder building (item 2), a possible Dorr-type thickener (item 3), and an ion-exchange building (item 4). Other production facilities are two possible final treatment buildings (items 5 and 6) and a possible preparation and packaging building (item 7).

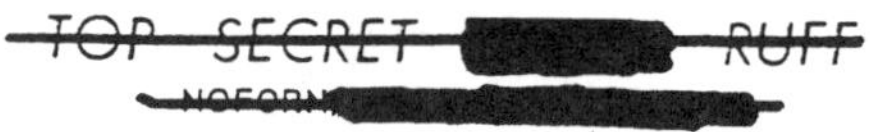

Table 1. Key To Annotations, Figure 2

Item No	Description	Approximate Dimensions (ft)	Approximate Roof Cover (sq ft)
1	Poss ore-receiving building	140 x 60	8,400
2	Poss ore classification, crusher, and grinder building	L-shaped	27,120
3	Poss Dorr-type thickener	140 diam	
4	Poss ion-exchange building	L-shaped	36,700
5	Poss final treatment building	300 x 80	24,000
6	Poss final treatment building	300 x 80	24,000
7	Poss preparation and packaging building	300 x 95	28,500
8	Prob storage and shipping building	160 x 85	13,600
9	Prob administration area (3 bldgs)	240 x 90(1) 65 x 60(2)	28,400
10	Thermal power plant, with 2 cooling towers, each 35 ft diam, and adjoining stack	300 x 85	25,500
		Total	216,220
11	Poss transformer yard	175 x 140	

Other facilities at the concentration plant include a probable storage and shipping building (item 8) and a probable administration area (item 9). A possible tailings area is located adjacent to the western edge of the plant area. An area of possible recovery ponds, with approximately 18 beds, is just north of the plant area. No pipelines are discernible on this photography.

With these facilities, the Steiu mill would appear to be a complex plant for the treatment of probably both uranium ore and concentrates. Both would be brought from the mining areas by truck to the ore-receiving and classification buildings (items 1 and 2).

16. *(Continued)*

~~TOP SECRET~~ ~~RUFF~~

~~NOFORN~~

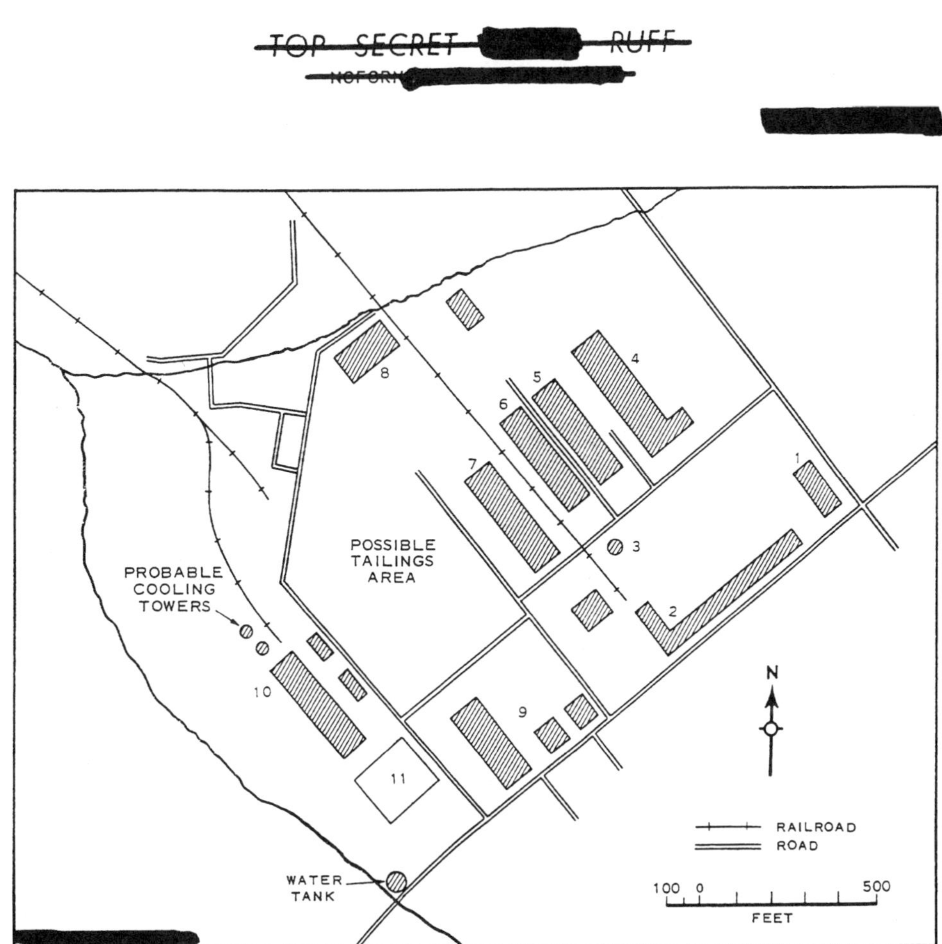

FIGURE 2. URANIUM ORE CONCENTRATION PLANT, STEIU, RUMANIA.

Blended ore would be passed through the crusher and grinder system (item 2), with some ores going to the possible thickener (item 3). All ores would then go to the ion-exchange building (item 4), and thence to the final treatment buildings (items 5 and 6). The waste material or slurry would be piped to the possible tailings pile. The possible recovery pond area is connected to the plant by a probable pipeline. It contains 18 possible settling or evaporation ponds, covering an area 600 by 500 feet.

Uranium concentrates could be shipped directly from the packaging building (item 7) or could be stored in the probable storage and shipping building (item 8) until shipment is made.

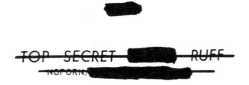

16. *(Continued)*

Since no solvent-extraction plant is visible in the vicinity of the concentration plant, it is possible that no further upgrading to green salt or metal takes place at the Steiu plant.

Production Estimates

If a plant of this size is treating mostly concentrates and small shipments of ore, its output could represent a considerable production of uranium concentrate.* There is no way of determining what portion of the mill feed is crude ore and what is concentrate from primary mills. The product of the Steiu plant probably is ammonium diuranate containing 75 to 90 percent uranium oxide.

It is very difficult even to attempt an estimate of the possible output of the Steiu plant because the scale of the satellite photography makes it impossible to determine the height of the possible tailings area and the volume of the possible recovery ponds.

TRANSPORTATION AND SECURITY

The Steiu uranium concentration plant is served by both road and rail. A single-track spur off the Oradea-Vascau single-track line serves the plant area, with spurs serving the thermal power plant, the possible research institute, and a U-shaped unidentified dead-end spur to the northeast of the built-up area (Figure 3).

A reported five-track holding yard, 1,730 feet long, is on the western edge of Steiu, with a large adjacent storage area parallel to the tracks. There are no rail facilities discernible between the uranium ore concentration plant and any of the mine areas. All transportation of the ores to the plant appears to be by road. Concentrated ores could be shipped to the Soviet Union for further processing through a rail transshipment point at Halmeu, Rumania, on the Soviet border.

* It is felt that, during its first years, the plant's input consisted largely of crude ore, but that the input of concentrates increased steadily, so that the input would now be high in concentrates and low in crude ore.

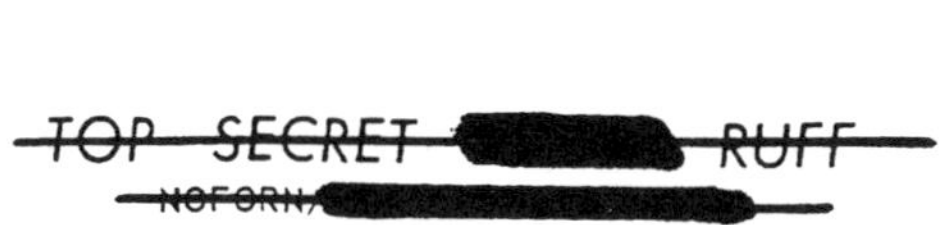

16. *(Continued)*

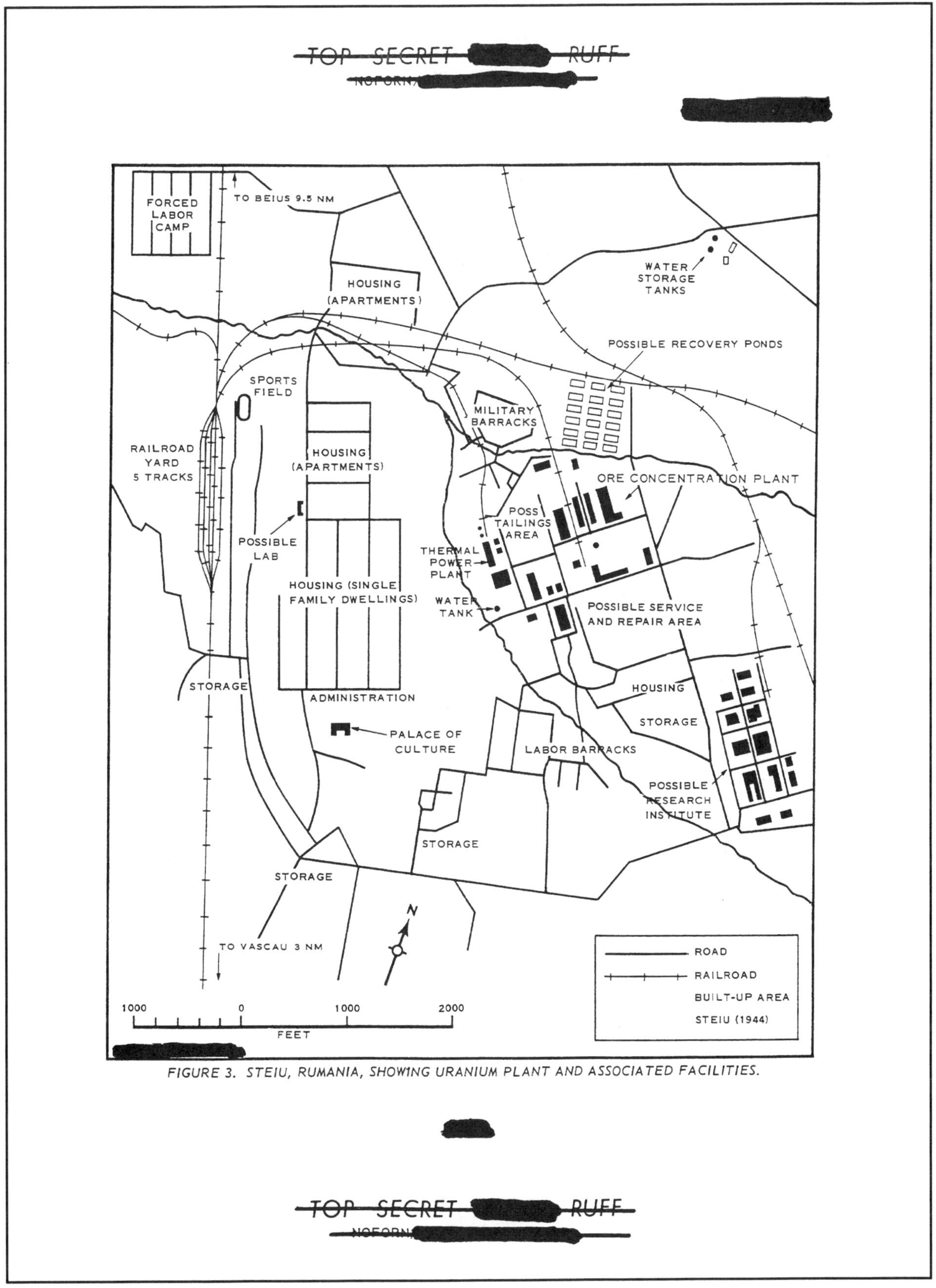

FIGURE 3. STEIU, RUMANIA, SHOWING URANIUM PLANT AND ASSOCIATED FACILITIES.

16. *(Continued)*

~~TOP SECRET~~ [redacted] ~~RUFF~~
~~NOFORN~~ [redacted]

No fences or walls can be seen surrounding the uranium concentration plant, but it is reported that the area is divided into three strictly divided zones of security, with limited access to each zone. People in the various zones are not permitted into all zones, and are very restricted in their movement within the area, as well as in their movement to other parts of Rumania.

SUPPORT FACILITIES

The extensive support facilities seen on the photography could be central facilities for servicing the tributary mines and plants in the area as well as the ore concentration plant.

The rail-served thermal power plant immediately west of the uranium ore concentration plant, contains a boilerhouse and generator hall (item 10, Figure 2), two probable cooling towers, a possible transformer yard (item 11), and a water tank.

A possible service and repair area, adjacent to the south edge of the plant, is probably for both the uranium ore concentration plant and the three mining areas. The area contains 13 buildings of various sizes.

The possible research institute 1/ is located on the southeast edge of the town area. Nothing is known of the work of the institute, [redacted] 1/, 7/

The town of Steiu has grown tremendously since the original Rumanian agricultural village was seen on German photography of 1944. On the photography of 1960-61, several large areas of multistory apartment buildings, single-family dwelling areas, reported military and labor barracks areas, [redacted] 8/ were viewed.

Water is supplied from the Crisul Negru, with water storage tanks located at various points throughout the built-up area. The small-scale satellite photography reveals no pipelines.

CONCLUSIONS

Satellite photography of August 1960 and July and September 1961 confirms the existence and operation of a uranium ore concentration plant, a storage and repair area, and a possible research institute at Steiu, approximately 35 nm southeast of Oradea and 49 nm west-southwest of Cluj, in the northwest part of Rumania.

The following conclusions may be drawn from the photography and from collateral information used:

1. The Steiu uranium ore concentration plant produces an unestimated amount of uranium oxide in the form of ammonium diuranate from ores [redacted] [redacted]. The plant product is estimated to have a grade of 75 to 90 percent uranium oxide.

2. The plant probably upgrades concentrates from preliminary processing plants located at three uranium mines. There are three nearby areas of uranium mining, located 8-10 miles southeast and connected by road to the Steiu uranium ore concentration plant.

3. A possible storage and repair area is probably associated with the plant and the mining areas.

4. A possible research institute, probably connected with the uranium ore concentration plant and the mining areas, is located on the southeast edge of Steiu.

16. *(Continued)*

REFERENCES

PHOTOGRAPHY

Mission	Date	Pass/Camera	Frames	Classification
GX 9664	7 Oct 44	SK	87, 88	U
9009	19 Aug 60	9	45	TSR
9019	8 Jul 61	9	114	TSR
9023	1 Sep 61	25	118	TSR

MAPS AND CHARTS

AMS. Series M 506 (GSGS4375), Sheet R-26, 3d ed, Mar 54, scale 1:250,000 (U)

AMS. Series M 606 (GSGS4417), Sheet 2065, 3d ed, Jan 54, scale 1:100,000 (U)

ACIC. USAF Operational Navigation Chart 251, 1st ed, Mar 59, scale 1:1,000,000 (U)

ACIC. US Air Target Chart, Series 200, Sheets 0251-8A and 13A, 1st ed, Mar and May 59, scale 1:200,000 (S)

DOCUMENTS

1. CIA. Free Rumanian Press, release #8, 20 Feb 57 (U)

REQUIREMENT

17. CIA/NPIC, Photographic Intelligence Report, "Regional Nuclear Weapons Storage Site Near Berdichev, USSR," May 1963

~~TOP SECRET~~

May 1963

CENTRAL INTELLIGENCE AGENCY

PHOTOGRAPHIC INTELLIGENCE REPORT

REGIONAL NUCLEAR WEAPONS STORAGE SITE NEAR BERDICHEV, USSR

~~Handle Via~~ [redacted] - TALENT - KEYHOLE ~~Controls Jointly~~

WARNING

This document contains classified information affecting the national security of the United States within the meaning of the espionage laws U. S. Code Title 18, Sections 793, 794, and 798. The law prohibits its transmission or the revelation of its contents in any manner to an unauthorized person, as well as its use in any manner prejudicial to the safety or interest of the United States or for the benefit of any foreign government to the detriment of the United States. It is to be seen only by personnel especially indoctrinated and authorized to receive [redacted] and TALENT-KEYHOLE information. Its security must be maintained in accordance with [redacted] and KEYHOLE and TALENT regulations. No action is to be taken on any [redacted] which may be contained herein, regardless of the advantages to be gained, unless such action is first approved by appropriate authority.

Published and Disseminated by

NATIONAL PHOTOGRAPHIC INTERPRETATION CENTER

~~TOP SECRET~~

GROUP 1
Excluded from automatic downgrading and declassification

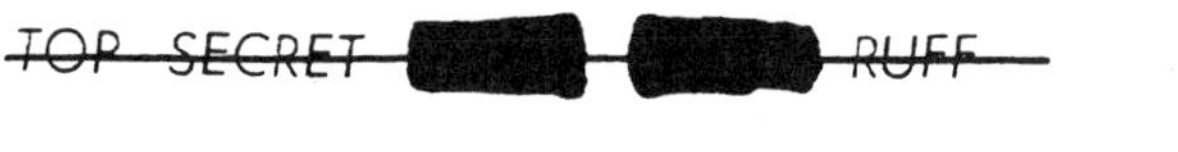

REGIONAL NUCLEAR WEAPONS STORAGE SITE

NEAR BERDICHEV, USSR

Good-quality photography providing details of the regional nuclear weapons storage site (49-56N 28-16E) located 2.1 nautical miles (nm) west of the Berdichev/Mikhaylenki Regional Military Storage Installation and 12 nm west of Berdichev, USSR, is available from several KEYHOLE missions, particularly from Mission 9037 of June 1962 (Figures 1 and 2). The cruciform buildings at the Berdichev site (Figure 3) are very similar to the cruciform building observed under construction at the Type III nuclear weapons storage site at Dolon Airfield (Figures 1 and 4). 1/ The layouts of the sites at Berdichev and Dolon are also similar.

DESCRIPTION OF BERDICHEV SITE

This site consists of a double-fenced area measuring approximately 4,900 by 1,650 feet and a small single-fenced support area adjoining the eastern side of the double-fenced area (Figure 2). All of the buildings appear completed although the cruciform buildings were not yet earth mounded.

The double-fenced area contains two cruciform buildings, a drive-through checkout building, and a small unidentified building. The cruciform buildings are located at the northwest and southeast ends of the double-fenced area. Each

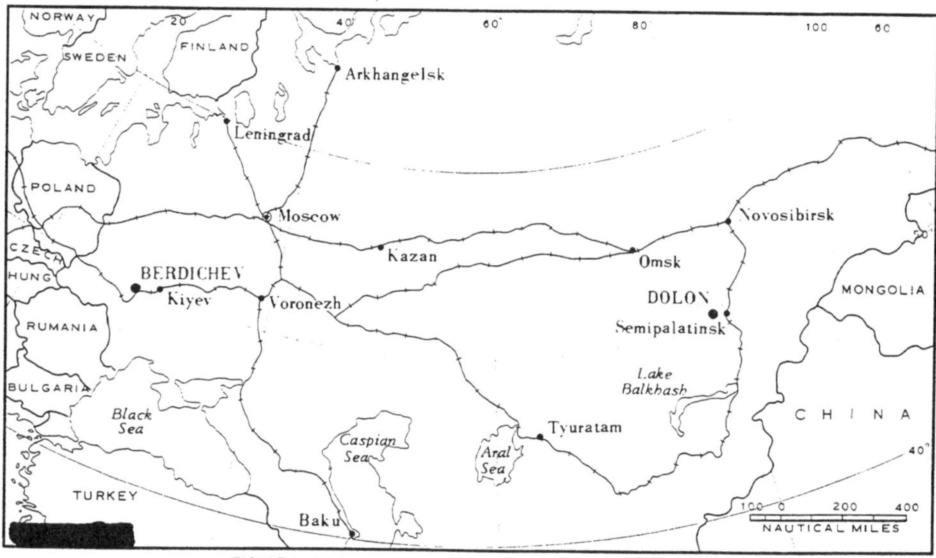

FIGURE 1. LOCATIONS OF NUCLEAR WEAPONS STORAGE SITES.

is a heavily constructed drive-through building and is encircled by a road. Photography of November 1962 revealed that the southeast cruciform building, which appeared under construction in June 1962, is complete. The cruciform buildings, located approximately 3,250 feet apart, are connected by road. A drive-through checkout building (190 by 85 feet) is located along this road approximately 1,400 feet from the southeast cruciform building. A road within the area parallels the inner fence and frames the area. A small unidentified building (55 by 30 feet) is located between the inner and outer fences on the southwestern side of the area.

The support area consists of an administration building and four support buildings. The administration building measures 210 by 80 feet, three of the support buildings measure 85 by 45 feet, and one support building measures 75 by 45 feet. Except for this small support area, the only transportation, communications, and other support facilities serving the site are located 2.1 nm west at the Berdichev Regional Military Storage Installation.

COMPARISON WITH DOLON SITE

The regional nuclear weapons storage site at Berdichev and the Type III nuclear weapons

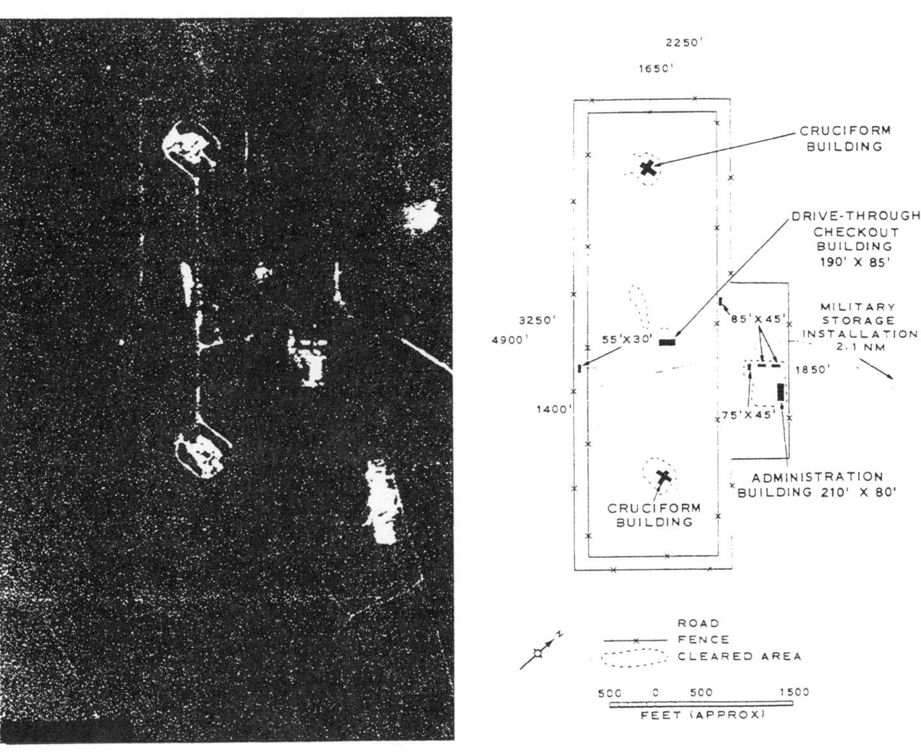

FIGURE 2. LAYOUT OF NUCLEAR WEAPONS STORAGE SITE NEAR BERDICHEV (JUNE 1962).

17. *(Continued)*

~~TOP SECRET~~ [redacted] ~~RUFF~~

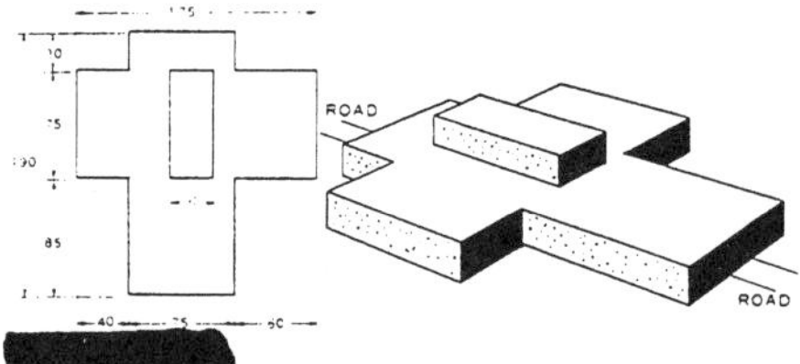

FIGURE 3. PLAN AND PERSPECTIVE VIEW OF NORTHWEST CRUCIFORM BUILDING AT REGIONAL NUCLEAR WEAPONS STORAGE SITE NEAR BERDICHEV, USSR.

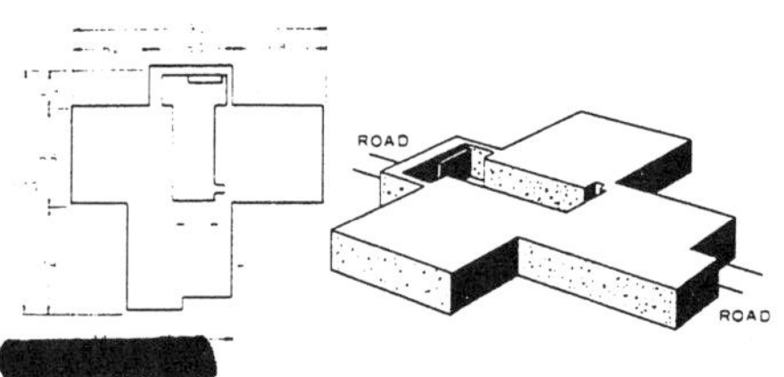

FIGURE 4. PLAN AND PERSPECTIVE VIEW OF CRUCIFORM BUILDING UNDER CONSTRUCTION AT TYPE III NUCLEAR WEAPONS STORAGE SITE AT DOLON AIRFIELD, USSR.

storage site at Dolon Airfield are generally similar. The measurements of the cruciform buildings at Berdichev (measurements for the northwest building are given in Figure 3) are close to the measurements of the cruciform building at Dolon (given in Figure 4). The measurements at Berdichev, based on KEYHOLE photography, are less precise than those at Dolon, based on TALENT photography. Minor details observed at Dolon are not discernible on the small-scale photography of Berdichev. The main difference between the cruciform buildings at the two sites is that the drive-through section of the northwest cruciform building at Berdichev is the longer section while the drive-through section of the cruciform building at Dolon is the shorter section.

The Berdichev and Dolon sites differ in the location of various buildings, particularly the checkout building, and in the layout of security fencing. Both sites have adjoining support areas, but there is variation in the number and dimensions of the buildings in the areas.

———— • ————

17. *(Continued)*

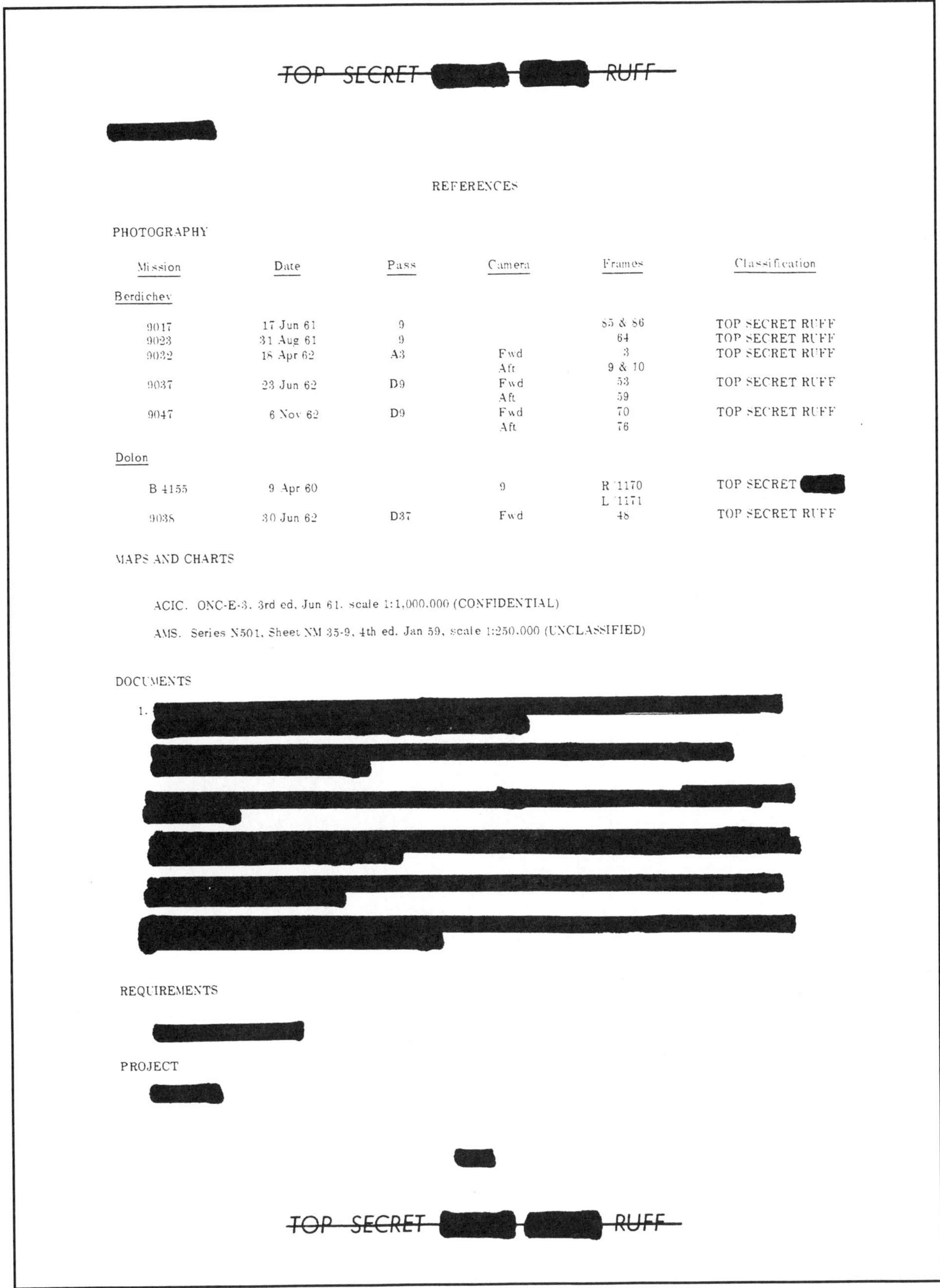

~~TOP SECRET~~ [redacted] [redacted] ~~RUFF~~

REFERENCES

PHOTOGRAPHY

Mission	Date	Pass	Camera	Frames	Classification
Berdichev					
9017	17 Jun 61	9		85 & 86	TOP SECRET RUFF
9023	31 Aug 61	9		64	TOP SECRET RUFF
9032	18 Apr 62	A3	Fwd	3	TOP SECRET RUFF
			Aft	9 & 10	
9037	23 Jun 62	D9	Fwd	53	TOP SECRET RUFF
			Aft	59	
9047	6 Nov 62	D9	Fwd	70	TOP SECRET RUFF
			Aft	76	
Dolon					
B 4155	9 Apr 60		9	R 1170	TOP SECRET [redacted]
				L 1171	
9038	30 Jun 62	D37	Fwd	48	TOP SECRET RUFF

MAPS AND CHARTS

ACIC. ONC-E-3. 3rd ed. Jun 61. scale 1:1,000,000 (CONFIDENTIAL)

AMS. Series N501. Sheet NM 35-9. 4th ed. Jan 59. scale 1:250,000 (UNCLASSIFIED)

DOCUMENTS

1. [redacted]

REQUIREMENTS

PROJECT

~~TOP SECRET~~ [redacted] [redacted] ~~RUFF~~

18. CIA/NPIC, Photographic Intelligence Report, "Search for Uranium Mining in the Vicinity of A-Ko-Su, China," August 1963

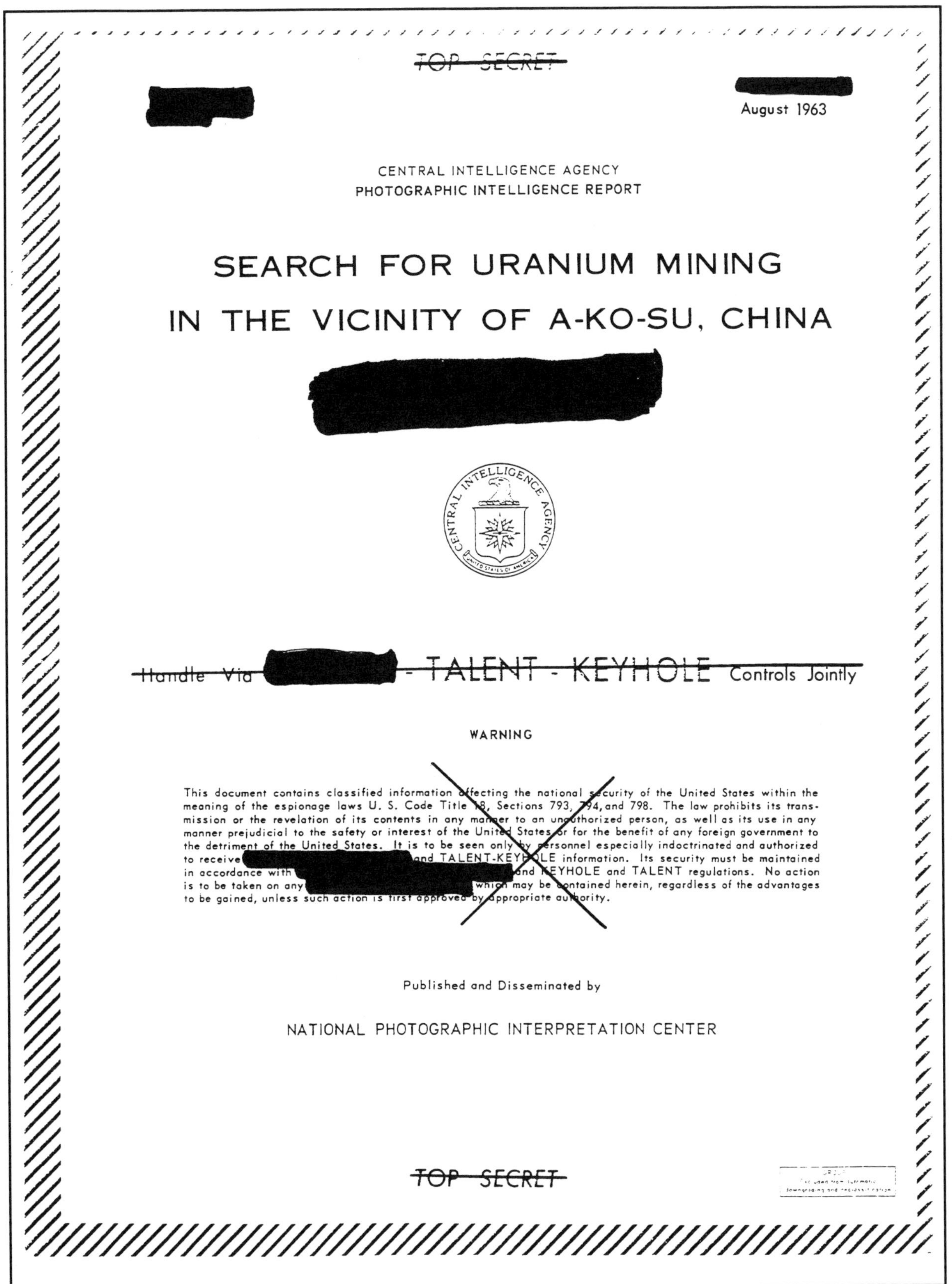

~~TOP SECRET~~

August 1963

CENTRAL INTELLIGENCE AGENCY
PHOTOGRAPHIC INTELLIGENCE REPORT

SEARCH FOR URANIUM MINING IN THE VICINITY OF A-KO-SU, CHINA

~~Handle Via~~ [redacted] ~~- TALENT - KEYHOLE~~ Controls Jointly

WARNING

This document contains classified information affecting the national security of the United States within the meaning of the espionage laws U. S. Code Title 18, Sections 793, 794, and 798. The law prohibits its transmission or the revelation of its contents in any manner to an unauthorized person, as well as its use in any manner prejudicial to the safety or interest of the United States or for the benefit of any foreign government to the detriment of the United States. It is to be seen only by personnel especially indoctrinated and authorized to receive [redacted] and TALENT-KEYHOLE information. Its security must be maintained in accordance with [redacted] and KEYHOLE and TALENT regulations. No action is to be taken on any [redacted] which may be contained herein, regardless of the advantages to be gained, unless such action is first approved by appropriate authority.

Published and Disseminated by

NATIONAL PHOTOGRAPHIC INTERPRETATION CENTER

~~TOP SECRET~~

SEARCH FOR URANIUM MINING IN THE VICINITY OF A-KO-SU, CHINA

SUMMARY AND CONCLUSIONS

This report is in response to a request for a search from photography for uranium mining or other activity related to atomic energy within a 50-nautical mile (nm) radius of A-ko-su (41-10N 80-16E), Hsin-chiang Sheng (Sinkiang Province), China (Figure 1). Photography from four KEYHOLE missions (December 1960, December 1961, November 1962, and December 1962) was examined. The search revealed two areas of mining and prospecting activity located approximately 30 to 45 nm northeast of A-ko-su in the valleys of the Tien-shan range on the Sino-Soviet border (Figure 2). A supply base for this activity was not definitely located.

The activity observed near A-ko-su is identified as the mining of coal which may possibly contain uranium. Lignite deposits are known to exist in the vicinity of A-ko-su. The total cumulative production of lignite for the period 1959 through 1961 is estimated from photographic evidence at 30,000 to 40,000 metric tons of lignite. If uranium is present, this output could yield from 15 to 30 metric tons equivalent of uranium oxide (U_3O_8).

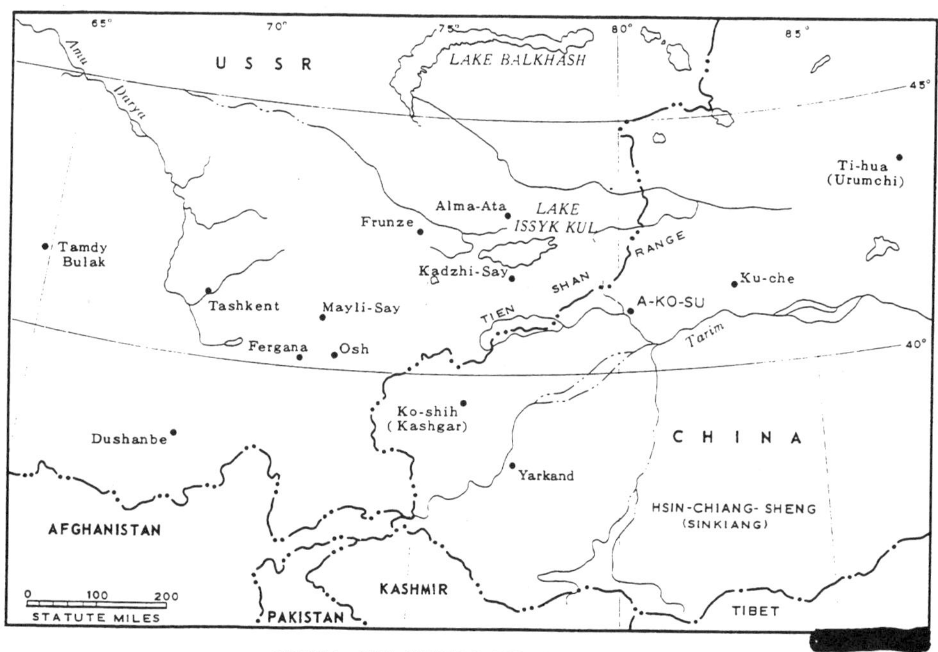

FIGURE 1. AREA OF SEARCH FOR URANIUM MINING.

TOP SECRET RUFF

~~TOP SECRET~~ ~~RUFF~~

Although photographic evidence of uranium processing was not observed, the possibility of uranium extraction cannot be discounted. Some evidence of extra security which is usually associated with uranium activity was observed at the mining sites. Observations of some activity at the mines during periods of snow cover is evidence of the priority that would be assigned to uranium extraction.

Commercial-grade uraniferous ore deposits are known to exist on the Soviet side of the Tien-shan range and the presence of uraniferous coal deposits in the A-ko-su region is suspected.

PHOTOGRAPHIC OBSERVATIONS

The activity observed in the vicinity of A-ko-su is located in two areas which are designated in this report as the Eastern Area and the Western Area (Figure 2). In the Eastern Area, five mining sites, one prospecting site, a treatment plant, and a possible explosives magazine were observed. In the Western Area, three prospecting sites were observed. For purposes of description, site numbers have been assigned to identify the locations of mines and prospects.

Evidence of Mining. The earliest photography (December 1960) of the mining sites (all in the Eastern Area) showed a cluster of five mines at Site 5, all apparently in production, and two mines--one at Site 1 and another at Site 2--apparently being readied for production. The December 1961 photography revealed all mines in production and the presence of a higher pile of coal refuse, although the pile covered approximately the same area as it had in 1960. Track patterns evident at times of snow cover indicated truck traffic on the roads serving the mines. This also indicated the continuing operation and development of the mines during winter.

Accumulation of coal in piles for possible reprocessing was observed at the treatment plant in the Eastern Area. The stockpiling of coal at the treatment plant may indicate the possibility that the coal is reprocessed for the extraction of a by-product. A by-product such as uranium could be produced in such small quantities that it would elude photographic observation. It could be transported to a center without perceptible traffic indications. Little or no accumulation of coal or ashes was observed in the towns and villages in the region.

Production Estimate. Based on the observed accumulation of coal refuse at the treatment plant, the total cumulative production of coal in the Eastern Area from 1959 (when digging probably began) to December 1961 is estimated at 30,000 to 40,000 metric tons. Coal production for the period December 1960-December 1961 is estimated at 25,000 metric tons. If the observed mining prospect at Site 2 in the Western Area is developed into a producing mine, the area's annual coal production could increase by an additional 10,000 metric tons. These estimates do not allow for some local consumption of coal.

For an estimate of possible uranium yield, the coal deposits of the A-ko-su area are assumed to resemble other weathered near-surface deposits of uranium-bearing Jurassic coals, such as those on the Soviet side of the border. These deposits may yield from 0.05 to

~~TOP SECRET~~ ~~RUFF~~

18. *(Continued)*

~~TOP SECRET~~ [redacted] [redacted] ~~RUFF~~

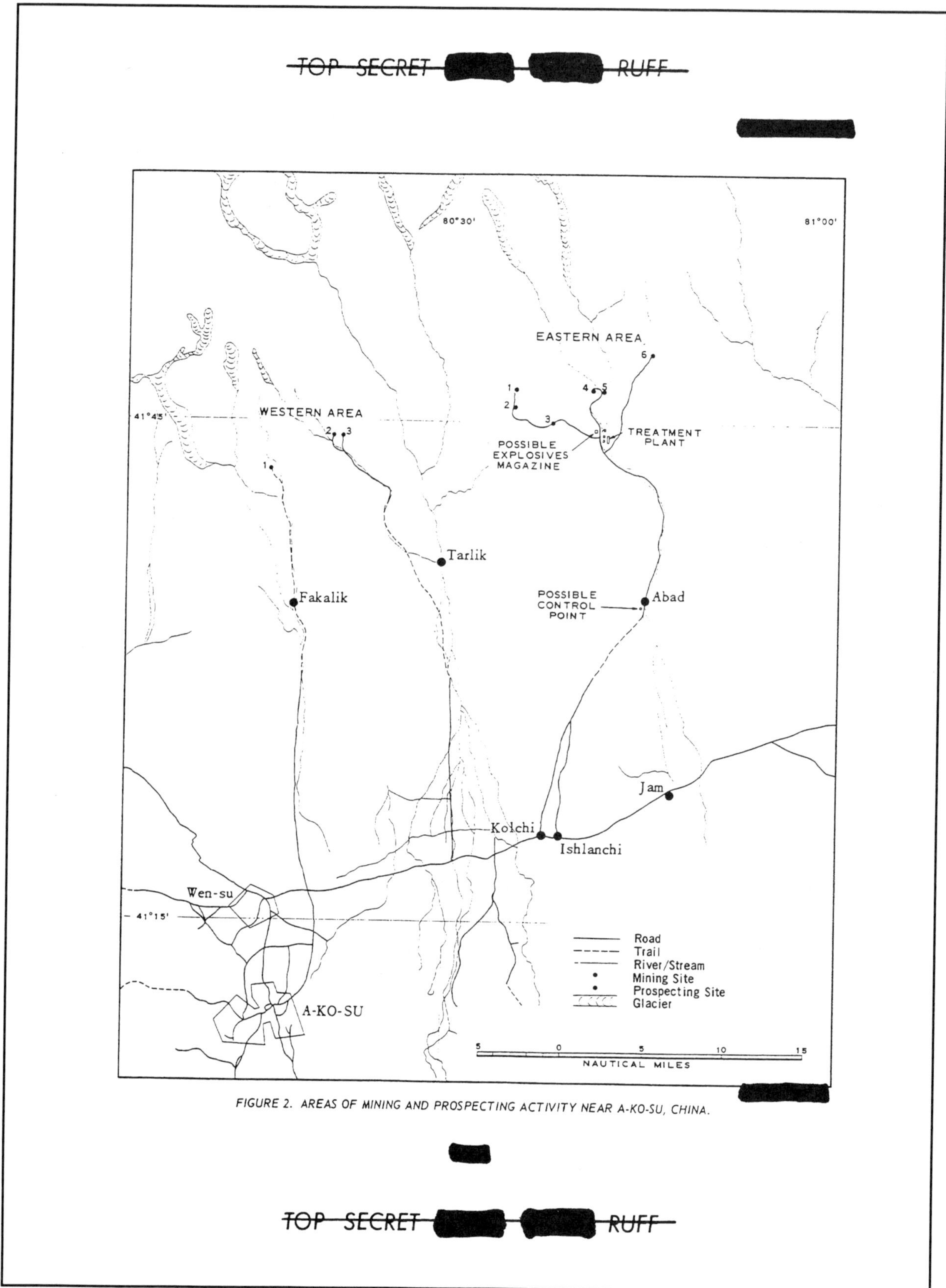

FIGURE 2. AREAS OF MINING AND PROSPECTING ACTIVITY NEAR A-KO-SU, CHINA.

~~TOP SECRET~~ [redacted] [redacted] ~~RUFF~~

~~TOP SECRET~~ [redacted] [redacted] ~~RUFF~~

0.75 percent U_3O_5 equivalent. This could place the area's total cumulative yield of U_3O_5 equivalent through December 1961 at 15 to 30 metric tons.

Evidence of Security. The areas of mining activity are located in remote valleys. To the north is an area of high glacier-covered mountains, and to the south, treacherous sands are found on the alluvial fans at the valley mouths. The valleys physiographically resemble those emptying on the Fergana Valley of the USSR in being deep and narrow at their lower ends. These characteristics permit easy control of entrance to and egress from a valley and allow compartmentalization of operations in a valley. Prison labor could be used at mine sites in these valleys with a minimum of control. Prisoners probably were brought to the mines in 1959 and 1960. Some security precautions were observed in the Eastern Area. A possible control point is located on the access road from Jam near the mouth of the first valley at Abad. Possible fences were observed at two of the mines (Sites 2 and 5).

Search for a Support Base. A support base for the observed mining activity was not definitely located. Such identifying features as piles of coal, ashes, or pit props in association with warehouses were not observed. No laboratory-type building was visible in the Eastern or Western Areas. Photography of the principal settlements--A-ko-su, Jam, and Wen-su--was examined closely. The activity at Wen-su, located in a valley with a steep eastern slope, is probably local in nature and not related to mining. A-ko-su is the most likely location for a general support base. Roads from the prospecting and mining sites converge on A-ko-su. Barnlike buildings on the eastern outskirts of the town and on the road to the mining areas may serve a support function. Traffic in the vicinity of these buildings, as indicated by patterns in the snow, seemed to exceed the level expected from local agricultural activity. However, winter shelter for grazing animals may account for the additional activity. Barracks-type buildings on the large farms (presumably state farms), 21 nm south-southeast of the mines near Jam, may house administrative support for the Eastern Area, and the farms, located on the northern edge of the irrigated plains, may supply provisions for the mining settlements.

EASTERN AREA

The Eastern Area, located approximately 39 nm north-northeast of A-ko-su, is the principal area of mining activity in the region (Figure 2). A treatment plant and a possible explosives magazine are centrally located with respect to the mining and prospecting sites in the area. Access to the area is by a road from Jam 21 nm to the south-southeast. A possible control point is located 10 nm to the south-southeast of the treatment plant on the access road and near the mouth of the valley. At the times of snow cover the access road appeared lightly traveled.

Treatment Plant (41-44N 80-42E). This plant is located at the junction of the access road from Abad with the road from the northwestern (Sites 1-3) and northern (Sites 4 and 5) sites and a road from the northeastern site (Site 6). The plant consists of two small adjoining square buildings identified as mills. A small pile of waste is located just south of each mill, a small rectangular building is located southeast of each mill, and a possible storage building is situated east of each mill. Water is piped to the plant from the river which flows from the northwest valley. However, dry-milling and manual removal of waste are apparently practiced during periods of freeze. The photography of November and December 1962

~~TOP SECRET~~ ~~RUFF~~

during snow cover indicated that the plant was probably operating at a low rate. Tracks in the snow indicated light traffic on the roads serving the treatment plant. Dust and water seepage were evident at the plant's coal pile.

Other features observed near the plant include the following: three medium-sized dormitory-type buildings located just west of the mills; a small square building identified as a possible control building located southwest of the plant on the west side of the access road; and a motor pool and/or equipment park, including a small rectangular building, located in a triangular area across the entrance-exit road.

Possible Explosives Magazine (41-44N 80-41E). This facility is located northwest of the treatment plant off the road to the western sites and near the junction with the road to the northwestern sites. Its location on the route between the mines and the treatment plant would allow trucks to carry return loads of explosives to the mines. The possible explosives magazine is secured and road served. Although this facility appeared on the November 1962 photography to be inactive, light activity indicating partially operating mines was observed on the December 1962 photography.

Site 1 (41-47N 80-35E). This site is located on the eastern side of a valley and contains an opencut mine. A village is located west of the mine in a valley. The December 1960 photography indicated that the mine was being readied for production. The site is the western terminus of a well-traveled road which also serves Sites 2 and 3. The road was not being used extensively in 1960. December 1962 photography revealed that the mine had probably been shut down, although tracks in the snow to the mine were observed.

Site 2 (41-46N 80-35E). This site contains an open-pit mine, the largest mine in the area. The terrain of a fenced area east of the mine appears broken, apparently caused by slumping from underground mining. A village is located south of the mine. The December 1960 photography showed the mine in production. At that time the road from this site to the treatment plant was well traveled. The December 1962 photography revealed a coal pile below the mine. Tracks in the snow indicating traffic activity at the mine were observed. The road from the site to the treatment plant was open, but the continuation of the road to Site 1 appeared to be inactive.

Site 3 (41-45N 80-38E). Site 3 consists of two small opencut prospects which are located halfway up the west side of a ridge. The site probably contains only limited reserves of coal.

Site 4 (41-47N 80-41E). This site contains a possible opencut mine and a small housing area. The site is located on a perched upper slope. It is served by a branch from the well-traveled road which also serves Site 5. The mine appeared to be inactive on the December 1962 photography, although tracks in the snow to the mine were discernible.

Site 5 (41-47N 80-42E). This site is the oldest and best developed mining site in the area. It consists of a large portal mine located on the eastern side of a valley and a cluster of four small opencut mines located on the broken western slope of the valley where faults probably limit the availability of reserves. The portal mine may have large reserves. A small pile, probably of coal, is observed on the floor of the narrow valley at the junction of a loop road serving these mines and the road to the treatment plant. A possible housing area is located in the center of the valley. A possible guard fence, with guard towers, crosses the valley below the mines and the possible housing area. A fence partially encloses the portal mine.

~~TOP SECRET~~ ~~RUFF~~

Lack of heavy traffic patterns on the road toward the treatment plant at the time of snow cover suggests that coal produced was being stockpiled at the site. The December 1962 photography revealed that the four opencut mines were inactive. Tracks in the snow to the mines showed maintenance activity was in progress.

Site 6 (41-49N 80-46E). Site 6 contains three small opencut mines located halfway up the eastern slope of a ridge. Each mine is served by a steep, well-defined trail. Scattered settlements are located 3 nm down the valley. The December 1962 photography indicated that the mines were inactive, although tracks in the snow to the mine were observed.

WESTERN AREA

The Western Area, located approximately 34 nm north of A-ko-su, contains three prospecting sites (Figure 2). A prospect at one of the sites (Site 2) is being developed for a mine. The sites are served by two separate trails. Routes suitable for vehicle use have been observed.

Site 1 (41-42N 80-15E). Site 1 contains a prospect located in a mountain meadow, and numerous trails leading to cliffs indicate other prospecting activity. Three rows of unidentified objects, possibly huts or stacks of supplies, were observed in a valley west of the prospect. Ten small settlements near the site serve as centers for farming and prospecting. The principal trail serving the site leads southward through the village of Fakalik where it becomes a secondary road leading to the east side of A-ko-su.

Site 2 (41-44N 80-20E). Site 2 contains an opencut prospect which is being cleared for an open-pit mine. This prospect is located at the foot of the western side of a low mountain. Trails lead up the broken slopes of the mountain to small prospects. On the November 1962 photography at the time of snow cover, the prospect appeared as a small dark area, and tracks connected it with a village around the mountain. The December 1962 photography revealed a much wider and darker area at the prospect.

Site 3 (41-44N 80-21E). Site 3 contains five irregularly shaped opencut prospects which are located halfway up the eastern side of the mountain. A trail connects this site with a small settlement in the valley.

BACKGROUND

According to a Soviet geologist, V. M. Sinitsyn, geological reconnaissance of the northwestern part of the Tarim Basin began in 1942-43. 3/ Geological field work continued intermittently until 1952-53 when localized detailed studies were carried out. In 1953 Sinitsyn prepared a geological map of the region as a guide to prospecting, and during 1955-56 he drafted a report on the region. 3/

Photography of December 1960 showed that roads had been built from A-ko-su northward to the mining areas and that opencut mining and treatment of coal had been started. These developments indicated that initial geological work and prospecting were probably in progress by 1958, if not earlier.

After prospecting during 1956-1957 the accumulation of coal shown by the 1960 reconnaissance indicates that miners were brought to the mines by 1959-1960. The presence of control points and fences in a mountainous

region indicates the miners probably are prisoners.

The usual prospecting practice of trenching, pitting, and drilling was not seen at the prospects near A-ko-su. In order to confirm the size and extent of the deposits exposed at a likely prospect, the exposure of the outcrop is widened by digging away the overburden.

GEOLOGY OF THE A-KO-SU REGION

The geology of the Sino-Soviet border region supports the possibility that uranium is present in the ores mined near A-ko-su. The valleys where mining is observed near A-ko-su are geologically contemporaneous and lithologically similar to those emptying on the Fergana Valley in the USSR where lignite coal of the Jurassic geological age is mined. Likewise, at Kadzhi-Say, USSR, northwest of A-ko-su, lignites also of Jurassic age have been described. 4/ The A-ko-su, Fergana Valley, and the Kadzhi-Say regions are seen on photography to have deeply eroded east/west throughgoing fault zones. Such fault or crush zones would facilitate the descent of uraniferous ground water. The broken or faulted lignite seams would provide a reducing environment for the precipitation of the uranium from the percolating ground water. The coal seams north of A-ko-su are broken by a series of north/south faults whose crushed-rock zones have been enlarged by swiftly flowing rivers. The broken and faulted character of the A-ko-su mining region limits a knowledge of the reserves and makes prospecting and mining costly and uncertain. Sinitsyn concluded that special work would show the location of the coal-bearing zones. He stated that the eastern or Kuche (Kuchar) coal basin which includes the A-ko-su region is the counterpart of the western or Yarkand-Fergana basin which extends from southern Hsin-chiang Sheng (Sinkiang) into the USSR. The appearance from photography of the region north of A-ko-su agrees with Sinitsyn's brief generalized geological description.

18. *(Continued)*

REFERENCES

PHOTOGRAPHY

Mission	Date	Pass	Camera	Frames	Classification
9013	9 Dec 60	22		157-160	TOP SECRET RUFF
9029	15 Dec 61	38		90	TOP SECRET RUFF
9048	29 Nov 62	70	Fwd Aft	190-192 194-196	TOP SECRET RUFF
9050	15 Dec 62	D54	Fwd Aft	86-89 90-93	TOP SECRET RUFF

MAPS OR CHARTS

AMS. Series ESPA-1, Sheet NK-44-8. 1st ed, Aug 62, scale 1:250,000 (TOP SECRET RUFF)

ACIC. US Air target Chart, Series 200, Sheet 0329-19A, Jun 59, scale 1:200,000 (SECRET)

ACIC. WAC 329, Jul 59, scale 1:1,000,000 (CONFIDENTIAL)

US Geol Surv. Geologic Map of China, Plate 4-2, Binder 2, Part IV (China and North Korea), Uranium and Thorium Resources of Communist Countries series, Jan 55 (SECRET)

US Geol Surv. Map of Coal Basins 21 and 22, Fergana Uraniferous Province, Plate 2-15, Binder 4, Part II (USSR), Uranium and Thorium Resources of Communist Countries series, Aug 54 (SECRET)

DOCUMENTS

3. Sinitsyn, V. M. Northwest Part of the Tarim Basin (Severo-Zapadnaya Chast Tarimskovo Basseyna), Geological Institute, Academy of Sciences of the USSR, Moscow, 1957 (UNCLASSIFIED)

REQUIREMENT

PROJECT NUMBER

19. CIA/NPIC, Photographic Intelligence Report, "Suspect CW Agent Production Plants, Dzerzhinsk, USSR, Changes Since 1962," August 1963

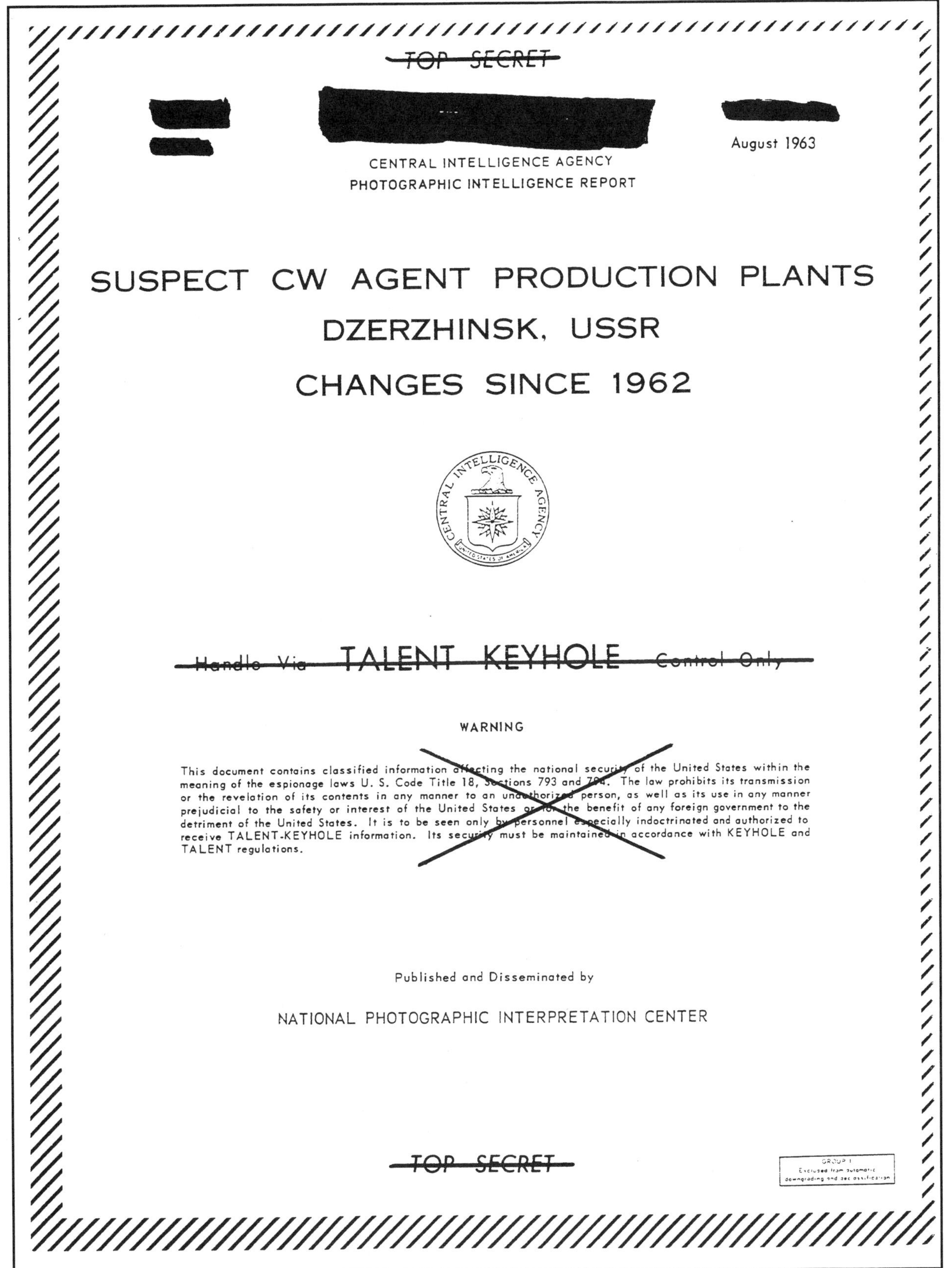

TOP SECRET

August 1963

CENTRAL INTELLIGENCE AGENCY
PHOTOGRAPHIC INTELLIGENCE REPORT

SUSPECT CW AGENT PRODUCTION PLANTS DZERZHINSK, USSR CHANGES SINCE 1962

Handle Via TALENT KEYHOLE Control Only

WARNING

This document contains classified information affecting the national security of the United States within the meaning of the espionage laws U. S. Code Title 18, Sections 793 and 794. The law prohibits its transmission or the revelation of its contents in any manner to an unauthorized person, as well as its use in any manner prejudicial to the safety or interest of the United States or for the benefit of any foreign government to the detriment of the United States. It is to be seen only by personnel especially indoctrinated and authorized to receive TALENT-KEYHOLE information. Its security must be maintained in accordance with KEYHOLE and TALENT regulations.

Published and Disseminated by

NATIONAL PHOTOGRAPHIC INTERPRETATION CENTER

TOP SECRET

GROUP 1
Excluded from automatic downgrading and declassification

19. *(Continued)*

~~TOP SECRET RUFF~~

SUSPECT CW AGENT PRODUCTION PLANTS
DZERZHINSK, USSR
CHANGES SINCE 1962

The two Dzerzhinsk Chemical Plants, Stroy 96 and Kalinin, which were described in [redacted], are further examined in this report. A third plant, Rulon 148, is also discussed. These three plants (Figure 1), which are part of the Dzerzhinsk Chemical Industrial Complex (56-14N 43-32E), were examined on good-quality KEYHOLE photography from Mission 9053 of 4 April 1963.

DZERZHINSK CHEMICAL PLANT STROY 96

The refinery section of this plant shows little change since March 1962. There has been a slight increase in the storage capacity of the area and one new building has been added (Figure 2). In the chemical production area of Stroy 96, numerous buildings are seen for the first time. However, these may not all be new. It is probable that some had been constructed previously, but were not visible until Mission 9053.

In the southwest corner of the plant, the appearance of new scars indicates possible new construction activity. A new waste heap or raw material storage pile can also be seen in this area.

Because of the small scale of the photography and the numerous buildings within the site, the walls which enclose different areas within

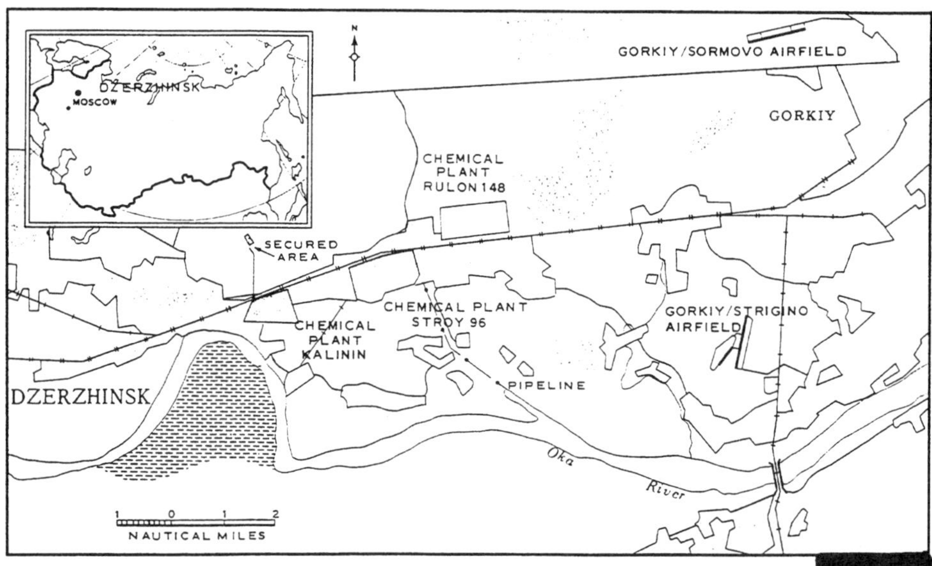

FIGURE 1. LOCATION OF DZERZHINSK CHEMICAL PLANTS.

~~TOP SECRET RUFF~~

19. *(Continued)*

~~TOP SECRET RUFF~~

Stroy 96 cannot be observed. However, in the relatively open area in the southern portion of the site, there appears to be a newly walled-in area.

DZERZHINSK CHEMICAL PLANT KALININ

The Kalinin plant appears very active. Interpretation of the photography of the western third of the plant was difficult because of the large amount of smoke issuing from factory chimneys and an apparent covering of soot. Only very vague outlines of some buildings are visible and an accurate detailed interpretation of this area is not possible.

Within the main secured area numerous new buildings are visible, but no walled-in areas can be seen (Figure 3). In the south-central portion of the plant area are two large new buildings and a number of smaller ones. In this same area there is also a large new scar. In the southwest section of the site there are five new buildings.

No changes can be seen in the small secured areas adjoining the eastern side of the Kalinin plant. As in the other areas, the security walls around these small sites are only partially visible.

One nautical mile north of Kalinin, in the area which was previously reported as a secured storage site, a large monitor-roofed building can be seen. In addition, there are two new buildings and a foundation for a third. The area is secured by a wall.

DZERZHINSK CHEMICAL PLANT RULON 148

It is now apparent that Chemical Plant Rulon 148 is more secured than either Stroy 96 or Kalinin (Figure 4). It is enclosed by a fence, a cleared strip, and a wall. Within the secured area there is no evidence of change since Mission 9031. However, just outside the southeastern corner of the secured area there are six areas of new earth scarring, and at least one new building to suggest recent construction activity.

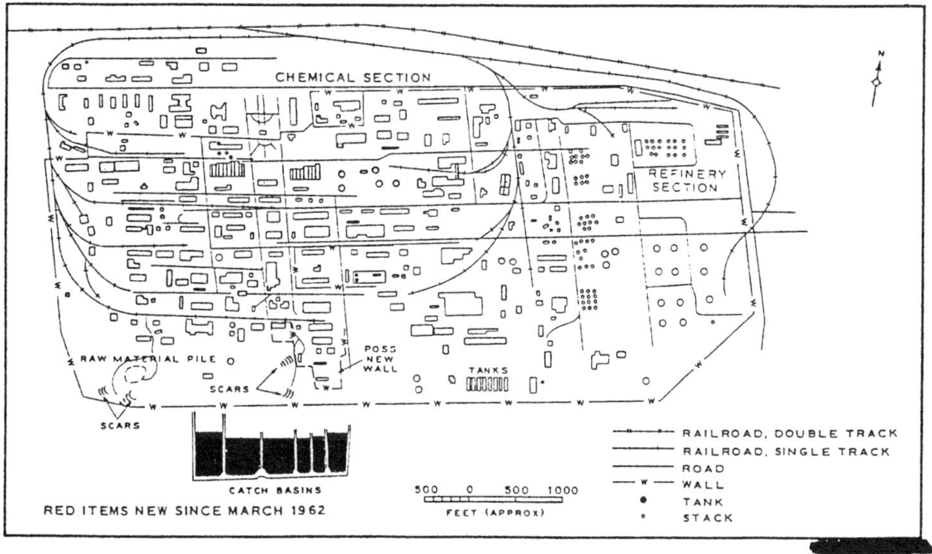

FIGURE 2. CHEMICAL PLANT STROY 96.

~~TOP SECRET RUFF~~

19. *(Continued)*

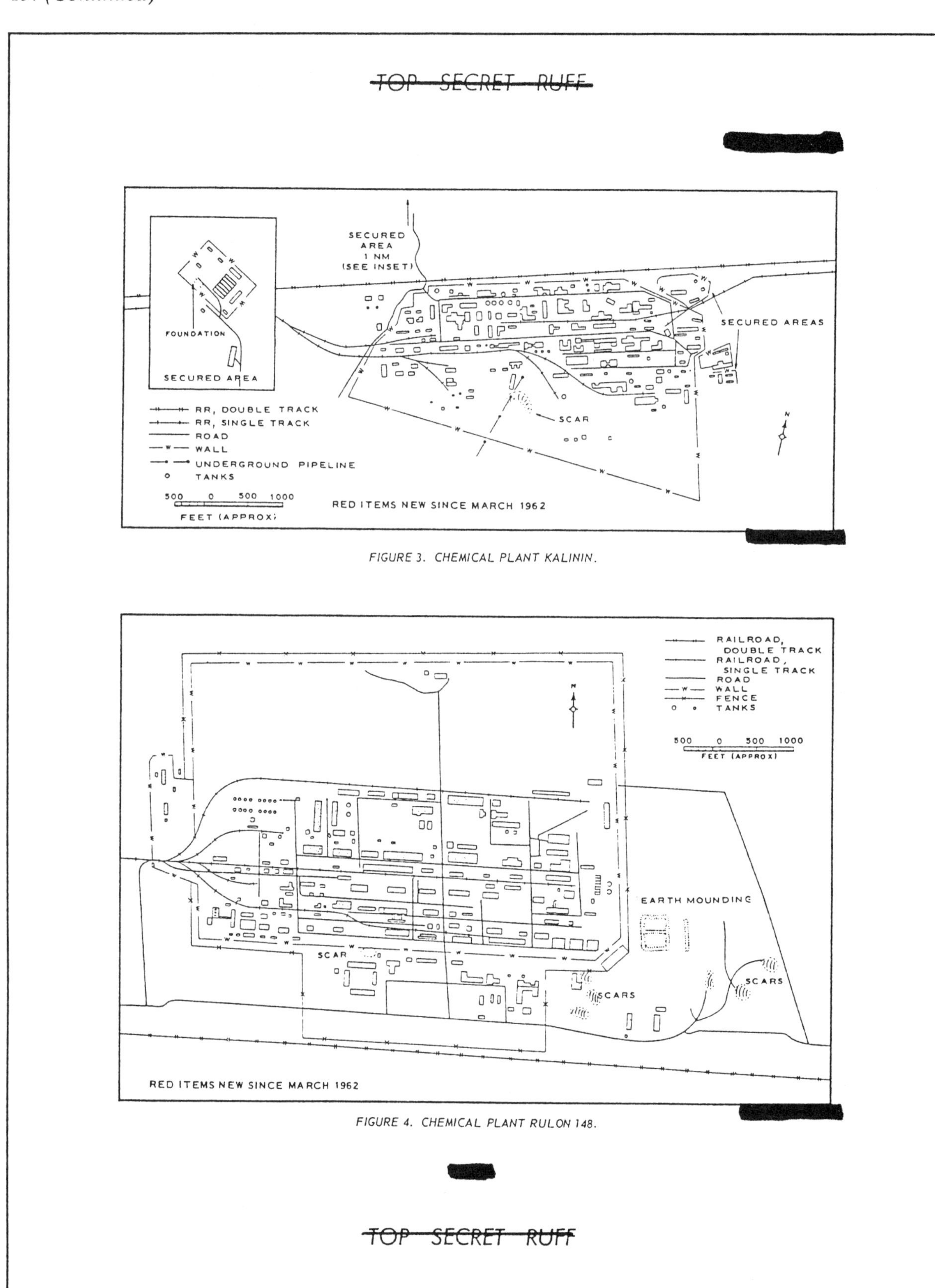

FIGURE 3. CHEMICAL PLANT KALININ.

FIGURE 4. CHEMICAL PLANT RULON 148.

19. *(Continued)*

~~TOP SECRET RUFF~~

REFERENCES

PHOTOGRAPHY

Mission	Date	Pass	Camera	Frames	Classification
9031	2 Mar 62	40D	Fwd	82	TOP SECRET RUFF
			Aft	90	
9053	2 Apr 63	8D	Fwd	101	TOP SECRET RUFF
			Aft	105	

MAPS AND CHARTS

ACIC. US Air Target Mosaic, Series 25, Sheet 0154-9983-2-25MA, 2d ed, Dec 56, scale 1:25,000 (CONFIDENTIAL)

ACIC. US Air Target Chart, Series 100, Sheet 0154-9983-100A, 2d ed, Dec 56, scale 1:100,000 (SECRET)

DOCUMENTS

1.

REQUIREMENT

PROJECT

~~TOP SECRET RUFF~~

20. CIA/NPIC, Photographic Intelligence Report, "Chronological Development of the Kapustin Yar/Vladimirovka and Tyuratam Missile Test Centers, USSR, 1957–1963," November 1963

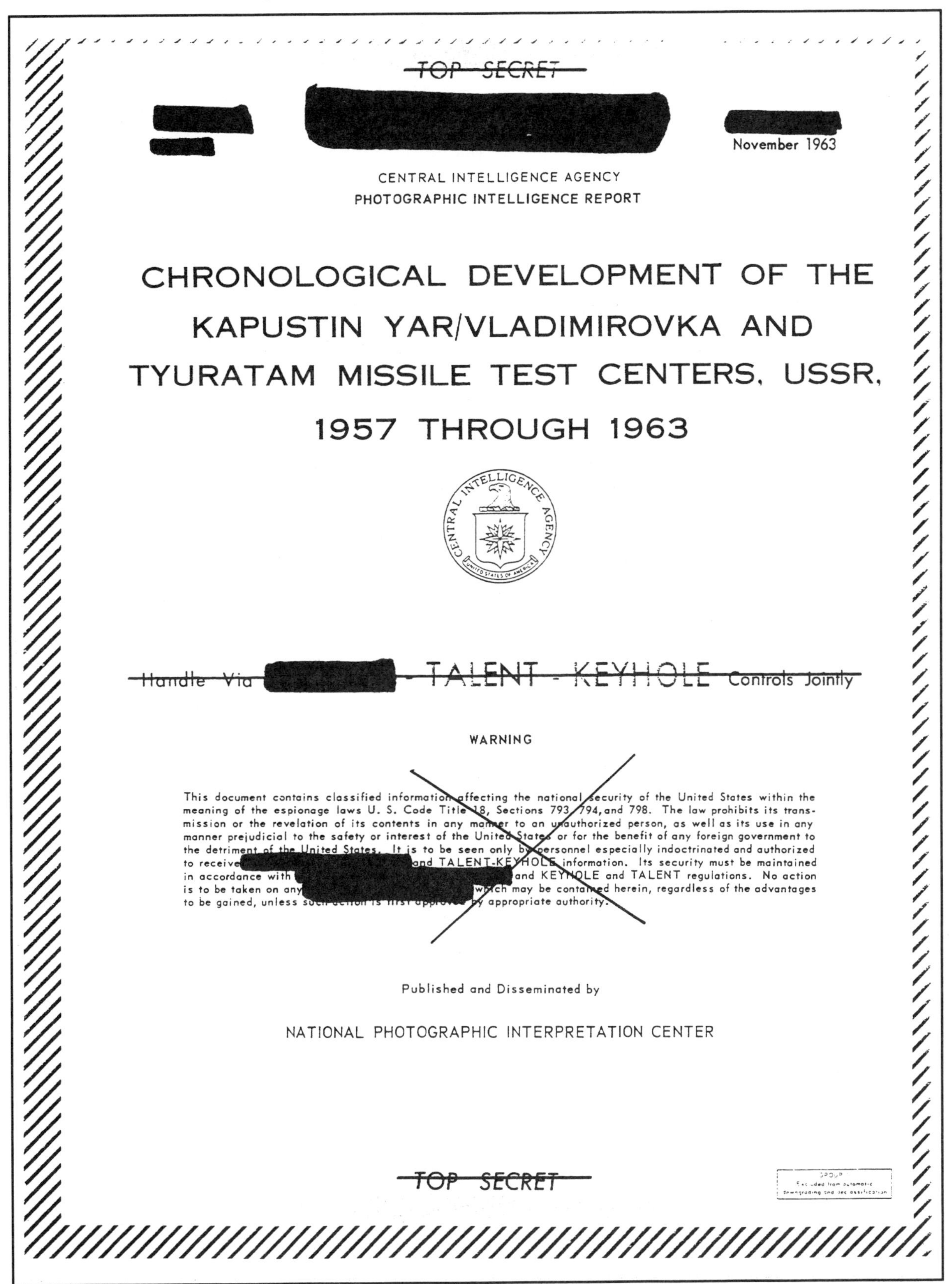

~~TOP SECRET~~

November 1963

CENTRAL INTELLIGENCE AGENCY
PHOTOGRAPHIC INTELLIGENCE REPORT

CHRONOLOGICAL DEVELOPMENT OF THE KAPUSTIN YAR/VLADIMIROVKA AND TYURATAM MISSILE TEST CENTERS, USSR, 1957 THROUGH 1963

~~Handle Via [redacted] - TALENT - KEYHOLE Controls Jointly~~

WARNING

This document contains classified information affecting the national security of the United States within the meaning of the espionage laws U. S. Code Title 18, Sections 793, 794, and 798. The law prohibits its transmission or the revelation of its contents in any manner to an unauthorized person, as well as its use in any manner prejudicial to the safety or interest of the United States or for the benefit of any foreign government to the detriment of the United States. It is to be seen only by personnel especially indoctrinated and authorized to receive [redacted] and TALENT-KEYHOLE information. Its security must be maintained in accordance with [redacted] and KEYHOLE and TALENT regulations. No action is to be taken on any [redacted] which may be contained herein, regardless of the advantages to be gained, unless such action is first approved by appropriate authority.

Published and Disseminated by

NATIONAL PHOTOGRAPHIC INTERPRETATION CENTER

~~TOP SECRET~~

GROUP 1
Excluded from automatic downgrading and declassification

20. *(Continued)*

CHRONOLOGICAL DEVELOPMENT OF THE KAPUSTIN YAR/VLADIMIROVKA AND TYURATAM MISSILE TEST CENTERS, USSR, 1957 THROUGH 1963

The sequence of growth at both Kapustin Yar/Vladimirovka and Tyuratam Missile Test Centers is depicted in color in Figures 1 and 2. Listings of all photographic coverages of the two test centers are given, showing the date of mission and camera system employed, in Tables 1 and 2. Not taken into account for the numerous coverages are weather conditions, cloud cover, or camera malfunctions. Significant developments at the launch complexes within the test centers are also summarized.

Of particular interest is the continued major expansion in the area of operations at Tyuratam between 1957 and 1963 compared to the minor areal expansion at Kapustin Yar/Vladimirovka during the latter part of the same period.

20. *(Continued)*

~~TOP SECRET~~ ~~RUFF~~

Table 1. Photographic Coverage of KY/Vlad MTC

Year	Month	Mission	System
1957	September	B 4059	TALENT A-2
1959	December	B 8005	TALENT B
1960	August	9009	KH-1
1961	June	9017	KH-2
	July	9019	KH-2
	August	9023	KH-3
1962	March	9031	KH-4
	April	9032	KH-4
	June	9035	KH-4
	July	9037	KH-4
	July	9038	KH-4
	July	9039	KH-4
	July	9040	KH-4
	August	9041	KH-4
	September	9044	KH-4
	October	9045	KH-4
	November	9047	KH-4
1963	March	9053	KH-4
	September	1002-1	KH-4J

KAPUSTIN YAR/VLADIMIROVKA MISSILE TEST CENTER

SAM Launch Facilities

- One SA-1 (herringbone) launch site
- Four SA-2 training launch sites
- Six SA-3 R & D and training launch sites
- Five deployed SA-2 launch sites
- Full or partial examples of all known SAM systems in R & D launch area

SSM Launch Facilities

- Launch Complex A (present system unknown)
  - Area 1A 2 launch pads; abandoned in 1959
  - Area 2A 2 large earth launch structures under construction in 1959; by 1963, 1 complete, 1 probably abandoned
  - Also, 7 inactive revetted field launch sites
- Launch Complex B (naval missile associated)
  - Area 1B 3 launchers: 2 inclined, 1 unidentified
  - Area 2B 2 launchers or ship simulators
  - Area 3B 2 launchers or ship simulators
- Launch Complex C (MRBM/IRBM associated)
  - Area 1C 1 launch pad, rail served since 1960, probable launch point for Cosmos satellites; prior to 1960, R & D and training for Sandal (SS-4) MRBM system; also 3 inactive revetted field launch sites
  - Area 2C 2 launch pads, probably associated with SS-5 IRBM system
  - Area 3C 3 launch pads; R & D and training for Shyster (SS-3) MRBM system
  - Area 4C 2 subareas, prototypes for deployed hard SS-4 and SS-5 launch sites
  - Area 5C 2 subareas, 2 launch pads each, for SS-4 and SS-5 troop training
- Launch Complex E (missile system unknown)
  - 1 large launch pad; no significant change since first observed in 1959
- Launch Complex F (troop training)
  - At least 65 revetted field-type training sites; abandoned by 1959
- Launch Complex G (troop training)
  - Area 1G 2 fixed launch pads
  - Area 2G 3 revetted field-type launch sites; 1 site active in 1959

Aerodynamic Facilities

- Launch Complex D
  - Area 1D 1 unusual rail-served launch structure in 1957 and 2 road-served probable launchers, 1962
  - Area 2D 1 launcher under construction in 1957; probably abandoned thereafter
  - Area 3D 1 launcher on pad; complete in 1960
  - Area 4D 1 launch pad; status and function undetermined

Vladimirovka Airborne Missile Loading Complex

- Air-to-air missile handling facilities
- Air-to-surface missile handling facilities

~~TOP SECRET~~ ~~RUFF~~

20. *(Continued)*

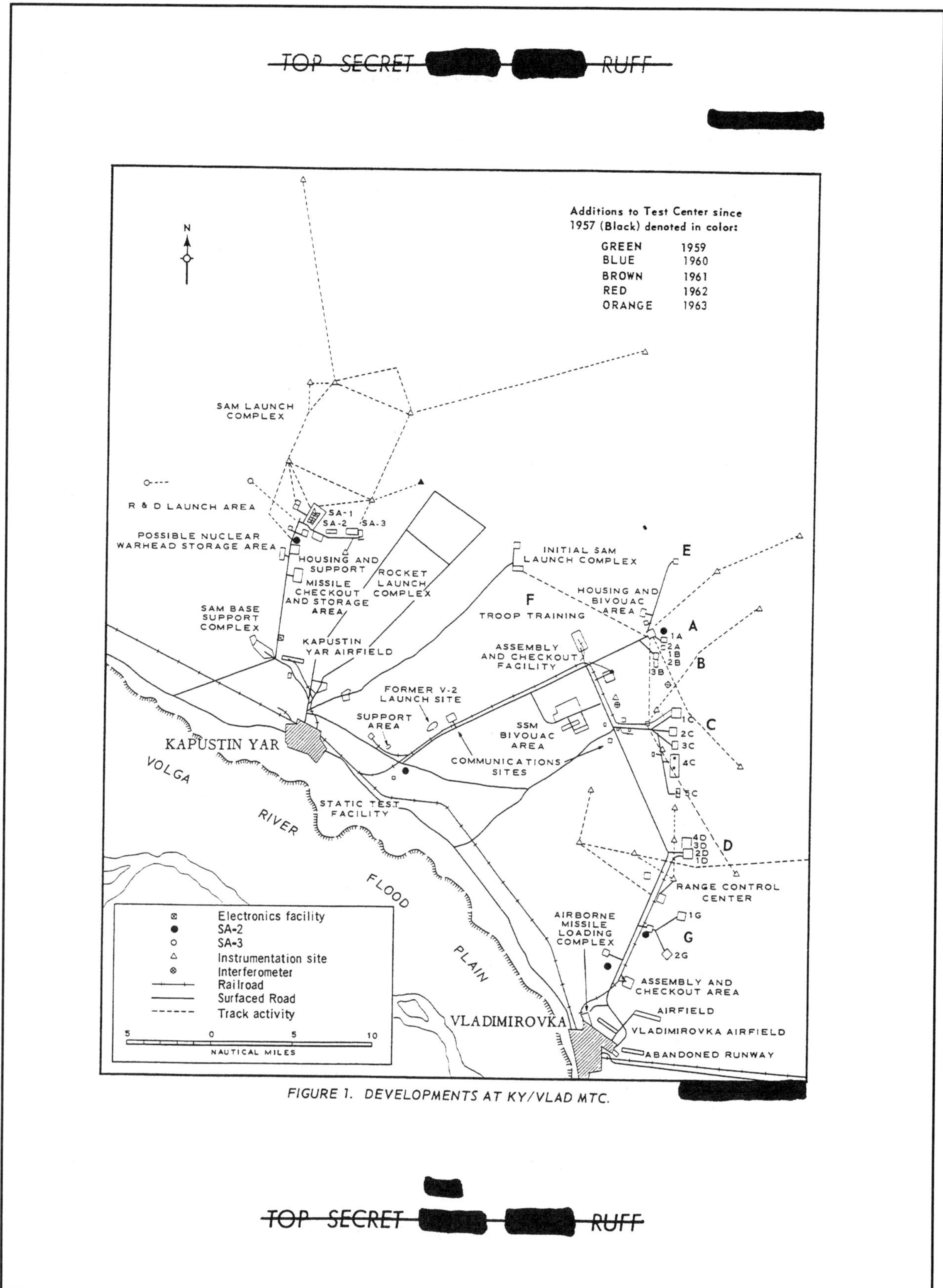

FIGURE 1. DEVELOPMENTS AT KY/VLAD MTC.

20. *(Continued)*

Table 2. Photographic Coverage of TTMTC

Year	Month	Mission	System
1957	August	B 4035	TALENT B
	September	B 4058	TALENT A-2
1959	July	B 4125	TALENT B
1960	April	B 4155	TALENT B
	December	9013	KH-2
1961	September	9022	KH-3
	December	9029	KH-3
1962	March	9031	KH-4
	June	9035	KH-4
	July	9038	KH-4
	August	9040	KH-4
	September	9044	KH-4
	October	9045	KH-4
	November	9048	KH-4
	December	9050	KH-4
1963	April	9053	KH-4
	June	9054	KH-4
	June	9056	KH-4
	August	1001-1	KH-4J
	September	1002-1	KH-4J

TYURATAM MISSILE TEST CENTER

Launch Complex A
- Pad A1 R & D for SS-6 missile system and probably for all space programs; large rail-served launch structure complete in 1957
- Pad A2 R & D for SS-8 missile system; complete in 1961; rail served; forerunner of pads at Complex E

Launch Complex B (prototype of SS-6 deployed sites)
- Large rail-served launch structures under construction, 1959; complete, 1960

Launch Complex C (R & D and prototype of soft SS-7 deployed sites)
- Pads C1 and C2 Pad separation 1,250 feet; complete in late 1960
- Pad C3 Combined with Pad C2, pad separation 950 feet; complete in mid-1962

Launch Complex D (prototype of hard SS-7 deployed sites)
- Site D1 Silo launch site with 2 or possibly 3 missile silos; probably complete in late 1962
- Site D2 Similar to Site D1; construction continuing in September 1963

Launch Complex E (prototype of soft SS-8 deployed sites)
- Pads E1 and E2 Pad separation 800 feet; complete in mid-1962
- Pad E3 Complete in 1963; resembles Pad A2

Launch Complex F (prototype of hard SS-8 deployed sites)
- Silo launch site with 2 or possibly 3 missile silos; probably complete in September 1963

Launch Complex G (space or ICBM associated)
- Pads G1 and G2 Rail served; pad separation 900 feet; started mid-1962; midstage of construction, September 1963
- Pads G3 and G4 Rail served; pad separation 1,800 feet; early to midstage in September 1963

Launch Complex H (missile system unknown)
- Pads H1 and H2 Road served; pad separation 580 feet; started in early 1963

Recent Construction
- At the rangehead, 2 road- and rail-served construction activities started in summer, 1963

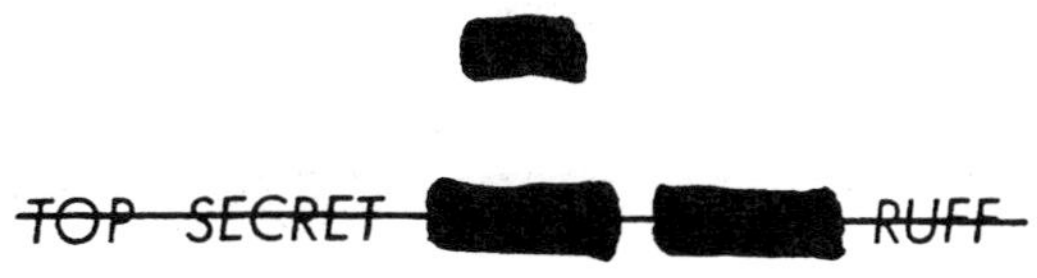

20. *(Continued)*

TOP SECRET [redacted] [redacted] RUFF

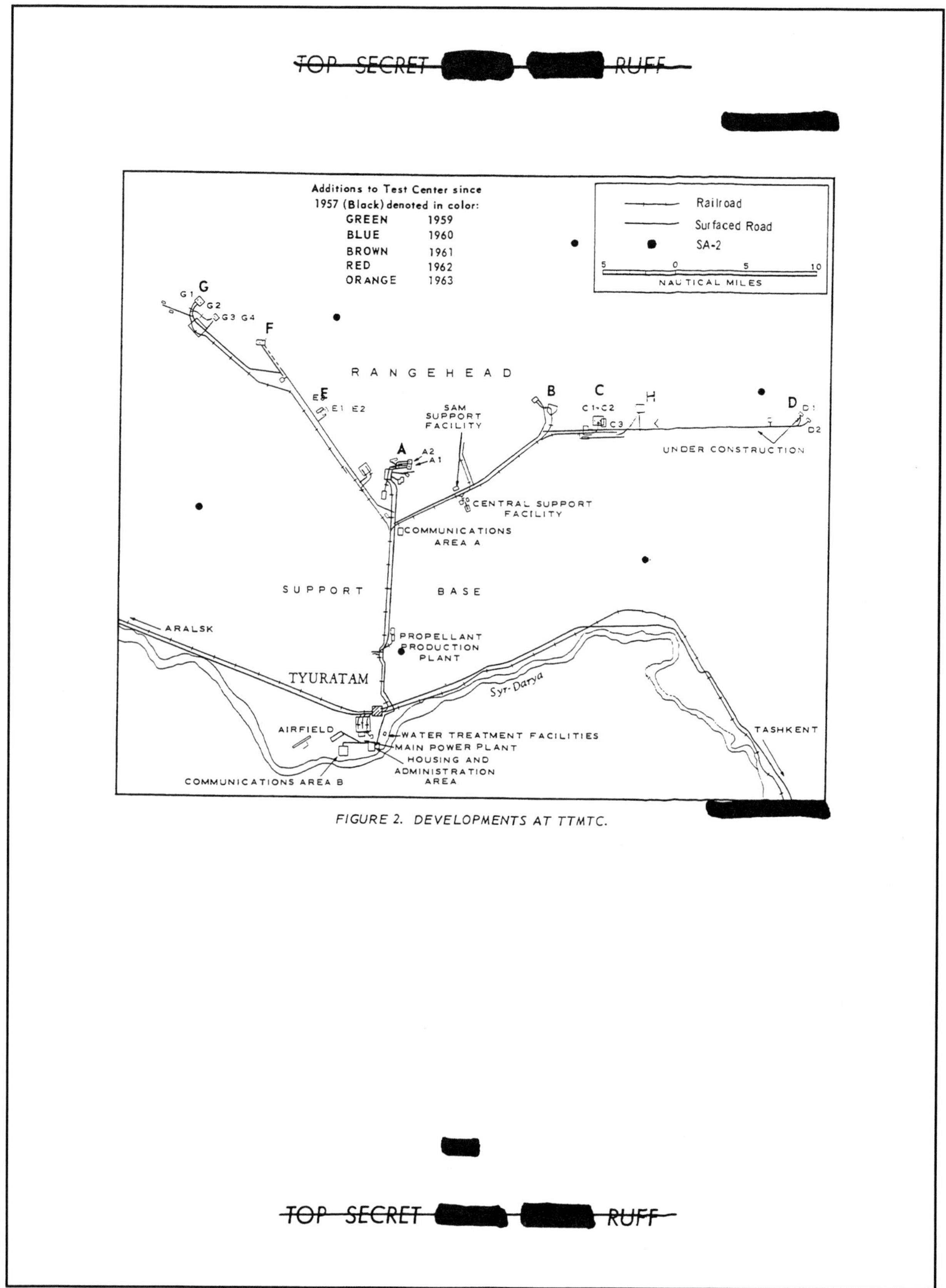

FIGURE 2. DEVELOPMENTS AT TTMTC.

TOP SECRET [redacted] [redacted] RUFF

21. CIA/NPIC, Photographic Intelligence Report, "Probable Solid Propellants Testing Facilities and Associated Explosives Plants in the USSR," December 1963

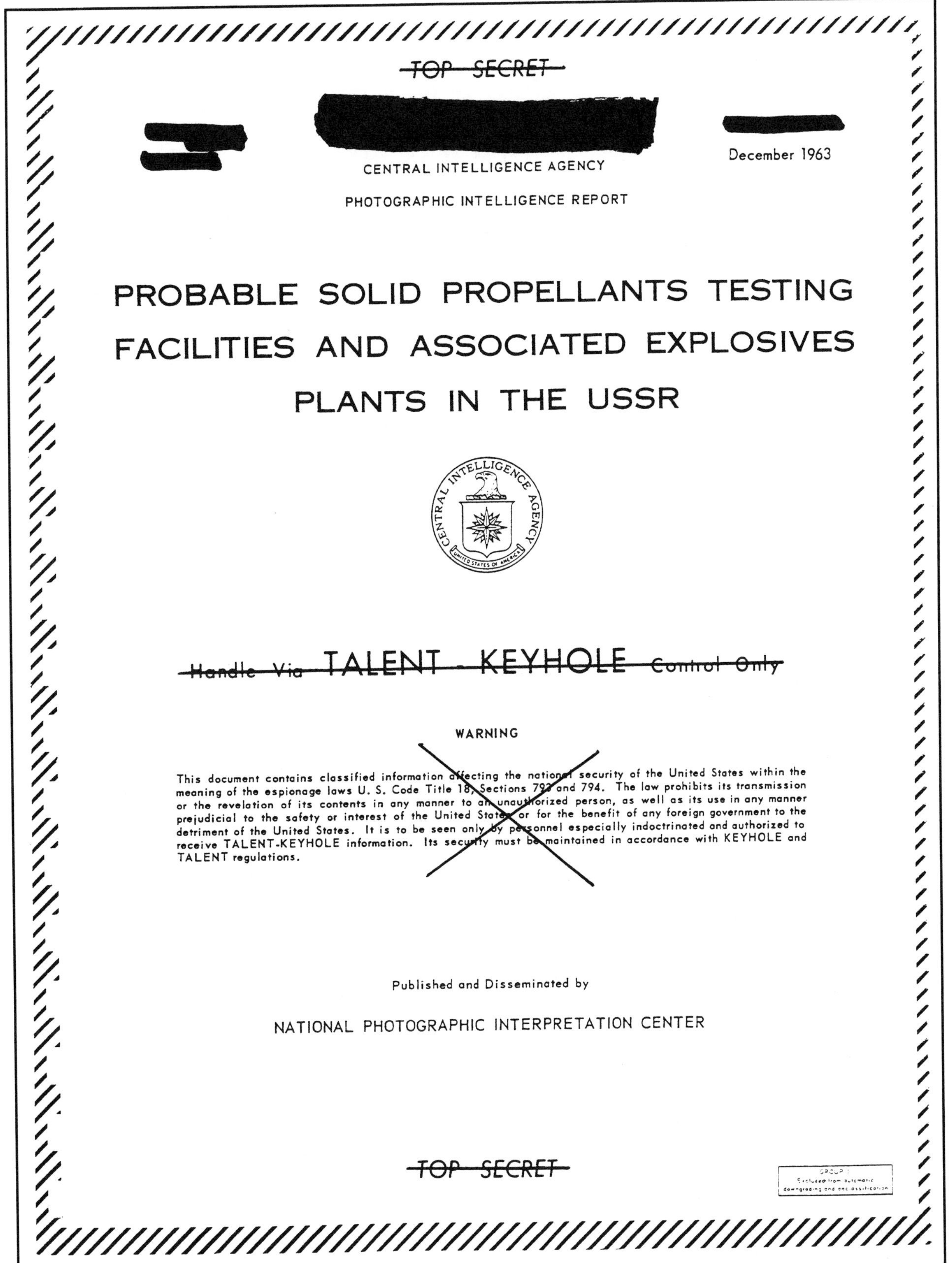

~~TOP SECRET~~

December 1963

CENTRAL INTELLIGENCE AGENCY

PHOTOGRAPHIC INTELLIGENCE REPORT

PROBABLE SOLID PROPELLANTS TESTING FACILITIES AND ASSOCIATED EXPLOSIVES PLANTS IN THE USSR

~~Handle Via~~ TALENT - KEYHOLE ~~Control Only~~

WARNING

This document contains classified information affecting the national security of the United States within the meaning of the espionage laws U. S. Code Title 18, Sections 793 and 794. The law prohibits its transmission or the revelation of its contents in any manner to an unauthorized person, as well as its use in any manner prejudicial to the safety or interest of the United States or for the benefit of any foreign government to the detriment of the United States. It is to be seen only by personnel especially indoctrinated and authorized to receive TALENT-KEYHOLE information. Its security must be maintained in accordance with KEYHOLE and TALENT regulations.

Published and Disseminated by

NATIONAL PHOTOGRAPHIC INTERPRETATION CENTER

~~TOP SECRET~~

GROUP 1
Excluded from automatic downgrading and declassification

21. *(Continued)*

~~TOP SECRET RUFF~~

CENTRAL INTELLIGENCE AGENCY
PHOTOGRAPHIC INTELLIGENCE REPORT

PROBABLE SOLID PROPELLANTS TESTING FACILITIES AND ASSOCIATED EXPLOSIVES PLANTS IN THE USSR

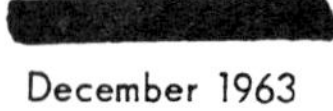

December 1963

Published and Disseminated by
NATIONAL PHOTOGRAPHIC INTERPRETATION CENTER

~~TOP SECRET RUFF~~

21. *(Continued)*

~~TOP SECRET RUFF~~

INTRODUCTION

Highly significant installations associated with the testing and production of unique explosives materials of a probable solid propellant nature have been identified at Biysk (52-31N 85-04E), Kamensk-Shakhtinskiy (48-19N 40-13E), Krasnoyarsk (56-02N 93-02E), Perm (57-58N 55-52E), and Sterlitamak (53-44N 56-00E), all in the USSR (Figure 1). These installations are identical to the extent that they are adjacent to or within explosives/munitions combines producing at least two explosives bases, and each has at least one test cell with a concrete-faced probable bunker/deflector. A total of eight test cells have been identified at the five sites.

Photography of these installations is provided by 20 KEYHOLE missions occurring between December 1960 and late August 1963. The quality and small scale of this photography preclude the determination of exact measurements and the assigning of definite functions to most of the buildings. Measurements of these facilities should be considered approximate, although in all cases scale factors were provided by TID/NPIC; where utilized, height factors were also provided by TID/NPIC.

For the purpose of this report, details on the Kamensk-Shakhtinskiy facility will not be included because of a lack of interpretable photography of the site.

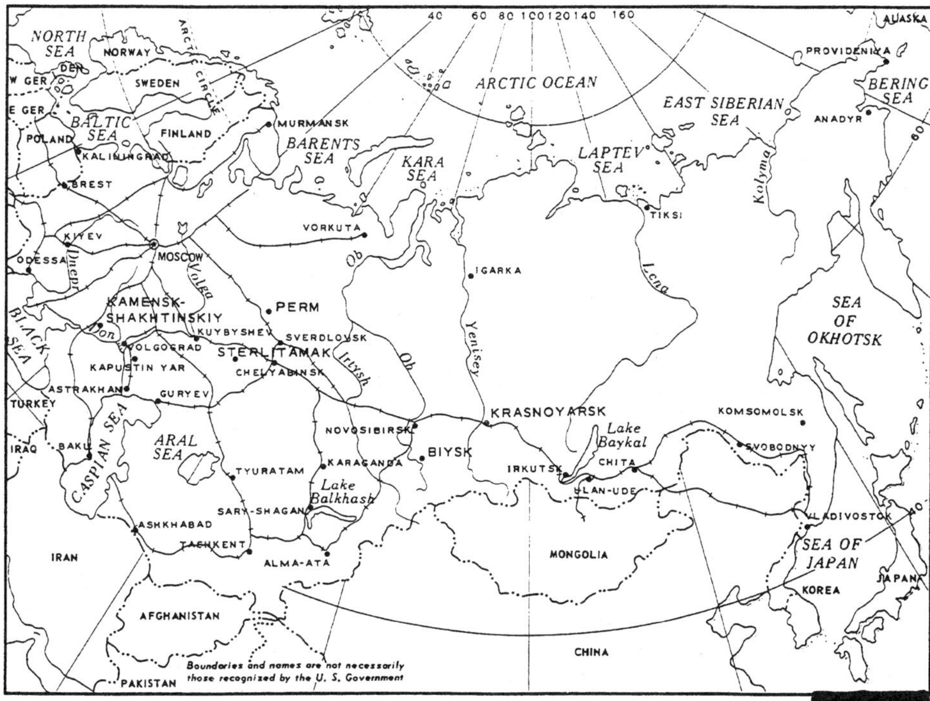

FIGURE 1. LOCATION OF PROBABLE SOLID PROPELLANTS TEST FACILITIES, USSR.

~~TOP SECRET RUFF~~

~~TOP SECRET RUFF~~

CHRONOLOGY

These facilities appear to be of recent construction, although only the Biysk facility can be negated on recent photography; it was not present in December 1960. The existence of the other four cannot be negated on any available KEYHOLE photography. The Sterlitamak and Kamensk-Shakhtinskiy facilities can be negated on captured German photography of 1942 and 1943, but the facilities at Perm and Krasnoyarsk cannot be negated on any available photography.

Confirmation of completion can be made at three of the facilities: Krasnoyarsk (September 1962), Perm (August 1962), and Sterlitamak (July 1963); as of June 1963, the test facilities at Biysk and Kamensk-Shakhtinskiy cannot be confirmed as complete. Criteria for confirmation would include: relative completion of the plant and storage facilities; paving of the large, concrete-faced probable bunker/deflectors; and completion of the support structures within the test facility.

TEST FACILITIES

BIYSK

The Biysk Probable Solid Propellants Test Facility (Figure 2) is located approximately 5 nautical miles (nm) west of the center of Biysk, USSR. The test facility is road and possibly rail served, and its area of 2,800 by 2,700 feet is secured by a single fence. The facility consists of two completed test cells and a probable third which appears under construction on photography of June 1963; the cells are annotated A, B, and C on Figure 2. A perspective sketch (Figure 3) presents an artist's concept of an oblique view of the test facility. Approximate dimensions of various structures at the Biysk facility can be found in Table 1 which is keyed to Figure 2.

One of the salient recognition features at the Biysk facility is a multi-level H-shaped building (item 1, Figure 2). This structure is similar in appearance and probably identical in function to H-shaped buildings at Krasnoyarsk and Sterlitamak. Because of the unusual configuration, it has been suggested that this building is possibly as many as six different buildings separated by possible blast walls. This structure was noted under construction in December 1961 and confirmed as complete by June 1962.

A second significant feature of the facility is the presence of the two completed test cells and the probable third under construction. Cells A and B are approximately 710 feet apart, and cells B and C are about 990 feet apart; the relative positions of the three within the facility can be seen on Figure 2.

Table 1. Associated Structures, Biysk Probable Solid Propellants Test Facility (item numbers are keyed to Figure 2)

Item	Dimensions (ft)	Item	Dimensions (ft)
1		10	230 x 65
a	310 x 85	11	230 x 65
b	95 x 90	12	240 x 65
c	140 x 75	13	145 x 65
d	160 x 75	14	85 x 45
e	120 x 70	15	65 x 65
2	125 x 40	16	40 x 40
3	125 x 125	17	90 x 55
4	80 x 80	18	230 x 65
5	90 x 75	19	75 x 30
6	85 x 60	20 (3)	90 x 55
7	320 x 85	21	60 x 60
8	55 x 55	22 (2)	55 x 40
9	290 x 45	23	210 x 100

~~TOP SECRET RUFF~~

21. *(Continued)*

~~TOP SECRET RUFF~~

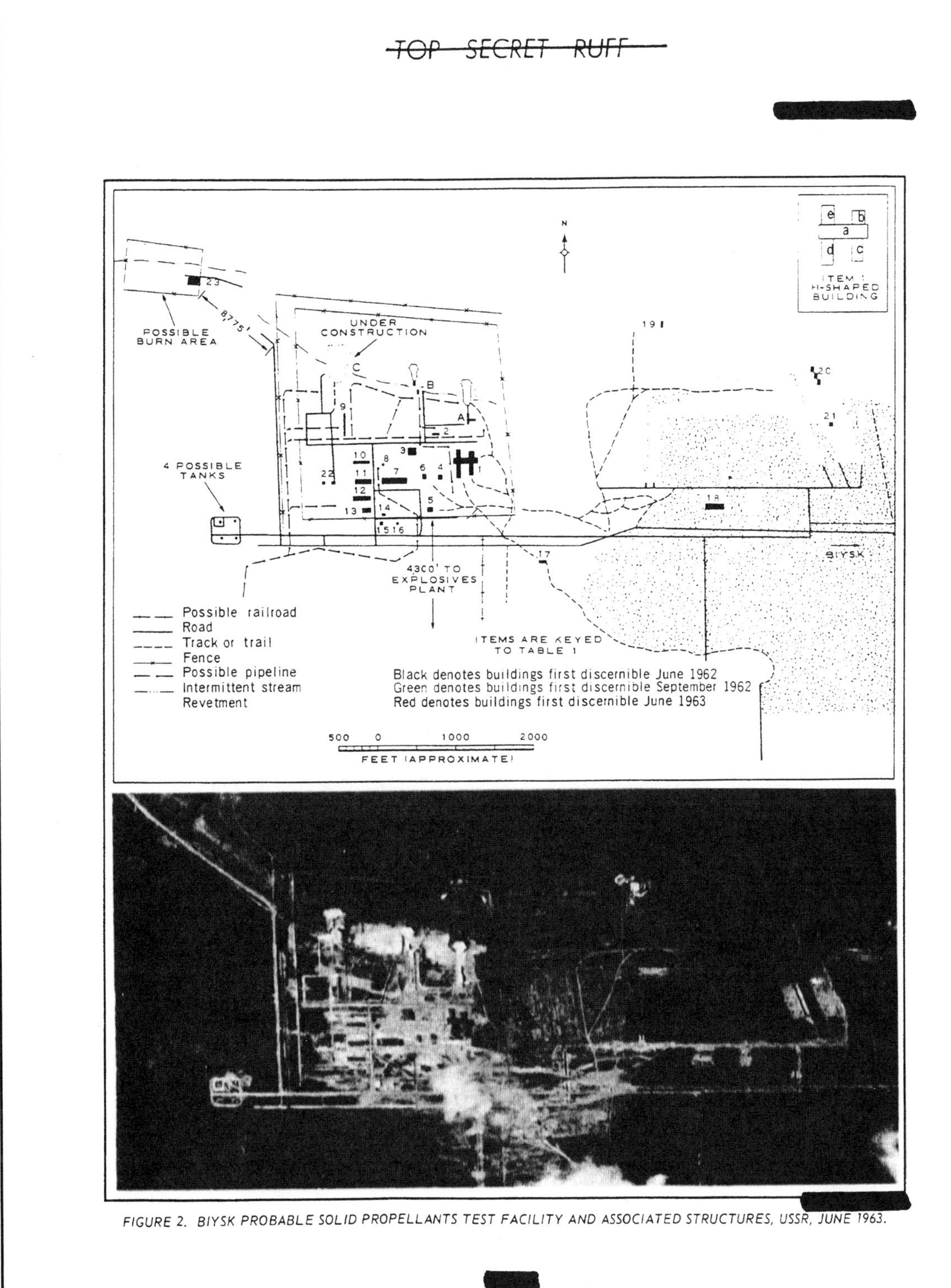

FIGURE 2. BIYSK PROBABLE SOLID PROPELLANTS TEST FACILITY AND ASSOCIATED STRUCTURES, USSR, JUNE 1963.

~~TOP SECRET RUFF~~

21. *(Continued)*

~~TOP SECRET RUFF~~

Test cell A was observed under construction on photography of December 1961; photographic limitations, however, did not permit a confirmation of the physical presence of the cell until June 1962. Cell A is road served from its rear or south end, is in at least three sections, and measures approximately 260 feet in overall length. A large revetment appears immediately to the east of the test cell. Test cell B can be identified as under construction on photography of June 1962 and complete by that of September 1962; it has several of the same features noted at cell A. Cell B is road served from the rear, is in three sections, and has an overall length of 170 feet. A large revetment appears about 25 feet west of cell B; this revetment and the one at cell A could serve instrumentation/safety functions. Probable test cell C can be identified as under construction in June 1963; no definitive statement or measurements can be made on cell C because of the construction status.

Another salient feature at the Biysk facility (and at every other facility identified thus far) is the concrete-faced probable bunker/deflector which is observed adjacent to each test cell; each is identified with a letter to correspond with the cell it serves. Line drawings of Biysk test cells A and B, their associated probable bunker/deflectors, detailed dimensions, and profile elevation sketches can be found on Figure 4.

Probable bunker/deflector A was observed under construction concurrently with test cell A; however, the concrete facing could not be confirmed until photography of September 1962. It measures about 235 feet from its base to the front end of the test cell; the distance from the nearest end of the H-shaped building to the rear of the probable bunker/deflector is approximately 950 feet. Probable bunker/deflector B, first noted under construction in June 1962, was faced with concrete by June 1963. It measures approximately 135 feet from the base to the front of the corresponding test cell, and the distance between the rear of cell B and the rear of probable bunker/deflector B is approximately 450 feet. Probable bunker/deflector C is visible under construction on photography of June 1963.

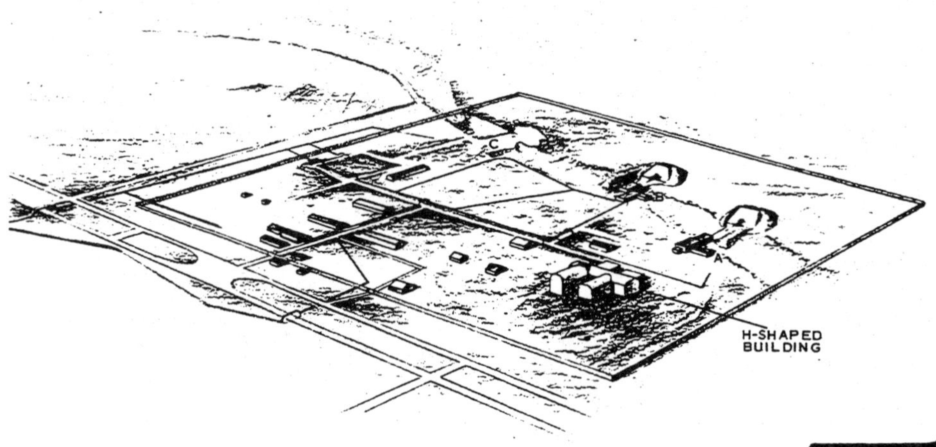

FIGURE 3. ARTIST'S CONCEPT OF TEST FACILITY, BIYSK, USSR.

~~TOP SECRET RUFF~~

~~TOP SECRET RUFF~~

A fourth feature at the Biysk facility is the group of three offset or staggered buildings (item 20, Figure 2) approximately 4,200 feet east of the test facility; they were first observed in June 1963. These buildings are similar in appearance and probably identical in function to comparable structures at Perm, Sterlitamak, and Kamensk-Shakhtinskiy.

Another significant item at Biysk is a secured area of 1,000 by 600 feet located approximately 8,775 feet northwest of the test facility; the area is road served and was first observed in the early stages of construction in June 1962. The purpose of this unidentified area cannot be adequately explained, although a single heavy revetment suggests a possible burn area where highly combustible material is handled. A similar area is found at the Perm test facility.

KRASNOYARSK

The Krasnoyarsk Probable Solid Propellants Test Facility (Figures 5 and 6) is located near Explosives Plant 580 (not to be confused with the new plant which serves the test facility) approximately 5 nm east of the center of Krasnoyarsk, USSR. Although this facility cannot be negated on available photography, it can be determined that it was in an early/mid stage of construction by June 1961. It consists of two test cells which are approximately 600 feet apart. The Krasnoyarsk facility is road served, and the area of approximately 2,500 by 1,000 feet is double secured; one of the fences is solid.

The Krasnoyarsk test facility has an H-shaped building (item 1, Figure 6) similar in appearance and probably identical in function to the irregular structures found at Biysk and Sterlitamak. The Krasnoyarsk building appeared to be in an early stage of construction in June 1961, and its completion can be confirmed by photography of September 1962. It is approximately 65 feet high at the highest point. Approximate dimensions of this building and other

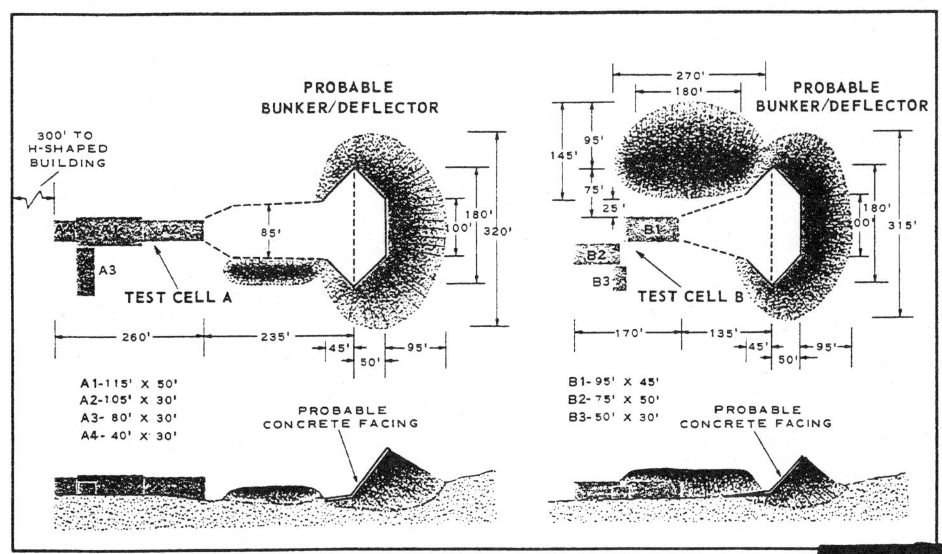

FIGURE 4. TEST CELLS A AND B AND PROBABLE BUNKER/DEFLECTORS, BIYSK, USSR.

~~TOP SECRET RUFF~~

21. *(Continued)*

~~TOP SECRET RUFF~~

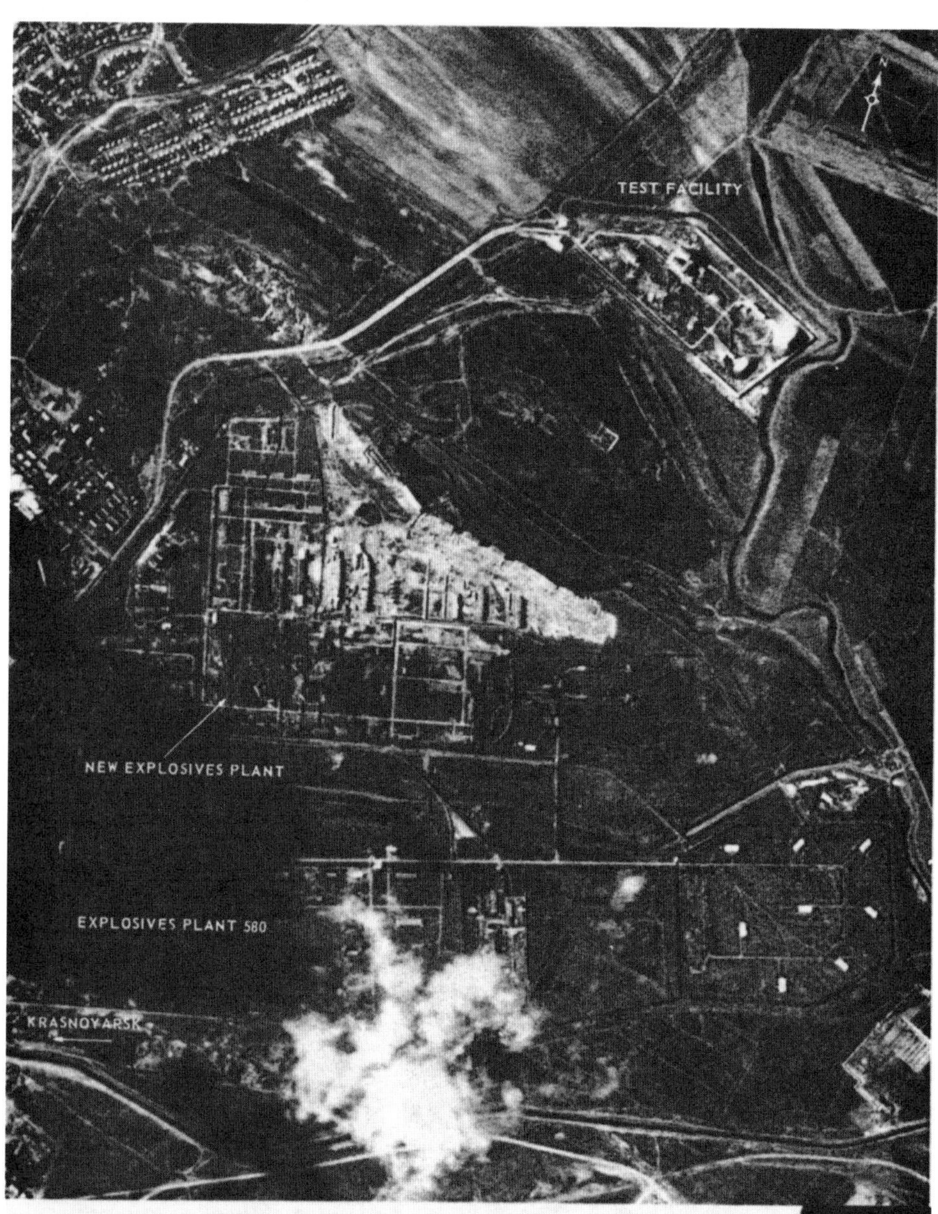

FIGURE 5. KRASNOYARSK PROBABLE SOLID PROPELLANTS TEST FACILITY, USSR, SEPTEMBER 1962.

~~TOP SECRET RUFF~~

~~TOP SECRET RUFF~~

structures at the test facility and explosives plant are given in Table 2 which is keyed to Figure 6.

The two test cells have been annotated A and B on Figure 6, which also illustrates their relative positions at the site. A perspective sketch (Figure 7) presents an artist's concept of an oblique view of the test facility.

Test cell A, the larger and newer of the two cells, was observed under construction and apparently essentially complete on photography of March 1962. The cell is in three sections, has an overall length of 250 feet, and appears to be road served from the rear. The cell is not revetted.

Test cell B, possibly the oldest of the test cells observed in the USSR thus far, appeared essentially complete in September 1961. It measures about 175 feet in overall length, is in three sections, and is connected to a revetted building approximately 300 feet to the rear of cell A by overhead piping or covered walkways. Test cell B is not revetted.

The concrete-faced probable bunker/deflector A (Figure 8) is at least 60 feet high and was first discernible under construction in June 1961; its completion can be confirmed on September 1962 photography. It measures approximately 260 feet from the base to the end of the test cell, and the distance from the nearest end of the H-shaped building to the rear of the probable bunker/deflector is about 1,400 feet. Probable bunker/deflector B (Figure 8) is approximately 45 feet high and can be confirmed as complete in June 1961. The base is about 95 feet from the front of test cell B; the distance from the rear of the cell to the rear of probable bunker/deflector B is approximately 440 feet.

A final feature at the Krasnoyarsk test facility is some scarring observed in front of test cell A on September 1962 photography. Although the funnel shape of this scar suggests

Table 2. Associated Structures, Test Facility and Explosives Plant, Krasnoyarsk (item numbers are keyed to Figure 6)

Item	Dimensions (ft)	Item	Dimensions (ft)
Test Facility			
1		2	316 x 95
a	300 x 75	3	200 x 210
b	100 x 95	4	80 x 80
c	145 x 80	5	115 x 55
d	160 x 80 x 65 (h)	6	125 x 75
e	125 x 70	7	35 x 70
		8	95 x 80
Explosives Plant			
1	355 x 75	28	215 x 65
2	330 x 50	29	185 x 65
3	300 x 75	30	300 x 115
4	190 x 80	31	340 x 100
5	205 x 80	32	220 x 65
6	195 x 80	33	395 x 60
7	330 x 65 x 20 (h)	34	310 x 60
8	330 x 65 x 20 (h)	35	480 x 50
9	165 x 105 x 45 (h)	36	365 x 75
10	45 x 35	37	645 x 80
11	775 x 30 x 15 (h)	38	525 x 80
12	690 x 75 x 25 (h)	39	380 x 75
13	695 x 45	40	460 x 65
	end sections (2)	41	300 x 65
	130 x 60 x 35 (h)	42	390 x 80
		43	405 x 85
14	790 x 50 x 40 (h)	44	215 x 60
15	825 x 75	45	220 x 60
16	205 x 75	46	120 x 50
17	185 x 75	47	315 x 45
18	100 x 75	48	250 x 45
19 (6)	150 x 60	49	410 x 50
20	220 x 85	50	200 x 35
21	225 x 85	51	55 x 35
22	135 x 85	52	100 x 55
23	290 x 110 x 55 (h)	53	105 x 40
24	130 x 65	54	90 x 40
25	280 x 55	55	155 x 60
26	125 x 65	56	120 x 50
	wing 115 x 80	57	230 x 35
27	120 x 80	58	85 x 40
	wing 160 x 70	59 (2)	120 x 50
		60 (8)	170 x 55

~~TOP SECRET RUFF~~

21. *(Continued)*

TOP SECRET RUFF

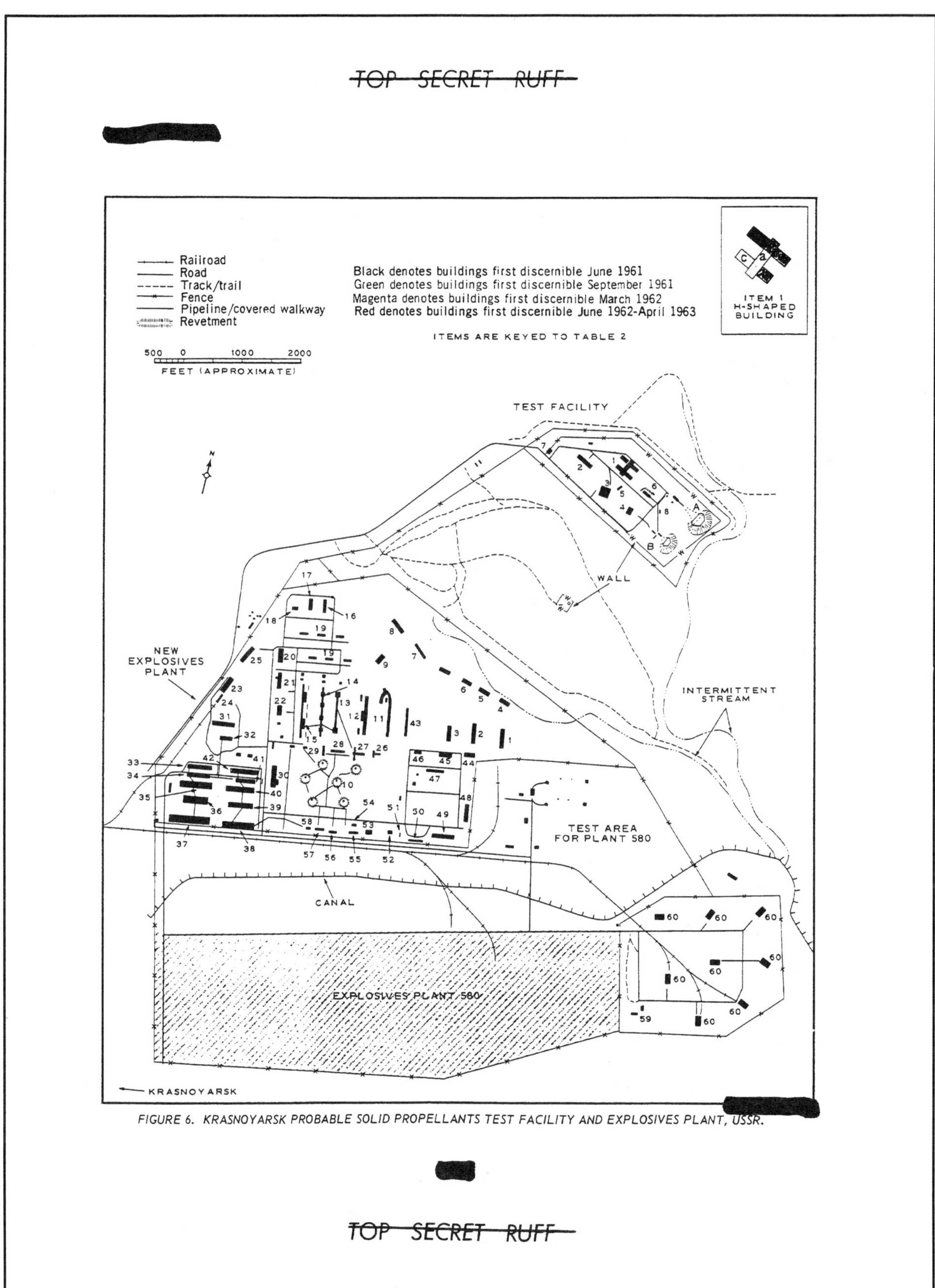

FIGURE 6. KRASNOYARSK PROBABLE SOLID PROPELLANTS TEST FACILITY AND EXPLOSIVES PLANT, USSR.

TOP SECRET RUFF

21. *(Continued)*

~~TOP SECRET RUFF~~

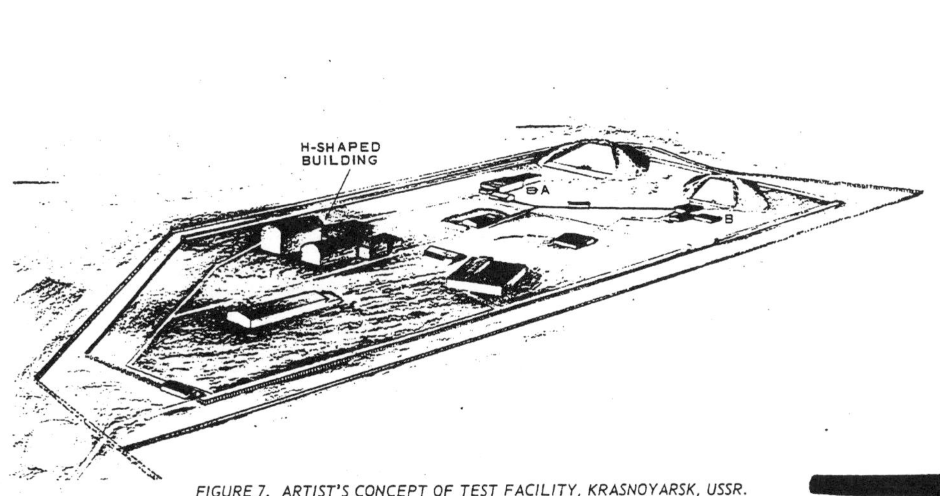

FIGURE 7. ARTIST'S CONCEPT OF TEST FACILITY, KRASNOYARSK, USSR.

a possible blast mark, no conclusive statement can be made on the basis of the photography available.

PERM

The Perm Probable Solid Propellants Test Facility (Figure 9) is located within the confines of the Perm Munitions and Chemical Combine K. Kirov 98, approximately 13 nm west of the center of Perm, USSR, along the Kama River. This facility cannot be negated on any available photography; when first observed in July 1961 it was in an undetermined stage of construction. The test facility is road and rail served, and the area of approximately 4,500 by 1,700 feet is secured by a single fence. Table 3, which is keyed

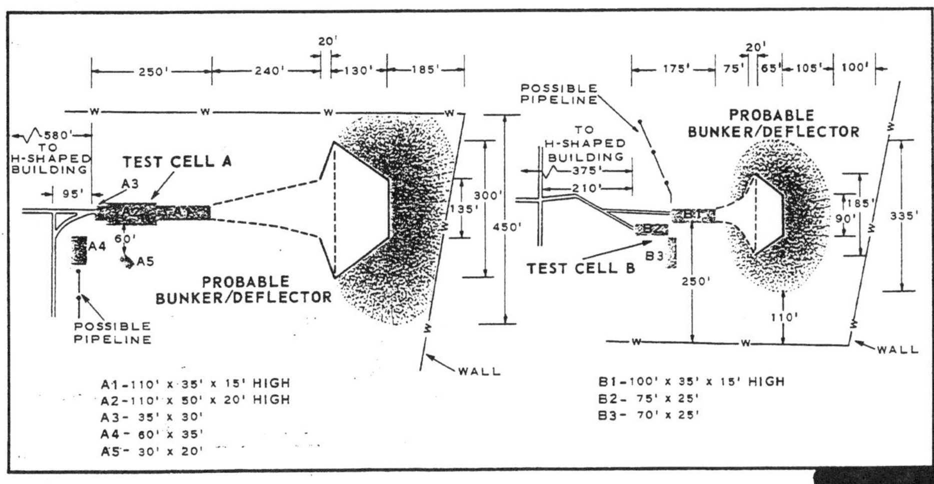

FIGURE 8. TEST CELLS A AND B AND PROBABLE BUNKER/DEFLECTORS, KRASNOYARSK, USSR.

~~TOP SECRET RUFF~~

~~TOP SECRET RUFF~~

Table 3. Associated Structures, Perm Probable Solid Propellants Test Facility (item numbers are keyed to Figure 9)

Item	Dimensions (ft)	Item	Dimensions (ft)
1	175 x 75	6	190 x 80
2	180 x 140	7	200 x 65
3	80 x 70	8	225 x 65
4	125 x 55	9	300 x 50
5	150 x 70	10 (4)	90 x 55
		11	120 x 55

to Figure 9, includes dimensions of structures at the Perm facility.

One test cell (annotated A on Figure 9) is within the secured area of the Perm test facility. This test cell is rail served to its front, and is the only cell identified thus far in the USSR which is served in this manner. Although the cell cannot be negated on available photography, it can be confirmed as complete in June 1962; it may

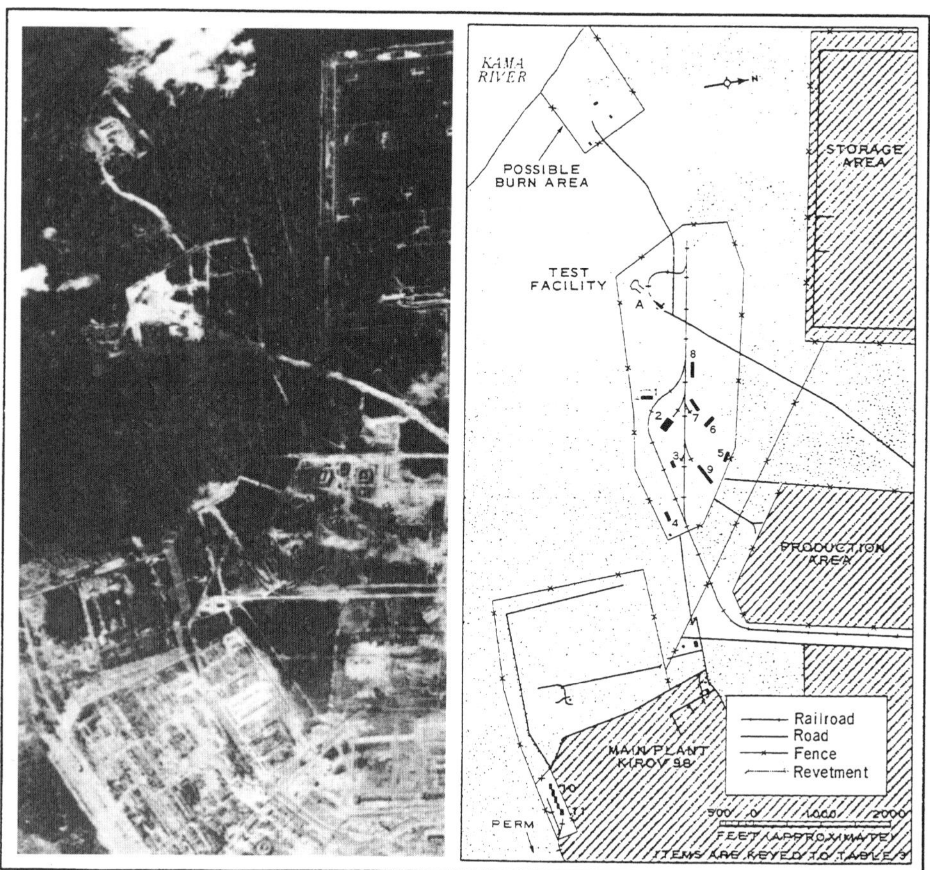

FIGURE 9. PERM PROBABLE SOLID PROPELLANTS TEST FACILITY AND PORTIONS OF PERM MUNITIONS AND CHEMICAL COMBINE K. KIROV 98, USSR, JUNE 1963.

~~TOP SECRET RUFF~~

21. *(Continued)*

~~TOP SECRET RUFF~~

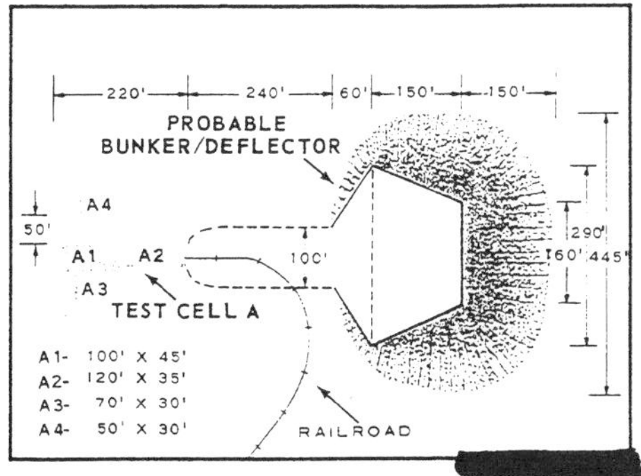

FIGURE 10. TEST CELL A AND PROBABLE BUNKER/DEFLECTOR, PERM, USSR.

be as old as cell B at Krasnoyarsk and therefore possibly one of the oldest in the USSR. The cell is made up of at least two sections and has an overall length of 220 feet.

The concrete-faced probable bunker/deflector (Figure 10) was present in July 1961. It is approximately 300 feet long from the base to the front of the test cell. The distance from the rear of the test cell to the rear of the probable bunker/deflector measures over 800 feet.

A group of five offset or staggered buildings (items 10 and 11, Figure 9) which appear to be separately secured from the rest of the associated explosives plant are located about 4,400 feet east-southeast of the test facility and appear to be rail served. These buildings were first discernible on August 1962 photography and are

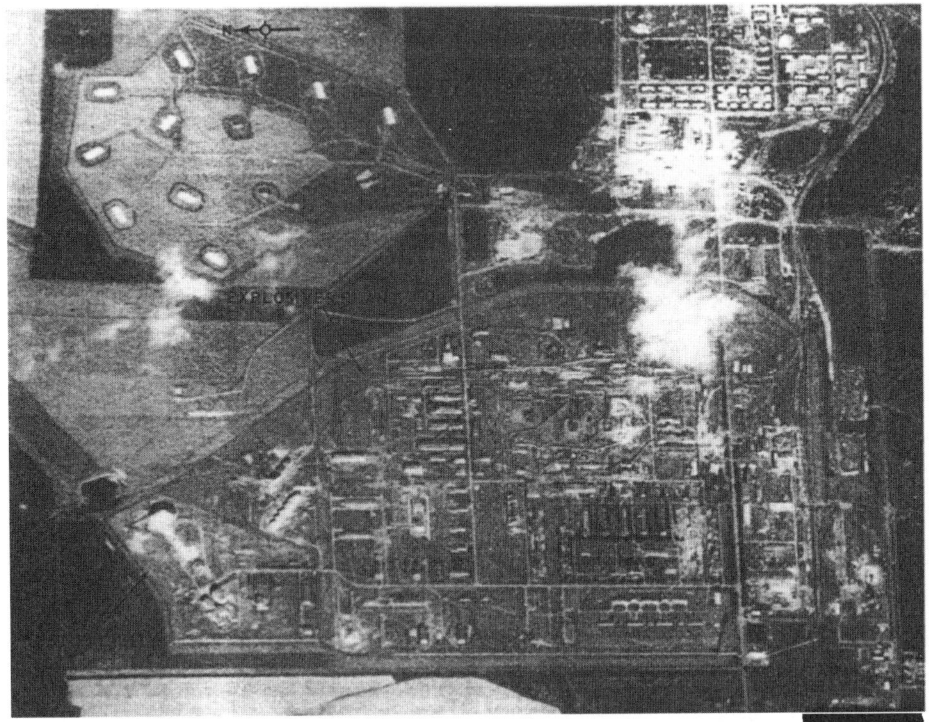

FIGURE 11. STERLITAMAK PROBABLE SOLID PROPELLANTS TEST FACILITY, USSR, JULY 1963.

~~TOP SECRET RUFF~~

~~TOP SECRET RUFF~~

similar in appearance and probably identical in function to those found at Biysk, Kamensk-Shakhtinskiy, and Sterlitamak.

A separately secured area about 1,800 feet west of the test facility measures approximately 1,300 by 1,150 feet. The area has three large, unexplainable, unoccupied revetments; its function may be that of a possible burn area, comparable to the similar area at Biysk.

STERLITAMAK

The Sterlitamak Probable Solid Propellants Test Facility (Figures 11 and 12) is adjacent to Explosives Plant 850, approximately 7 nm north of Sterlitamak, USSR, and about 3 nm west of the Belaya River. Although the test facility can be negated on captured German photography of July 1942, dating of the initial construction at the facility by photography is not possible. The test facility is road served, and its area of approximately 1,800 by 1,300 feet is partially double secured. It is possible that the outer fence is solid; only a single fence separates the test facility from the explosives plant.

The Sterlitamak test facility has an H-shaped building (item 1, Figure 12) which is similar in appearance and probably identical in function to those at Biysk and Krasnoyarsk. It was first discerned in April 1962, and confirmation as complete was possible on July 1963 photography; it is believed that construction of this structure was nearly complete in April 1962. Approximate dimensions of this building and other structures at the test facility and explosives plant can be found in Table 4 which is keyed to Figure 12.

The test cell at Sterlitamak (annotated A on Figure 12) is served from the front by a wide turn radius road; this is the only facility identi-

Table 4. Associated Structures. Test Facility and Explosives Plant, Sterlitamak (item numbers are keyed to Figure 12)

Item	Dimensions (ft)	Item	Dimensions (ft)	Item	Dimensions (ft)	Item	Dimensions (ft)
1		21	125 x 40	47	380 x 60	69	250 x 70
a	280 x 85	22	165 x 90	48	175 x 65	70	310 x 75
b	95 x 85	23	175 x 75	49	130 x 80	71, 74	310 x 105
c	150 x 80	24	340 x 80	50	170 x 110	72, 73	310 x 90
d	165 x 80	25	525 x 80	51	145 x 60	75	110 x 45
e	125 x 80	26	385 x 70	52	180 x 50		wing
2	110 x 110	27	285 x 70	53	200 x 75		90 x 60
3	110 x 70	28	100 x 30	54	245 x 110	76	365 x 85
4	265 x 90	29	115 x 50	55	255 x 60	77	460 x 110
5 (7)	90 x 60	30	320 x 50	56	250 x 90	78	160 x 90
6	120 x 60	31	110 x 100	57	265 x 60	79	105 x 80
7	100 x 70	32	105 x 40	58	525 x 100	80	80 x 75
8	140 x 70	33	125 x 60	59	180 x 70	81	115 x 85
9	395 x 60	34	175 x 65	60	150 x 50	82	75 x 60
10, 11	170 x 70	35	190 x 90	61 (4)	175 x 45	83	130 x 50
12	105 x 30	36	115 x 90	62 (3)	165 x 40	84	70 x 50
13	220 x 50	37	190 x 40	63 (4)	165 x 70	85	170 x 45
14, 15	170 x 70	38	165 x 80	64	825 x 160	86	85 x 60
16 (2)	each in 2		wing 45 x 45	65	580 x 110	87	200 x 50
	sections:	39	225 x 100	66	280 x 50	88 (12)	210 x 80
	90 x 80,	40	265 x 105	67	2 wings:	89	100 x 60
	90 x 60	41	300 x 140		140 x 20		
17	135 x 65	42	185 x 105		(each)		
18	400 x 50	43	375 x 150		center section:		
19	395 x 60	44, 45	180 x 90		145 x 25		
20 (2)	205 x 40	46	380 x 105	68	390 x 85		

~~TOP SECRET RUFF~~

~~TOP SECRET RUFF~~

fied thus far in the USSR which is served in this manner. The cell is not revetted or mounded and appeared essentially complete in April 1962. It has at least three sections and measures 250 feet in overall length. The relative position of the cell and other structures at the facility can be seen on Figure 12.

The probable bunker/deflector (Figure 13) can be observed on April 1962 photography; the concrete facing, however, cannot be confirmed until photography of July 1963. The structure is about 260 feet long from the base to the front of the test cell. From the rear of the probable bunker/deflector to the H-shaped building the distance is approximately 1,350 feet. A road serves both the front and the rear of the probable bunker/deflector; this is the only facility in the USSR which has this particular road pattern.

Eight staggered or offset buildings (items 5 and 6, Figure 12) which are possibly rail served are located within the explosives plant and adjacent to the test facility. Three of these buildings were discernible in May 1962, and the others could first be observed on June

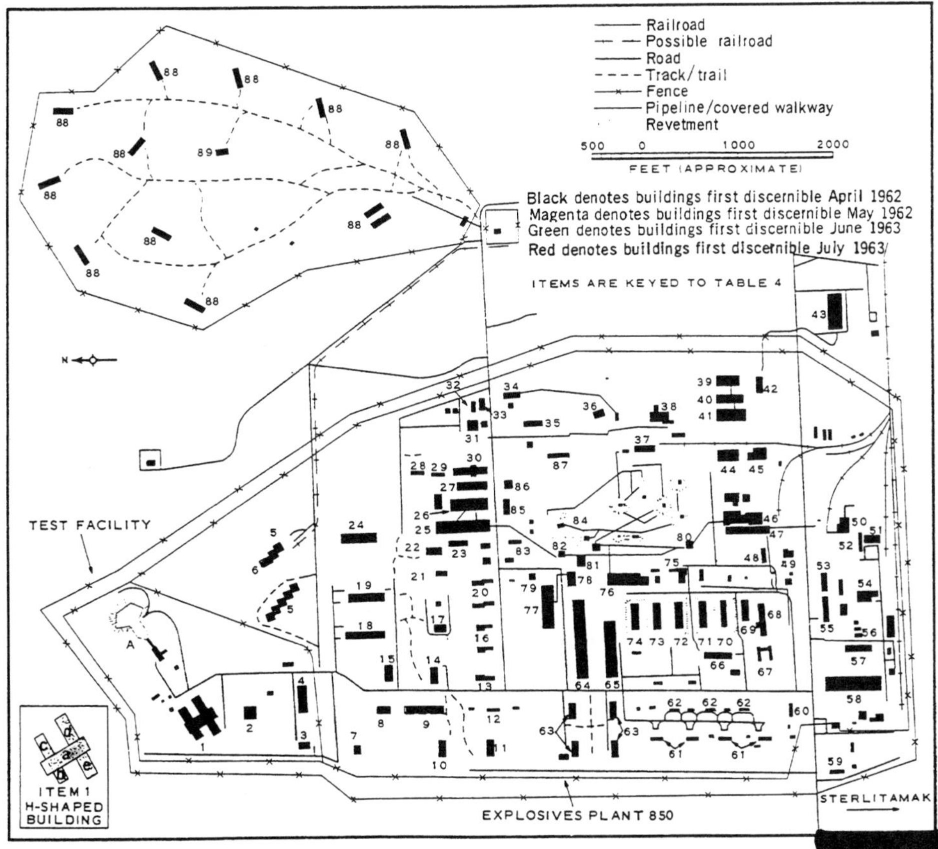

FIGURE 12. STERLITAMAK PROBABLE SOLID PROPELLANTS TEST FACILITY AND EXPLOSIVES PLANT 850, USSR.

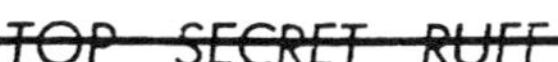

~~TOP SECRET RUFF~~

~~TOP SECRET RUFF~~

1963 photography. The buildings are similar in appearance to buildings at Biysk, Kamensk-Shakhtinskiy, and Perm and probably have an identical function.

THE PLANTS

Each of the four test facilities described is adjacent to a plant which has every evidence of producing at least two explosives bases. Although the buildings in the plants have not been identified as to type, it is apparent that mixing, casting, batching, and related functions could be carried on at each facility. All of the plants are road and rail served and are at least single secured; the plants at Biysk and Sterlitamak are double secured.

With the exception of the Perm plant, each explosives plant has shown significant construction since first photographic observation. The Biysk plant has been expanded by the addition of at least two explosives lines and has at least tripled in storage capacity since December 1960. The Krasnoyarsk plant (Figure 6), built adjacent to Explosives Plant 580, cannot be confirmed as completely constructed as of April 1963. The Sterlitamak plant (Figure 12), though showing less construction activity than the Biysk and Krasnoyarsk plants, has had evidences of construction since it was first observed on photography of April 1962.

The proximity of facilities to one another within the explosives plants precludes an unqualified, detailed analysis. Once an explosives line is constructed, it can often be used to work on new explosives bases.

CONCLUSIONS

1. There are test facilities at the following five cities in the USSR: Biysk, Kamensk-Shakhtinskiy, Krasnoyarsk, Perm, and Sterlitamak; the five facilities have a total of eight test cells.

2. Because these test facilities are within or adjacent to explosives plants capable of producing at least two explosives bases, the facilities can be considered probable solid propellants test facilities.

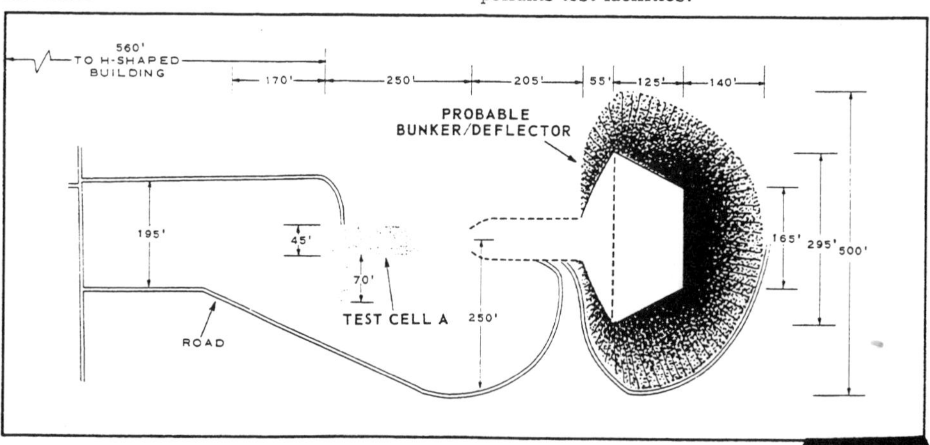

FIGURE 13. TEST CELL A AND PROBABLE BUNKER/DEFLECTOR, STERLITAMAK, USSR.

~~TOP SECRET RUFF~~

~~TOP SECRET RUFF~~

3. The probable bunker/deflectors at these sites are concrete faced, suggesting the function of deflector. Road service to both the front and rear of the bunker/deflector at Sterlitamak suggests the possibility of an instrumentation role; Sterlitamak, however, is the only facility at which this road characteristic could be noted.

4. These five facilities could be involved in research and development or production or both. The associated plants appear to have the capacity to produce explosives, while the test facilities at each of the installations have slight differences suggesting the possibility of research and development.

REFERENCES

PHOTOGRAPHY

Mission	Date	Pass	Camera	Frames	Classification
Biysk					
9056	30 Jun 63	54D	Fwd	43	TOP SECRET RUFF
			Aft	48, 49	
9044	1 Sep 62	53D	Fwd	42	TOP SECRET RUFF
			Aft	49	
9040	28 Jul 62	1A	Fwd	25	TOP SECRET RUFF
			Aft	32	
9038	1 Jul 62	52D	Fwd	24	TOP SECRET RUFF
			Aft	29	
	1 Jul 62	46A	Fwd	25	TOP SECRET RUFF
			Aft	33	
	30 Jun 62	31A	Fwd	56	TOP SECRET RUFF
9035	1 Jun 62	48A	Fwd	24, 25	TOP SECRET RUFF
			Aft	30	
9029	14 Dec 61	22	--	29	TOP SECRET RUFF
9013	10 Dec 60	37	--	67	TOP SECRET RUFF
Krasnoyarsk					
9057	21 Jul 63	38D	Fwd	28	TOP SECRET RUFF
			Aft	34	
9053	2 Apr 63	6D	Fwd	26	TOP SECRET RUFF
			Aft	31	
9047	9 Nov 62	53D	Fwd	20	TOP SECRET RUFF
			Aft	26	
9043	18 Sep 62	6D	Fwd	132, 133	TOP SECRET RUFF
			Aft	138	
9041	3 Aug 62	22D	Fwd	31	TOP SECRET RUFF
			Aft	37	
9040	30 Jul 62	32A	Fwd	14	TOP SECRET RUFF
			Aft	20	
9037	24 Jun 62	16A	Fwd	14	TOP SECRET RUFF
			Aft	20	
9031	1 Mar 62	22D	Fwd	21	TOP SECRET RUFF
			Aft	28	
9022	13 Sep 61	6D	--	81	TOP SECRET RUFF
9017	18 Jun 61	22	--	84	TOP SECRET RUFF
Perm					
1001-1	27 Aug 63	39D	Aft	23	TOP SECRET RUFF
	25 Aug 63	2A	Fwd	17	TOP SECRET RUFF
			Aft	22	

~~TOP SECRET RUFF~~

21. *(Continued)*

~~TOP SECRET RUFF~~

REFERENCES (Continued)

PHOTOGRAPHY

Mission	Date	Pass	Camera	Frames	Classification
9054	16 Jun 63	55D	Fwd	46	TOP SECRET RUFF
			Aft	51	
9041	3 Aug 62	17A	Fwd	16	TOP SECRET RUFF
			Aft	22	
9038	28 Jun 62	2A	Fwd	52, 53	TOP SECRET RUFF
			Aft	60, 61	
9035	30 May 62	2A	Fwd	42, 43	TOP SECRET RUFF
			Aft	48	
9019	8 Jul 61	1A	--	94, 95	TOP SECRET RUFF
Sterlitamak					
1001-1	27 Aug 63	39D	Aft	48	TOP SECRET RUFF
9057	19 Jul 63	8D	Fwd	49, 50	TOP SECRET RUFF
			Aft	55	
9054	14 Jun 63	17A	Fwd	1	TOP SECRET RUFF
			Aft	6	
9035	30 May 62	2A	Fwd	29	TOP SECRET RUFF
			Aft	35	
9032	18 Apr 62	2A	Fwd	22	TOP SECRET RUFF
			Aft	29	
GX 1570	22 Jul 42	--	--	65 - 69	CONFIDENTIAL

MAPS OR CHARTS

AMS. Series N501, Sheet No 40-7, 1st ed, Nov 57, scale 1:250,000 (UNCLASSIFIED)
AMS. Series N501, Sheet No 40-4, 1st ed, Nov 57, scale 1:250,000 (UNCLASSIFIED)
8RTS. US Air Target Chart, Series 200, Sheet 0159-23HL, 2d ed, May 63, scale 1:200,000 (SECRET)
2RTS. US Air Target Chart, Series 200, Sheet 0161-21HL, 2d ed, Aug 62, scale 1:200,000 (SECRET)
ACIC. US Air Target Chart, Series 200, Sheet 0165-15A, 1st ed, Sep 58, scale 1:200,000 (SECRET)
2RTS. US Air Target Chart, Series 200, Sheet 0234-24HL, 2d ed, Jul 63, scale 1:200,000 (SECRET)

REQUIREMENT

PROJECT

~~TOP SECRET RUFF~~

22. CIA/NPIC, Photographic Intelligence Report, "Comparison of Possible AMM–Associated Facilities at Moscow and Sary-Shagan, USSR, August and September 1963," May 1964

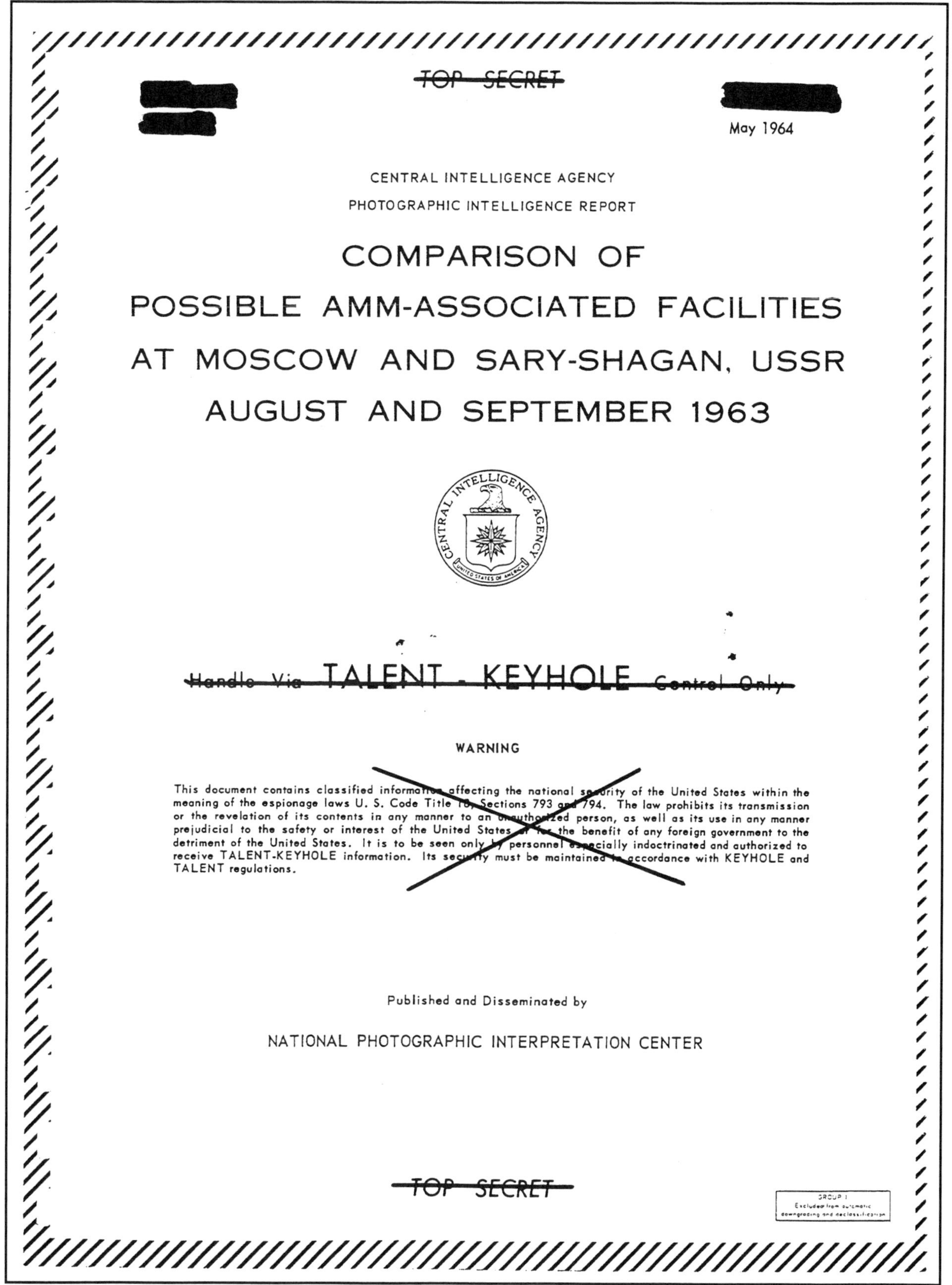

TOP SECRET

May 1964

CENTRAL INTELLIGENCE AGENCY

PHOTOGRAPHIC INTELLIGENCE REPORT

COMPARISON OF POSSIBLE AMM-ASSOCIATED FACILITIES AT MOSCOW AND SARY-SHAGAN, USSR AUGUST AND SEPTEMBER 1963

Handle Via TALENT - KEYHOLE Control Only

WARNING

This document contains classified information affecting the national security of the United States within the meaning of the espionage laws U. S. Code Title 18, Sections 793 and 794. The law prohibits its transmission or the revelation of its contents in any manner to an unauthorized person, as well as its use in any manner prejudicial to the safety or interest of the United States or for the benefit of any foreign government to the detriment of the United States. It is to be seen only by personnel especially indoctrinated and authorized to receive TALENT-KEYHOLE information. Its security must be maintained in accordance with KEYHOLE and TALENT regulations.

Published and Disseminated by

NATIONAL PHOTOGRAPHIC INTERPRETATION CENTER

TOP SECRET

GROUP 1
Excluded from automatic downgrading and declassification

22. *(Continued)*

~~TOP SECRET~~ [redacted] ~~RUFF~~

PHOTOGRAPHIC INTELLIGENCE REPORT

COMPARISON OF POSSIBLE AMM-ASSOCIATED FACILITIES AT MOSCOW AND SARY-SHAGAN, USSR AUGUST AND SEPTEMBER 1963

May 1964

NATIONAL PHOTOGRAPHIC INTERPRETATION CENTER

~~TOP SECRET~~ [redacted] ~~RUFF~~

~~TOP SECRET~~ [redacted] ~~RUFF~~

ERRATA

On page 1, the reference to Figure 1 in paragraph 3 should be deleted.

On page 15, in the data for August 1963, Mission 1001-1, the second reference to item 5 should be changed to item 6.

~~TOP SECRET~~ [redacted] ~~RUFF~~

22. *(Continued)*

~~TOP SECRET~~ [redacted] ~~RUFF~~

PREFACE

This report compares developments and configurations at the possible antimissile missile (AMM) associated facilities at Launch Complex B, Sary-Shagan Antimissile Test Center (SSATC), and at one of the four modified SA-1 SAM sites in the Moscow area. The information presented is derived from a study of TALENT and KEYHOLE photography available as of 25 September 1963. The report is prepared under project [redacted] in answer to CIA requirement [redacted]

~~TOP SECRET~~ [redacted] ~~RUFF~~

22. *(Continued)*

~~TOP SECRET~~ [redacted] ~~RUFF~~

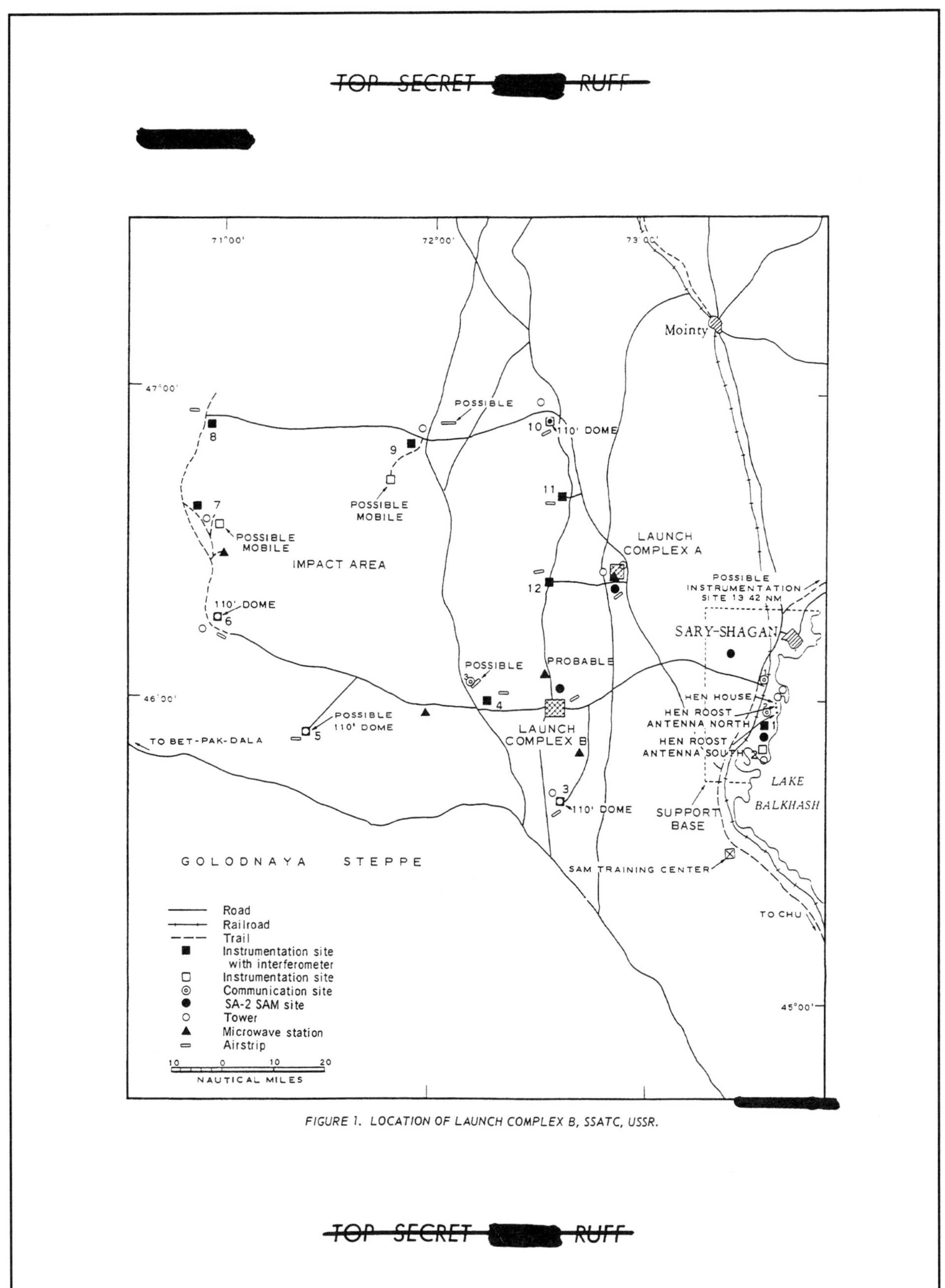

FIGURE 1. LOCATION OF LAUNCH COMPLEX B, SSATC, USSR.

~~TOP SECRET~~ [redacted] ~~RUFF~~

~~TOP SECRET~~ [redacted] ~~RUFF~~

INTRODUCTION

A preliminary analysis of the similarities between the possibly AMM-associated, modified SA-1 SAM sites near Moscow and the similar facilities at SSATC as of June 1963 was published in [redacted] 1/ This report supplements that analysis; it covers changes introduced at the Moscow sites between April 1962 and September 1963, and, particularly, changes discernible in the facilities at SSATC between April 1960 and August 1963.

Construction of all the possible AMM-associated facilities (modified SA-1 SAM sites) near Moscow follows the same general pattern; it probably commenced at approximately the same time and progressed at similar rates. For this comparison, SAM Site E33-1 has been selected as an illustration of all four possible AMM-associated sites in the Moscow area, and it is compared with the relatively new area containing a triad of buildings at Launch Complex B, SSATC. This new area is designated as Facility C. Selected features of the Sary-Shagan facility and the Moscow site correspond to the prototype herringbone pattern at the Kapustin Yar/Vladimirovka Missile Test Center (KY/Vlad MTC), USSR. A chronology of construction activity at Site E33-1 and Facility C is given in Table 3.

POSSIBLE AMM SITE IN THE MOSCOW AREA
APRIL 1962 - SEPTEMBER 1963

Moscow SAM Site E33-1 [redacted] is located at 56-20N 36-48E (Figure 1). Site preparation for the triad of buildings (items 1, 2, and 3, Figure 2) was first discernible on photography of August 1962. Building construction in the support area was observed on photography of April 1962. The buildings constructed in the triad and in the support area at Site E33-1 are typical of the possible AMM sites in the Moscow area. Item numbers are keyed to Figure 2 and Table 1.

BUILDING TRIAD AREA

Sites E33-1, E15-1, and possibly E05-1 have a group of three small rectangular structures (item 4) located approximately 500 feet from the large rectangular building (item 1). Sites E33-1 and E24-1 have two small rectangular structures with a possible third under construction (item 5). These three form a T-shaped pattern approximately 800 feet from the large building (item 1). Two structures probably similar to these (item 5) at Site E33-1 are located in a corresponding position at Site E05-1. Although these structures (items 4 and 5) may all be temporary construction support buildings, it should be noted that at Sites E33-1, E15-1, and E05-1 the three small rectangular structures (item 4) occupy the same position relative to the large building (item 1) and to each other. The consistency of this pattern suggests that these may be integral components of the facility as opposed to temporary construction support facilities.

A raised structure, approximately 70 feet long and possibly square, is under construction on the roof of the large building (item 1). This was observed for the first time on photography of September 1963.

New earth scarring within and near the

~~TOP SECRET~~ [redacted] ~~RUFF~~

22. *(Continued)*

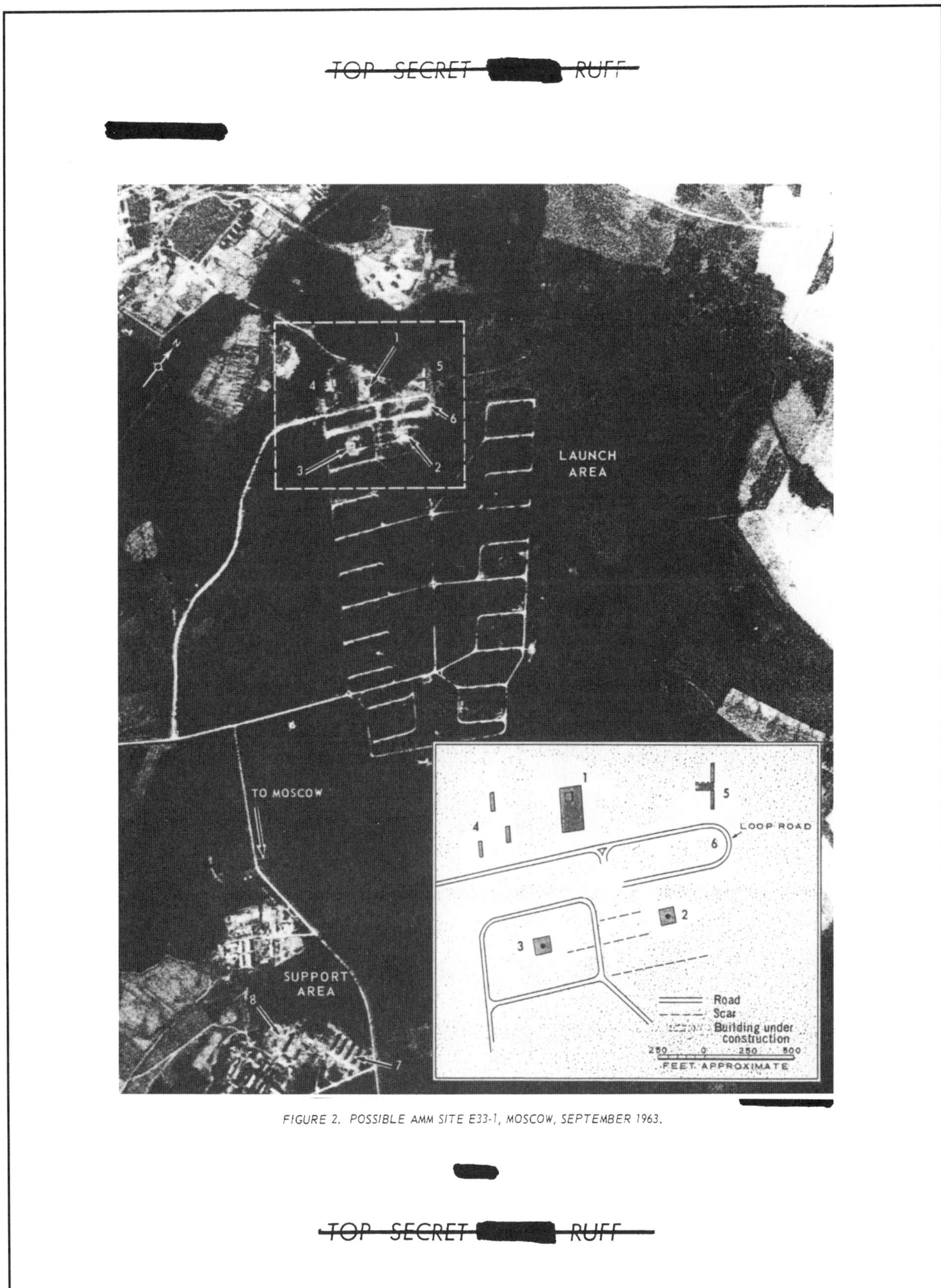

FIGURE 2. POSSIBLE AMM SITE E33-1, MOSCOW, SEPTEMBER 1963.

22. *(Continued)*

~~TOP SECRET [redacted] RUFF~~

Table 1. Key to Items on Figure 2

Item	Description
1	Large rectangular bldg with u/i structure on roof
2	Small square bldg
3	Small square bldg
4	3 rectangular structures
5	2 small rectangular structures and a possible 3d u/c forming T-shaped pattern
6	Loop road, extension of the access road
7	Row of 5 possible housing units
8	Bldg similar to 5 bldgs in item 7

herringbone pattern at all the possible AMM sites in the Moscow area and at a number of other SA-1 sites appeared between June and September 1963. (Figure 8 depicts the new activity at Site E33-1.)

The extension of the access road to the building triad area passes south of the three small rectangular structures (item 4, Figure 2) and the large rectangular building (item 1); it extends as far as the T-shaped pattern, then it loops in a wide-radius 180-degree turn (item 6) and returns to a cleared area adjacent to one of the rib roads of the herringbone launch area. The distance between the outside edges of the loop is approximately 260 feet. The road is about 25 feet wide and may be concrete surfaced. A strip 70 to 100 feet wide was cleared for the construction of the road; this includes a strip contiguous to a section of the herringbone road between buildings (items 1 and 3) of the triad. The loop road and cleared areas form a distinctive pattern among the possible AMM facilities in the Moscow area.

SUPPORT AREA

New construction activity has been observed in the support areas at all of the possible AMM sites in the Moscow area. In each case, construction of four or five multistory buildings was apparently planned to begin prior to the commencement of construction on the triad of buildings. The new buildings are approximately 200 by 50 feet and may be three- to five-story structures. The location and appearance of these new buildings and the roads serving them suggest that they may be housing units. At each site four or five buildings are constructed in a row, with another building, possibly serving a different function, located nearby. These new buildings at Site E33-1 (items 7 and 8, Figure 2) have an estimated total floorspace of 180,000 to 300,000 square feet. If they were troop barracks, allowing 60 square feet per person, each could house between 500 and 840 personnel; however, if they were designed as family apartments, the number accommodated would be much smaller.

FACILITY C, LAUNCH COMPLEX B, SSATC
APRIL 1960 - AUGUST 1963

Major components of the four possible AMM facilities at the SSATC follow the same general pattern, with variations probably the result of system development and testing. Significant variations observed at Facility C, Launch Complex B (46-01N 72-29E; [redacted] presented the first evidence of a possible launch facility directly associated with the triad of buildings (Figure 3). Item numbers are keyed to Figure 4 and Table 2.

BUILDING TRIAD AREA

Site preparation for the triad of buildings (items 1, 2, and 3) was first discernible on pho-

22. *(Continued)*

~~TOP SECRET~~ ~~RUFF~~

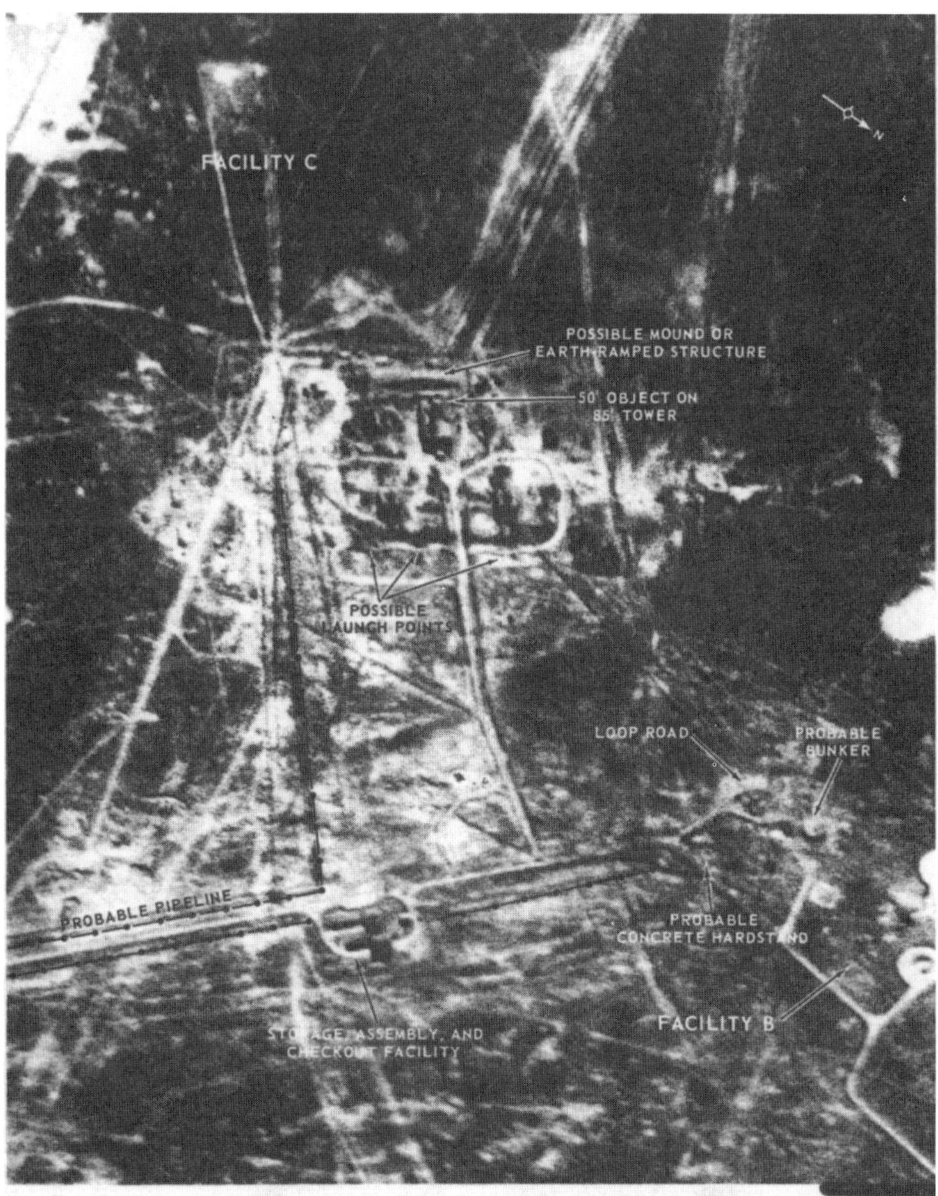

FIGURE 3. FACILITY C, LAUNCH COMPLEX B, SSATC, AUGUST 1963.

~~TOP SECRET~~ ~~RUFF~~

~~TOP SECRET~~ ~~RUFF~~

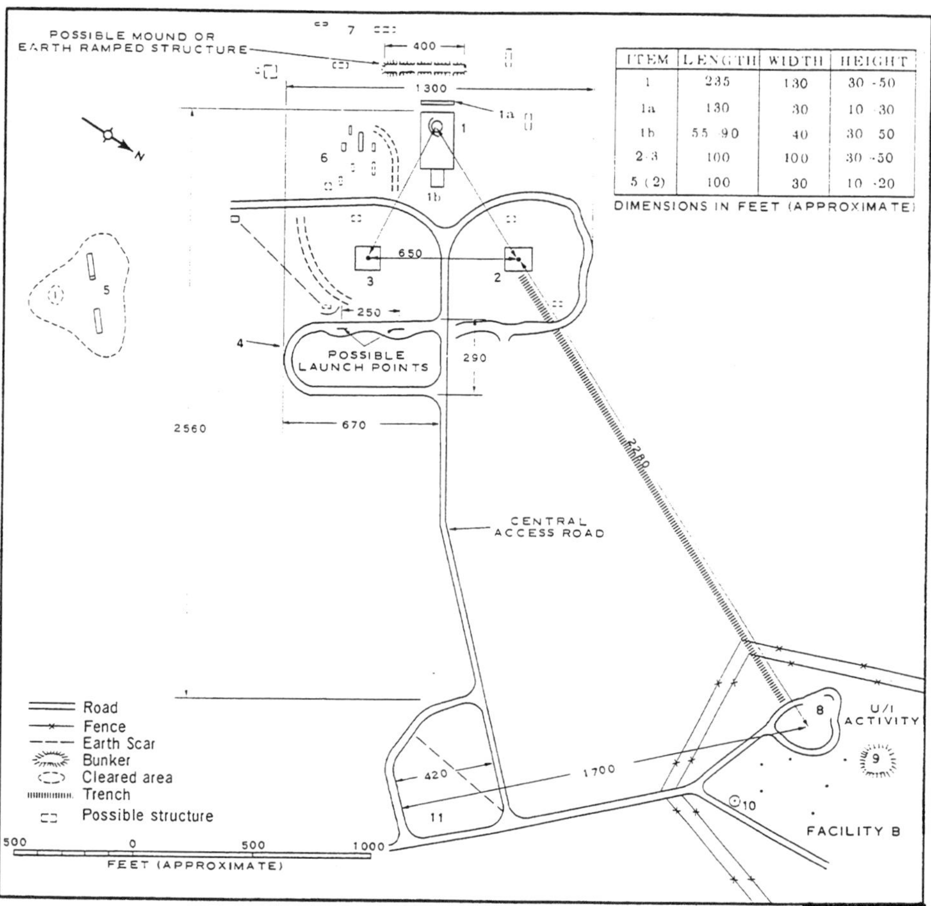

ITEM	LENGTH	WIDTH	HEIGHT
1	235	130	30 -50
1a	130	30	10 -30
1b	55 -90	40	30 50
2-3	100	100	30 -50
5 (2)	100	30	10 -20

FIGURE 4. RECTIFIED LINE DRAWING OF BUILDING TRIAD AREA, FACILITY C, SSATC.

Table 2. Key to Item Numbers on Figure 4

Item	Description
1	Large rectangular bldg with circular dish-shaped object on tower, 85 feet above roof level
a	U/I structure associated with large bldg
b	U/I structure associated with large bldg
2	Small square bldg with u/i object centered on roof
3	Small square bldg with u/i object centered on roof
4	Loop road
5	2 rectangular structures with a third u/i object
6	Area of u/i activity with one large object and several smaller objects
7	Possible bldgs
8	Loop road and u/i activity
9	Probable bunker
10	Circular probable concrete hardstand
11	Loop road at intersection of access roads

~~TOP SECRET~~ ~~RUFF~~

22. *(Continued)*

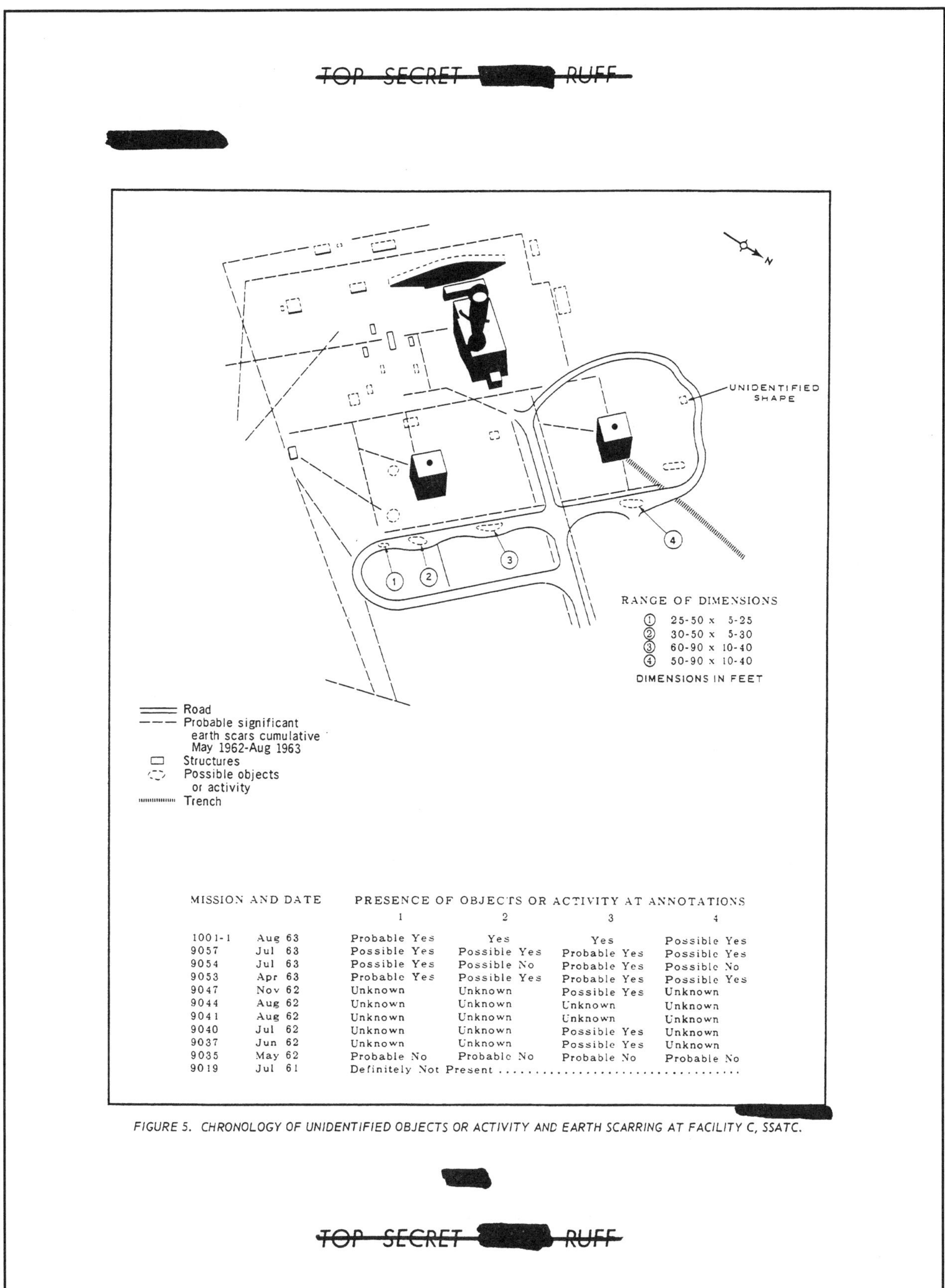

~~TOP SECRET~~ [redacted] ~~RUFF~~

MISSION AND DATE		PRESENCE OF OBJECTS OR ACTIVITY AT ANNOTATIONS 1	2	3	4
1001-1	Aug 63	Probable Yes	Yes	Yes	Possible Yes
9057	Jul 63	Possible Yes	Possible Yes	Probable Yes	Possible Yes
9054	Jul 63	Possible Yes	Possible No	Probable Yes	Possible No
9053	Apr 63	Probable Yes	Possible Yes	Probable Yes	Possible Yes
9047	Nov 62	Unknown	Unknown	Possible Yes	Unknown
9044	Aug 62	Unknown	Unknown	Unknown	Unknown
9041	Aug 62	Unknown	Unknown	Unknown	Unknown
9040	Jul 62	Unknown	Unknown	Possible Yes	Unknown
9037	Jun 62	Unknown	Unknown	Possible Yes	Unknown
9035	May 62	Probable No	Probable No	Probable No	Probable No
9019	Jul 61	Definitely Not Present			

FIGURE 5. CHRONOLOGY OF UNIDENTIFIED OBJECTS OR ACTIVITY AND EARTH SCARRING AT FACILITY C, SSATC.

~~TOP SECRET~~ [redacted] ~~RUFF~~

22. *(Continued)*

~~TOP SECRET~~ [redacted] ~~RUFF~~

tography of July 1961. Construction of an oblong loop road in the triad area was first detected on photography of June 1962; by April 1963 the road was essentially as shown in Figure 4. Two possible launch points, located approximately 250 feet east-northeast of the small building (item 3), may have been completed by June 1962. Unidentified objects or activity on or near the possible launch points has been detected over a period of 18 months. A chronology of these observations including metrical data is given in Figure 5. A possible mound or earth-ramped structure, approximately 400 feet long, is located southwest of the large building (item 1). This was first observed on photography of April 1963.

A circular object which appears to be a parabolic dish is mounted on an 85-foot tower constructed on the roof of the large building (item 1), and is interpreted as a suspect radar. The crescent-shaped shadow depicted in Figure 5 could only occur with a parabolic-type dish;

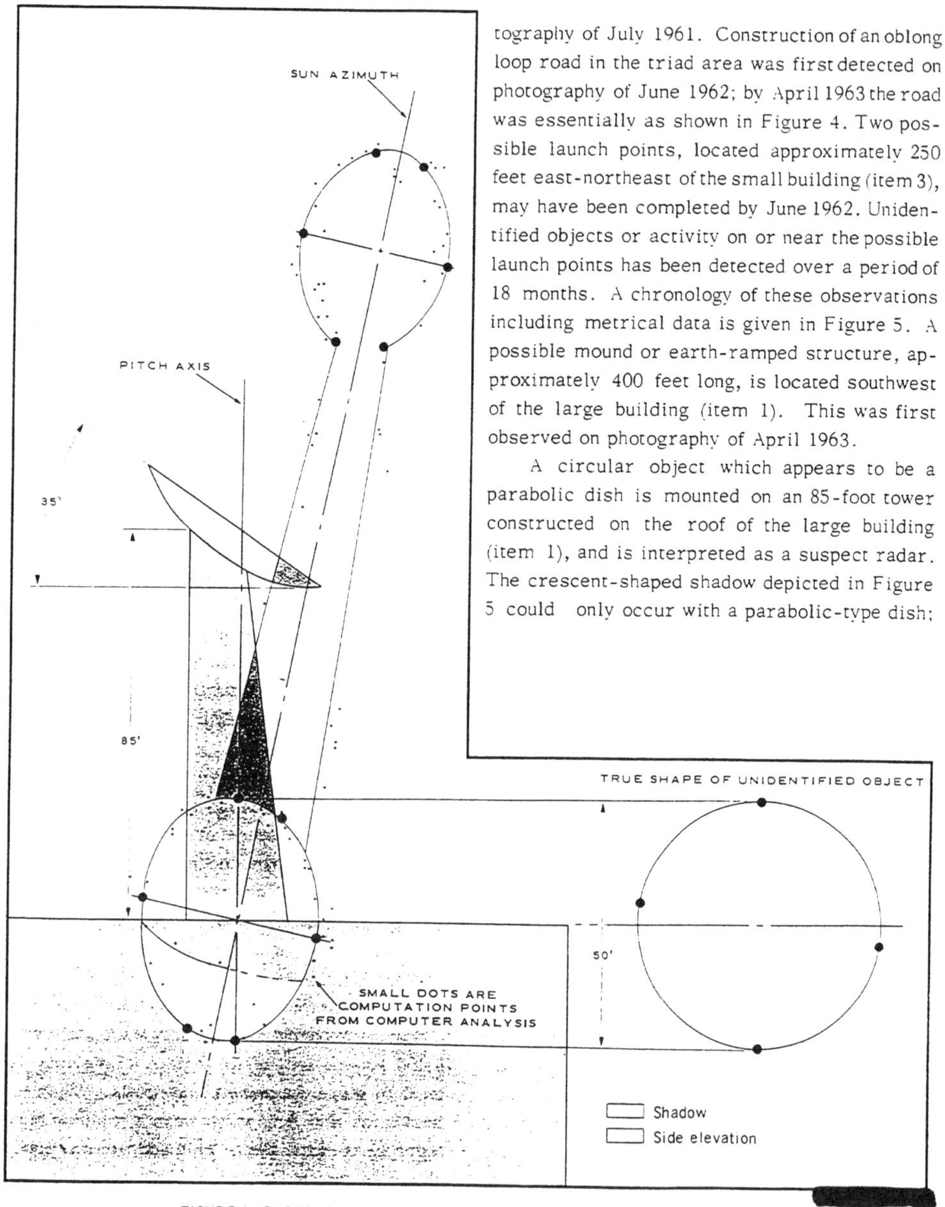

FIGURE 6. PHOTOGRAMMETRIC PROJECTION OF SUSPECT RADAR ANTENNA.

~~TOP SECRET~~ [redacted] ~~RUFF~~

if the object were spherical, a crescent-shaped highlight would have resulted along the top of the object. A side elevation and plan view projection of the object with metrical data are given in Figure 6. Possibly the same dish was seen on or near the ground and alongside the large building (item 1) in June 1963. The tower was probably under construction at this time.

A possibly raised section on the roof of the large building (item 1) is indicated by the fact that the shadow of the tower breaks as it crosses the roof of the building. The southeastern side of the building appears to curve outward slightly on photography of June and August 1963. This feature (shown in Figure 5) has not been detected at any of the corresponding structures at either SSATC or Moscow.

Several structures are separated from Facility C by a fence and are located within Facility B; however, they are probably associated with the more recently constructed Facility C. A probable bunker (item 9, Figure 4) may have been present in July 1961. A circular probable concrete hardstand (item 10) was present before construction began on Facility C. A possible electronic device was observed on the hardstand on photography of April 1960. A loop road (item 8) is within an area of unidentified activity. A trench connects this loop road with a building (item 2) in the triad area.

An area of unidentified activity (item 6), approximately 200 feet southeast of the large building (item 1) contains several small structures. One low rectangular structure is approximately 90 by 20 feet. Grouped around this structure are three to six 50- by 20-foot suspect footings. Activity has been observed in this area since construction was first identified in the building triad area. A possible structure could be seen in the area on photography of May 1962, and a possible cable or pipeline was visible entering the area. June 1962 photography revealed an apparent extension of this possible cable or pipeline from the area of unidentified activity to the large building (item 1). Figure 5 depicts all significant earth scars in cumulative portrayal.

A loop road (item 11, Figure 4) at the intersection of access roads, approximately 2,000 feet northeast of the building triad and 400 feet west of an assembly and checkout facility (shown on Figure 3), has been present since May 1962. No structure or activity has been observed in association with this loop road, other than an earth scar running through the enclosed area.

SUPPORT AREA, LAUNCH COMPLEX B

Additions to the Launch Complex B support area (not on graphics) since April 1960 are as follows.

(1) Storage, Assembly, and Checkout Area: Three buildings which are probably single story and have an estimated total floorspace of 4,500 square feet. These buildings were probably in place in December 1960.

(2) Housing Facility: Five two-story and two single-story housing units having an estimated total floorspace of 39,700 square feet. These additions could accommodate 660 persons if they were used as troop barracks, allowing 60 square feet per man. Most of these new buildings were added between April and December 1960.

(3) Possible Technical and Laboratory Facility: A probable single-story warehouse with 8,400 square feet of floorspace which may be used to store flammable materials. The slab foundation and fire walls were visible on photography of April 1960. This addition was probably complete in December 1960.

22. *(Continued)*

TOP SECRET [redacted] RUFF

SIMILARITIES BETWEEN THE POSSIBLE AMM FACILITIES MOSCOW AND SARY-SHAGAN

The shape and spacing of two adjacent possible launch points (item 4, Figure 4) at Facility C, SSATC, are similar to the shape and spacing of the launch points at the prototype herringbone launch site at KY/Vlad MTC (Figure 7). The launch points at Moscow Site E33-1 are similar to the launch points at Kapustin Yar in size, shape, and spacing, although the curved prepared areas appear to be less prominent at Moscow than at Kapustin Yar. Figures 7 and 8 are similarly scaled line drawings of these sites with Facility C, Launch Complex B, SSATC superimposed for comparison.

A tower with a parabolic dish-shaped object is mounted on the roof of the large building (item 1) in the triad at Facility C, SSATC, and a raised structure is under construction at approximately the same location on the large building (item 1, Figure 2) at Moscow Site E33-1.

One of the small buildings (item 3, Figure 2) occupies the same relative position with regard to adjacent launch points at Site E33-1 as the corresponding building at Facility C (item 3, Figure 4) occupies with reference to nearby possible launch points. The site of this building at Moscow Site E33-1 was previously occupied by an SA-1 site control bunker.

An oblong loop road with wide-radius turns is located within the triad areas at both the Moscow and SSATC facilities, although the placement of the road with reference to the buildings of the triad is different.

Approximately the same amount of unoccupied terrain surrounds the small buildings (items 2 and 3) at both the Moscow and SSATC facilities.

New construction has been observed in the support areas at the possible AMM sites near Moscow and at Facility C. Approximately 180,000 to 300,000 square feet of possible housing space has been added at Site E33-1, and approximately 39,700 square feet of probable housing space has been added to the support area at Launch Complex B, SSATC. However, there is no apparent correlation of size or shape between the buildings constructed at these two facilities.

A trench extending from one of the triad buildings (item 2, Figure 4) to a road-served area of unidentified activity (item 8, Figure 4) at Facility C, SSATC, is comparable to a possible trench at Site E33-1 which extends from the corresponding small building (item 2, Figure 2) to an area of unidentified activity as shown in Figure 8.

DIFFERENCES BETWEEN THE POSSIBLE AMM FACILITIES MOSCOW AND SARY-SHAGAN

The large building (item 1, Figure 2) at Moscow Site E33-1 does not yet have any associated structures which would correspond to those at Facility C, SSATC (items 1a and 1b, Figure 4). Structures similar to one of these (item 1a) have been observed at all the possible AMM facilities at SSATC; however, they have not yet been constructed at any of the possible AMM sites near Moscow. The other structure (item 1b) at Facility C is unique.

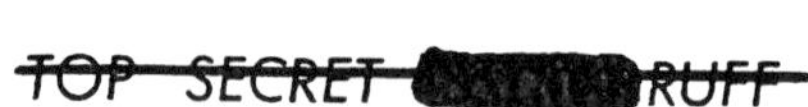

22. *(Continued)*

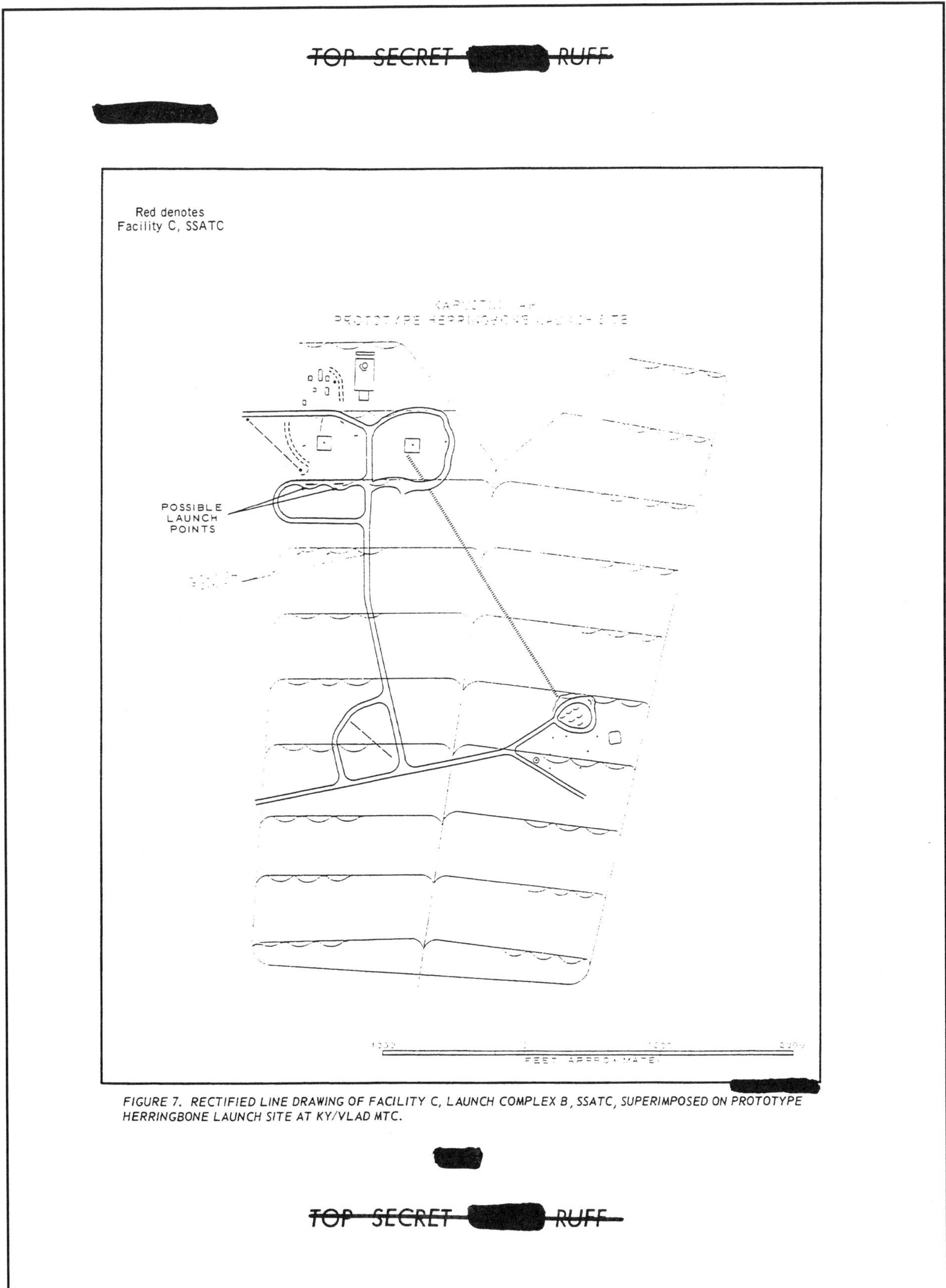

FIGURE 7. RECTIFIED LINE DRAWING OF FACILITY C, LAUNCH COMPLEX B, SSATC, SUPERIMPOSED ON PROTOTYPE HERRINGBONE LAUNCH SITE AT KY/VLAD MTC.

22. *(Continued)*

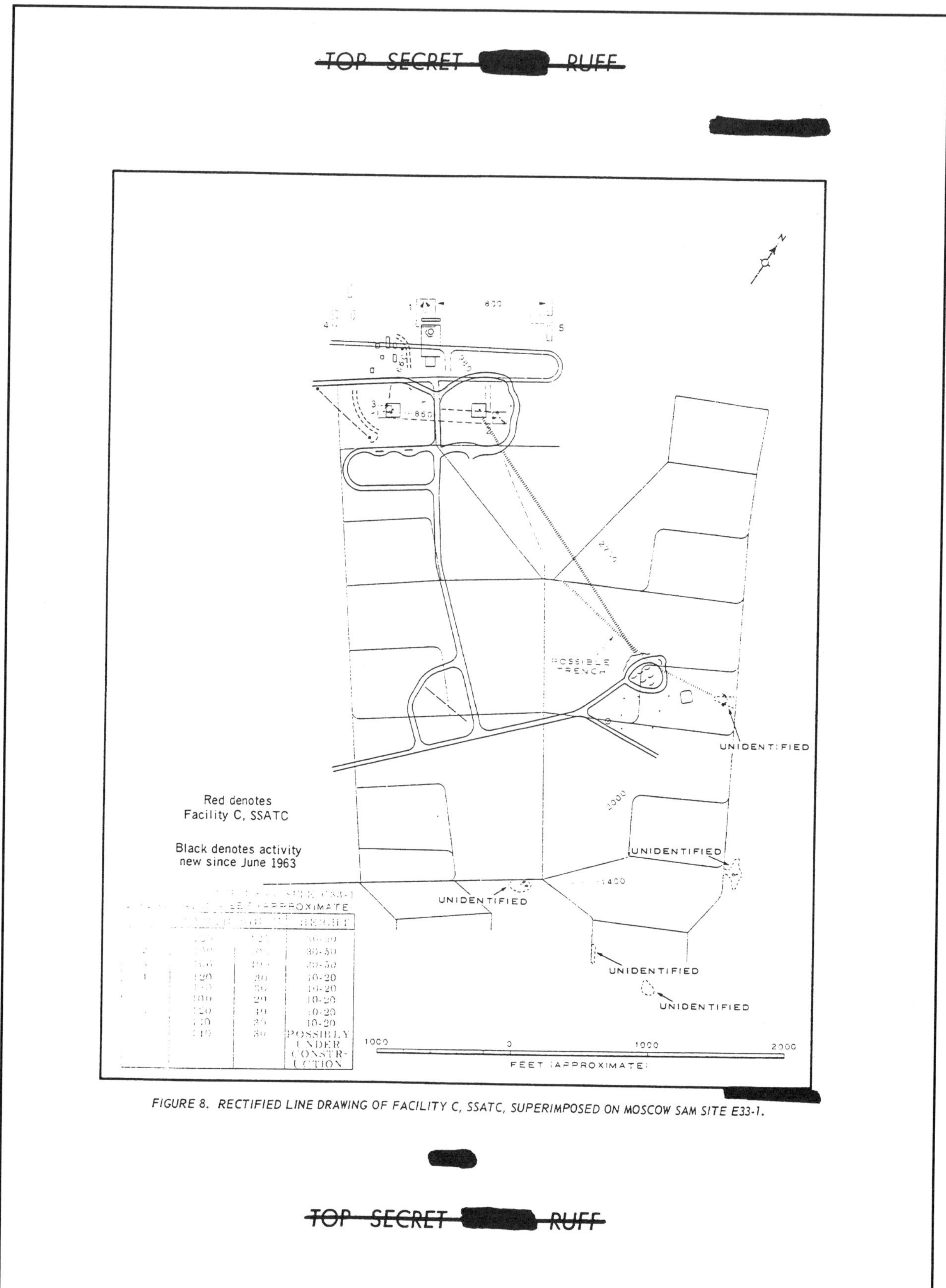

FIGURE 8. RECTIFIED LINE DRAWING OF FACILITY C, SSATC, SUPERIMPOSED ON MOSCOW SAM SITE E33-1.

22. *(Continued)*

~~TOP SECRET~~ ~~RUFF~~

There is no apparent correlation between the position of the loop road (item 6, Figure 2) with reference to the triad of buildings at Moscow Site E33-1 and the position of the loop road (item 4, Figure 4) within the building triad area at Facility C.

One side of the large building (item 1) at Facility C appears to curve outward, probably as shown in Figure 5. This feature is not discernible at any of the corresponding buildings at the other possible AMM facilities at either Moscow or SSATC.

Structures (items 4 and 5, Figure 2) which are possibly integral components of the triad facility at E33-1 are not discernible at any of the possible AMM facilities at SSATC.

The position of new earth scars within the herringbone pattern at the Moscow facility, as shown in Figure 8, does not appear to correlate with any significant features or activity observed at Facility C, SSATC.

Unidentified activity or objects have been observed at one or more of the SSATC possible launch points near the two small buildings (items 2 and 3) since June 1962. No similar features have been visible at Moscow Site E33-1.

The large possible mound or earth-ramped structure (Figure 3) at SSATC has no counterpart at any of the other possible AMM facilities at either Moscow or SSATC.

The small rectangular unidentified objects (item 6, Figure 4) located approximately 250 feet from the large building at SSATC have no counterpart at any of the other possible AMM facilities at either Moscow or SSATC.

DISCUSSION

Since this analysis is based primarily on KEYHOLE photography, there may be significant details at both the Moscow and SSATC facilities which escape detection. The launch points at Moscow Site E33-1 may differ considerably in detail from the possible launch points at Facility C, SSATC. Herringbone launch areas, clearly photographed from low altitude near Moscow, consist of regular, concrete, saw-toothed extensions of the access road; each extension is bounded on one side by a prepared area on which possible checkout equipment is located. These prepared areas give the launch points the curved shape which is particularly prominent at the KY/Vlad MTC prototype herringbone site. At Moscow Site E33-1 and KY/Vlad MTC only the road and the launch point extensions appear to have a concrete surface, whereas at Facility C, the entire area appears to be surfaced with concrete.

The areas at Facility C cannot be positively identified as launch points. Since the facility is probably concerned with electronics, another possibility is that they are used as hardstands for electronics equipment. However, if this were accepted as a hypothesis it would be difficult to explain the purpose of a road system with such wide-radius turns. The road pattern at Facility C suggests the necessity for repeated access to the possible launch areas by motor vehicles which cannot readily negotiate turns and require these elaborate provisions. The herringbone road pattern serves such a purpose. If the facilities at SSATC are neither launch points nor hardstands for electronics equipment, there is no apparent explanation for them.

The intersections of a loop road (item 11, Figure 4) and the access roads at Facility C have Y-shaped wide-radius turns. This area could serve as a hold area for transporters

███

with loads to be delivered to the possible launch points near the two small buildings (items 2 and 3).

The area of unidentified activity (item 6) appears to be an integral part of Facility C, SSATC. The unidentified objects in this area were suspected of being stacked crates and boxes used to transport supplies and equipment during construction of the building triad area; however, a careful analysis of the following photographic evidence tends to discount this possibility:

1. The placement of the large object and the smaller objects grouped around it forms an orderly pattern which does not resemble a dump.

2. These objects cast little or no shadow, as would stacks of crates and boxes.

3. The large central object (approximately 90 by 20 feet) was clearly visible on photography of April and August 1963, and there was no perceptible change in its size, position, or rectangular shape even though activity was apparent throughout the area. Furthermore, the object was possibly in this location as early as May 1962.

4. A prominent earth scar entering this area, probably a cable or pipeline, was observed on photography of May 1962. An apparent extension of this earth scar, visible on photography of June 1962, appears to connect it with the large building in the triad.

5. A possible construction crane was visible in this area on photography of April 1963.

Table 3. Chronology of Construction Activity at Moscow SAM Site E33-1 and Facility C, SSATC

	SAM SITE E33-1, MOSCOW Item Numbers are Keyed to Figure 2	FACILITY C, COMPLEX B, SSATC Item Numbers are Keyed to Figure 4
Apr 60 Mission B 4155	No photo coverage.	No evidence of construction activity.
Dec 60 Mission 9013	No photo coverage.	No change can be detected.
Jul 61 Mission 9019	Very poor photo quality -- identification only.	Early stages of site preparation in the triad area with extensive earth scarring, but no evidence of building foundations. A loop road (item 8) is possibly under construction.
Apr 62 Mission 9032	This is the first mission which produced interpretable photography. None of the buildings of the triad are present, nor is there any sign of foundations or footings. However, ground scarring can be detected between the large building (item 1) and the T-shaped pattern (item 5). Five 200- by 50-foot buildings are present in the support area. Of these, four are arranged in a row while the fifth is located nearby. The road leading south from the northwestern corner of the herringbone site is hardly visible and possibly lightly used.	No photo coverage.
May-Jun 62 Mission 9035	Portions of the herringbone pattern are cloud covered. The possible construction activity (scarring) noted on Mission 9032 cannot be seen; however, this is possibly due to poor photo quality. A portion of the road leading south from the northwestern corner of the herringbone site is visible. The road is more prominent, indicating a possible increase in its use for access to construction activity in what is now referred to as the building triad area.	Construction on the triad of buildings is apparently in advanced stages, with all walls and probably the roofs in place. The small structure (item 1a) southwest of the large building cannot be seen. Unidentified construction activity adjacent to the large building is noted. Trenches in the triad area are visible and a loop road (item 8) is possibly under construction.

███

22. *(Continued)*

Table 3. (Continued)

	SAM SITE E33-1, MOSCOW Item Numbers are Keyed to Figure 2	FACILITY C, COMPLEX B, SSATC Item Numbers are Keyed to Figure 4
Jun 62 Mission 9037	No photo coverage.	Construction activity is in evidence throughout the area of the triad installation; however, the structure (item 1a) southwest of the large building is probably still missing. Evidence of construction activity on an oblong loop road pattern northeast of the triad of buildings is noted for the first time. The loop road (item 8) is probably complete. Trench work can be seen in the area. A road loop around one of the small buildings (item 2) is possibly under construction.
Jul 62 Mission 9040	No photo coverage.	No change can be detected since the last mission.
Aug 62 Mission 9041 Mission 9044	The quality of this photography is poor; however, possible construction activity in the building triad area can be detected. The access road to the building triad area is more prominent, possibly indicating increased use since April 1962. Photo quality precludes analysis of support area structures. (9044, no coverage.)	9041: A structure (item 1a) is seen under construction for the first time. Other construction activity continues. 9044: Small buildings (items 2 & 3) appear complete with objects on the roofs. Construction activity can be seen on the roof of the large building (item 1).
Nov 62 Mission 9047	Partial cloud cover and poor quality. Possible construction activity can be seen in the building triad area. A probable sixth large building is present in the support area; it is similar to and in line with the four buildings previously observed.	Construction activity continues. A trench to the large building appears to have been enlarged.
Apr 63 Mission 9053	No photo coverage.	The small structure (item 1b) in back of the large building is seen for the first time. It is not possible to tell if this structure extends to the edge of the large building, as it is partly hidden by shadow. Three buildings may be under construction in a new area of activity (item 5). A possible mound or earth-ramped structure has been added in front of the large building. Two small possible buildings (item 7) can be seen approximately 300 feet south of this feature. The activity approximately 200 feet southeast of the large building continues; a small, low rectangular structure with two or more smaller shapes nearby can be seen. A possible crane is nearby. Two possible launch points can be seen off the loop road approximately 250 feet east-northeast of the small building (item 3). A suspect third launch point is u/c east-northeast of the other small building of the triad (item 2).
Jun 63 Mission 9054 Mission 9056	No photo coverage on 9054. 9056: Two buildings (items 1 and 2) appear complete; however, possible construction activity continues on the top rear portion of the large building (item 1). Another building (item 3) appears to be incomplete. An un-	Construction activity continues in the area southeast of the large building and on the loop road (item 4) in back of the triad of buildings. A circular dish-shaped object, approximately 50 feet in diameter, can be seen on or near the ground alongside the large building. It may be

~~TOP SECRET~~ ~~RUFF~~

Table 3. Continued

	SAM SITE E33-1, MOSCOW Item Numbers are Keyed to Figure 2	FACILITY C, COMPLEX B, SSATC Item Numbers are Keyed to Figure 4
	identified small object is centered on the top of one of the small buildings (item 2). Possible footings or foundations for the structures in a T-shaped pattern (item 5) are in place. Three buildings (item 4) are possibly in place. The row of five large buildings is prominent in the support area, approximately 9,000 feet southeast of the building triad area. The sixth building, similar in size, is located about 700 feet to the southwest. A loop road with a wide-radius turn is prominent between the large building and the two small buildings in the triad.	the dish which later photography reveals atop the tower, yet to be completed on the roof of the large building. Construction of a trench between a small building (item 2) and the loop road is in progress. A possible trench or ditch can be seen on the south side of the loop road (item 8). 9056: No cover.
Jul 63 Mission 9057	No photo coverage.	Construction activity in the area continues. The large building (item 1) is probably complete. A tower is being constructed on the front portion of the roof of the large building. The tower is now approximately 40 feet above roof level.
Aug 63 Mission 1001-1	Partial cloud cover and haze limit interpretation. No changes can be seen.	Construction activity continues. The loop pattern of roads in the triad area is now more distinct. Two probable buildings (item 5) are now visible. The loop road encircling the building at item 2 bends outward, and an unidentified shape can be seen just inside this bend. Earth scarring can be detected at the loop road (item 8). A trench from the building at item 2 to this loop road is complete. Activity in the area southeast of the large building continues (item 5). The central structure in this area is approximately 90 by 20 feet in size. Grouped around this long narrow structure are three to six 50- by 20-foot structures. All structures are low, casting little or no shadow. The tower on the roof of the building is complete. It is 85 feet tall, 40 to 50 feet wide at the base, and has a 50-foot circular dish-shaped object at its top. The dish is elevated at an angle of 55 degrees from the horizontal, on an azimuth of 85 degrees.
Sep 63 Mission 1002-1	All three buildings in the triad appear to be complete. The small buildings (items 2 & 3) have an unidentified small object centered on each roof. A raised structure, about 70 feet long and possibly square, is being constructed on the rear portion of the large building. The top of the raised structure is 15 to 30 feet above the level of the roof. New track activity and earth scarring are noted in the herringbone pattern since the last mission. This new activity is depicted on Figure 8.	No photo coverage.

~~TOP SECRET~~ ~~RUFF~~

REFERENCES

PHOTOGRAPHY

Mission	Date	Pass	Camera	Frames	Classification
Moscow SAM Site E33-1					
1002-1	25 Sep 63	24D	Fwd Aft	40 47	TOP SECRET RUFF
1001-1	28 Aug 63	50A	Fwd Aft	56 61	TOP SECRET RUFF
9056	27 Jun 63	02A	Fwd Aft	6 12	TOP SECRET RUFF
9047	7 Nov 62	24D	Fwd Aft	43 49	TOP SECRET RUFF
9041	2 Aug 62	02A	Aft	29	TOP SECRET RUFF
9035	1 Jun 62	40D	Fwd Aft	35 41	TOP SECRET RUFF
9032	18 Apr 62	03A	Fwd Aft	34 40	TOP SECRET RUFF
9019	9 Jul 61	17A	-	52, 53	TOP SECRET RUFF
SSATC, Launch Complex B					
1001-1	25 Aug 63	07D	Fwd Aft	60 64	TOP SECRET RUFF
9057	19 Jul 63	23D	Fwd Aft	132 137	TOP SECRET RUFF
9054	14 Jun 63	23D	Fwd Aft	137 141	TOP SECRET RUFF
9053	2 Apr 63	07D	Fwd Aft	77 82	TOP SECRET RUFF
9047	6 Nov 62	07D	Aft	76	TOP SECRET RUFF
9044	31 Aug 62	33A	Fwd Aft	10 16	TOP SECRET RUFF
9041	3 Aug 62	23D	Fwd Aft	67 73	TOP SECRET RUFF
9040	28 Jul 62	07D	Fwd Aft	80 87	TOP SECRET RUFF
9037	24 Jun 62	23D	Fwd Aft	82 86	TOP SECRET RUFF
9035	31 May 62	23D	Fwd Aft	108 113	TOP SECRET RUFF
9019	8 Jul 61	7	-	174	TOP SECRET RUFF
9013	8 Dec 60	7	-	130	TOP SECRET RUFF
B 4155	9 Apr 60	-	9	L2189	TOP SECRET

22. *(Continued)*

~~TOP SECRET [redacted] RUFF~~

[redacted]

REFERENCES (Continued)

MAPS AND CHARTS

Moscow

SAC. US Air Target Chart, Series 200, Sheet 0154-23HL, 2d ed, Apr 63, scale 1:200,000 (SECRET)

SAC. US Air Target Chart, Series 200, Sheet 0167-5HL, 2d ed, Apr 63, scale 1:200,000 (SECRET)

SAC. US Air Target Chart, Series 200, Sheet 0154-22HL, 2d ed, Mar 63, scale 1:200,000 (SECRET)

SAC. US Air Target Chart, Series 200, Sheet 0167-4HL, 2d ed, Mar 63, scale 1:200,000 (SECRET)

SAC. US Air Target Chart, Series 200, Sheet 0167-10HL, 2d ed, Feb 63, scale 1:200,000 (SECRET)

SAC. US Air Target Chart, Series 200, Sheet 0167-9A, 1st ed, Jan 59, scale 1:200,000 (SECRET)

SSATC

AMS. Series DESPA-1, Sheet NL43-4, 1st ed, Jun 62, scale 1:250,000 (TOP SECRET RUFF)

AMS. Series DESPA-1, Sheet NL43-7, 1st ed, Jun 62, scale 1:250,000 (TOP SECRET RUFF)

AMS. Series DESPA-1, Sheet NL43-5, 1st ed, May 62, scale 1:250,000 (TOP SECRET RUFF)

SAC. US Air Target Chart, Series 200, Sheet 0245-15AL, 2d ed, Mar 63, scale 1:200,000 (SECRET)

SAC. US Air Target Chart, Series 200, Sheet 0245-9-200AL, 2d ed, May 61, scale 1:200,000 (SECRET)

SAC. US Air Target Chart, Series 200, Sheet 0245-14AL, 3d ed, Mar 61, scale 1:200,000 (SECRET)

SAC. US Air Target Chart, Series 200, Sheet 0245-10AL, 2d ed, Jun 60, scale 1:200,000 (SECRET)

DOCUMENTS

1. [redacted]

RELATED DOCUMENTS

[redacted]

REQUIREMENT

[redacted]

PROJECT

[redacted]

[redacted]

~~TOP SECRET [redacted] RUFF~~

23. Special National Intelligence Estimate 13-4-64, "The Chances of an Imminent Communist Chinese Nuclear Explosion," 26 August 1964

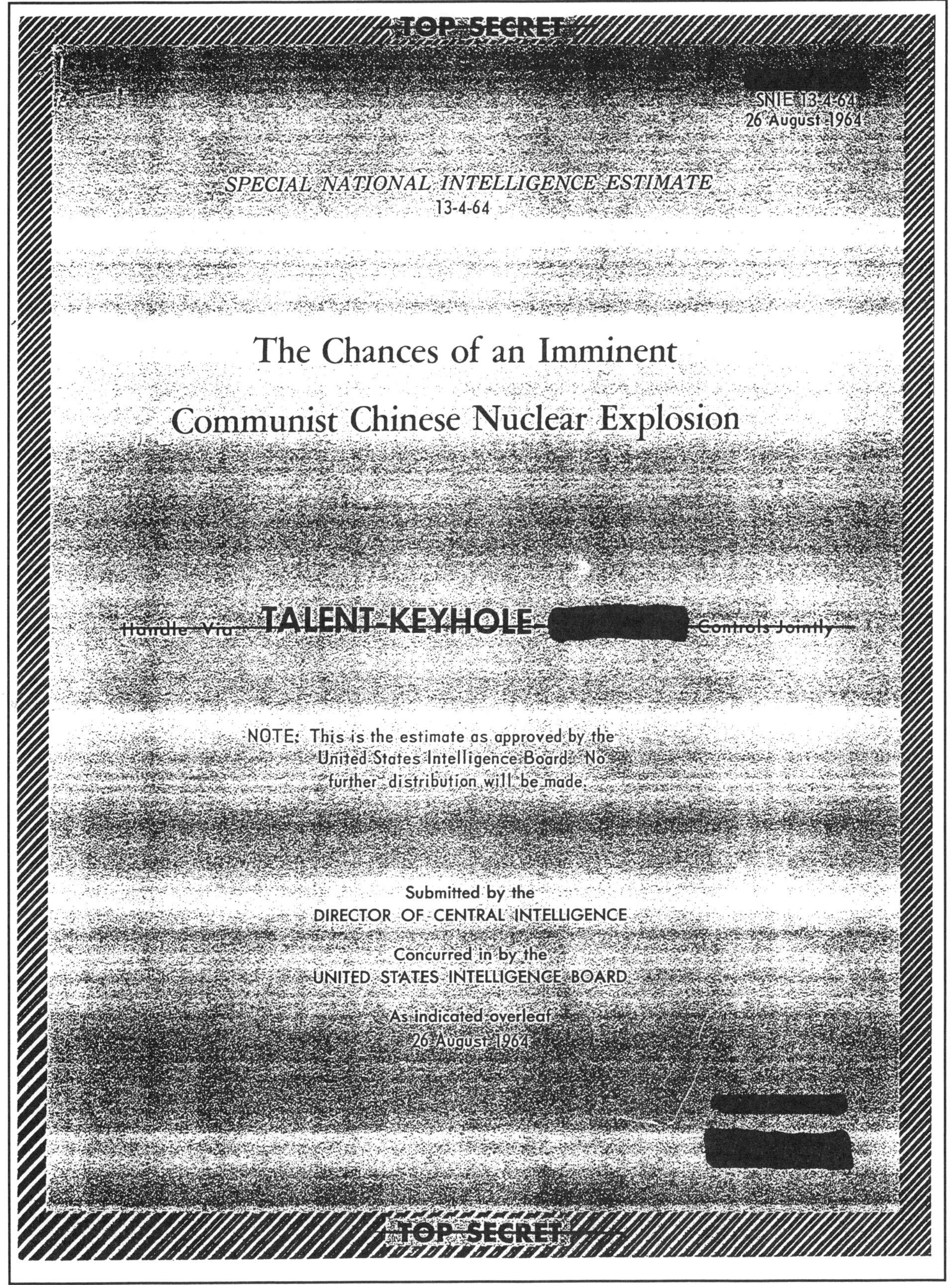

TOP SECRET

SNIE 13-4-64
26 August 1964

SPECIAL NATIONAL INTELLIGENCE ESTIMATE
13-4-64

The Chances of an Imminent Communist Chinese Nuclear Explosion

~~Handle Via~~ TALENT-KEYHOLE- [redacted] ~~Controls Jointly~~

NOTE: This is the estimate as approved by the United States Intelligence Board. No further distribution will be made.

Submitted by the
DIRECTOR OF CENTRAL INTELLIGENCE

Concurred in by the
UNITED STATES INTELLIGENCE BOARD

As indicated overleaf
26 August 1964

TOP SECRET

23. *(Continued)*

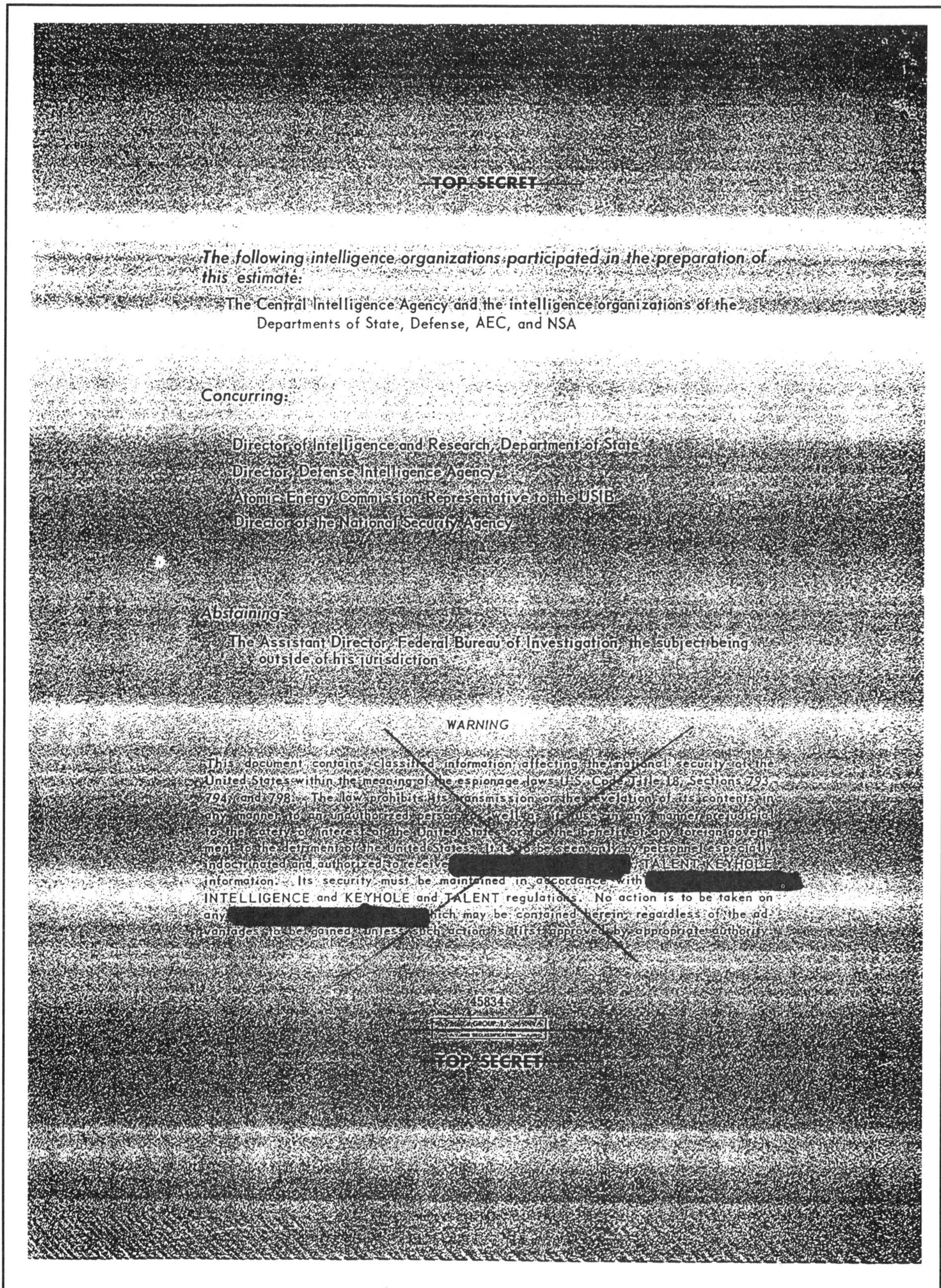

~~TOP SECRET~~

The following intelligence organizations participated in the preparation of this estimate:

The Central Intelligence Agency and the intelligence organizations of the Departments of State, Defense, AEC, and NSA

Concurring:

Director of Intelligence and Research, Department of State
Director, Defense Intelligence Agency
Atomic Energy Commission Representative to the USIB
Director of the National Security Agency

Abstaining:

The Assistant Director, Federal Bureau of Investigation, the subject being outside of his jurisdiction

WARNING

This document contains classified information affecting the national security of the United States within the meaning of the espionage laws U.S. Code Title 18, Sections 793, 794, and 798. The law prohibits its transmission or the revelation of its contents in any manner to an unauthorized person, as well as its use in any manner prejudicial to the safety or interest of the United States or for the benefit of any foreign government to the detriment of the United States. It is to be seen only by personnel especially indoctrinated and authorized to receive [redacted] TALENT-KEYHOLE information. Its security must be maintained in accordance with [redacted] INTELLIGENCE and KEYHOLE and TALENT regulations. No action is to be taken on any [redacted] which may be contained herein, regardless of the advantages to be gained, unless such action is first approved by appropriate authority.

45834

~~TOP SECRET~~

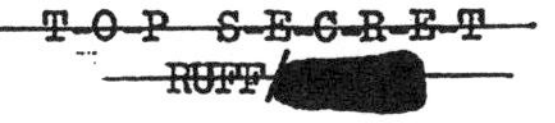

C E N T R A L   I N T E L L I G E N C E   A G E N C Y

26 August 1964

SUBJECT: SNIE 13-4-64: THE CHANCES OF AN IMMINENT COMMUNIST CHINESE NUCLEAR EXPLOSION

THE PROBLEM

To assess the likelihood that the advanced stage of construction at a probable nuclear test site in Western China indicates that the Chinese Communists will detonate their first nuclear device in the next few months.

CONCLUSION

On the basis of new overhead photography, we are now convinced that the previously suspect facility at Lop Nor in Western China is a nuclear test site which could be ready for use in about two months. On the other hand the weight of available evidence indicates that the Chinese will not have sufficient fissionable material for a test of a nuclear device in the next few months. Thus, the evidence does not permit a very confident estimate of the chances of a Chinese Communist nuclear detonation in the next few

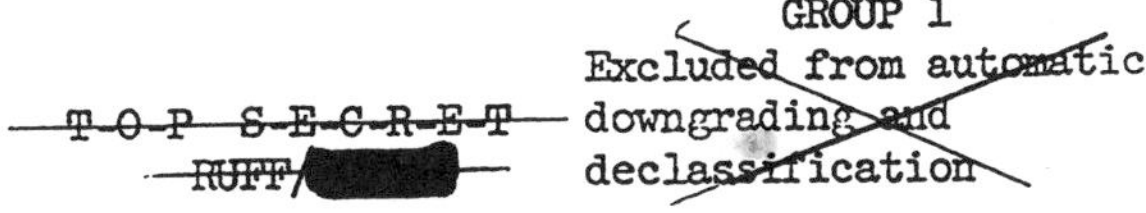

GROUP 1
Excluded from automatic downgrading and declassification

months. Clearly the possibility of such a detonation before the end of this year cannot be ruled out -- the test may occur during this period. On balance, however, we believe that it will not occur until sometime after the end of 1964.

DISCUSSION

1. Overhead photography of 6-9 August shows that the previously suspect facility near Lop Nor in Sinkiang is almost certainly a nuclear testing site. Developments at the facility include a ground scar forming about 60 percent of a circle 19,600 feet in diameter around a 325-foot tower (first seen in April 1964 photography), and work on bunkers near the tower and instrumentation sites at appropriate locations is underway. the outward appearance and apparent rate of construction indicate that the site could be ready for a test in two months or so. The characteristics of the site suggest that it is being prepared for both diagnostic and weapon effect experiments.

2. Analysis of all available evidence on fissionable material production in China indicates -- though it does not prove -- that the Chinese will not have sufficient material for a test of a nuclear device in the next few months. The only Chinese production reactor identified to date is

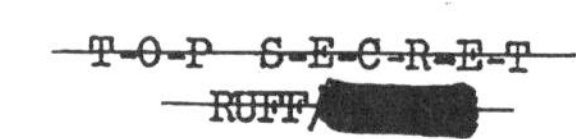

23. *(Continued)*

~~T-O-P S-E-C-R-E-T~~
~~RUFF/~~

the small, air-cooled reactor at Pao-t'ou. As of September 1963, [redacted] Construction was continuing throughout the site, including some fairly substantial work around the building which houses the reactor. Photography of March 1964 indicated that major construction at the site -- including service roads, [redacted], and additional security provisions -- had apparently been completed. Thus we believe the reactor went into operation possibly in the latter part of 1963 but more probably in early 1964. We estimate that, even if no major obstacles were encountered, it would take at least 18 months, and more likely two years, after the starting up of the Pao-t'ou reactor before a nuclear device would be ready for testing. Thus, if the Pao-t'ou reactor started operation no earlier than late 1963 and if it is China's only operating production reactor, the earliest possible date for testing is mid-1965.

3. It is, of course, possible that the Chinese have another source of fissionable material. Such a facility might have been started with Soviet aid as a result of the 1957 Soviet-Chinese aid agreement, probably about the same time as the Lanchou gaseous diffusion building. We would expect this reactor to be a fairly large water-cooled production reactor. There are areas, particularly parts of Szechwan, which are suitable for such a reactor and have not been photographed. Since it is doubtful that

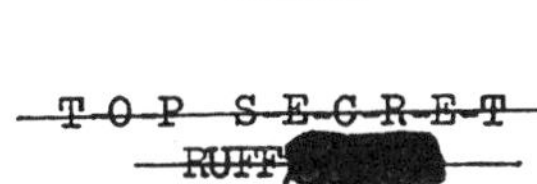

~~T-O-P S-E-C-R-E-T~~
~~RUFF/~~

~~TOP SECRET~~
~~RUFF~~/[redacted]

a reactor of this type could have been finished before the withdrawal of Soviet technicians in 1960, its completion would have depended on a native Chinese effort, a difficult but not impossible task. Such a reactor might have started operations in 1962 or 1963, thus making available sufficient plutonium for a test by the end of this year.

4. On the other hand we have photographed much of the area around virtually all locations where A-E activity is indicated [redacted] about half of all locations that might be geographically suitable for reactor sites. Apart from Pao-t'ou, no operating production reactor or isotope separation plant has been found. We believe it unlikely -- though clearly not impossible -- that such an operating facility exists.

5. It is also possible that the Chinese may have acquired fissionable material from a foreign source,

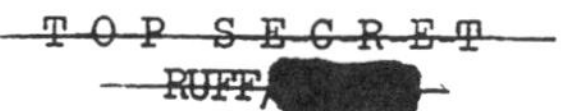

~~TOP SECRET~~
~~RUFF~~/[redacted]

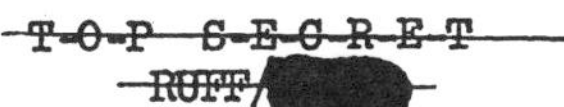

T-O-P S-E-C-R-E-T
RUFF/

As for the Soviets, we do not believe that in the past they have transferred appreciable amounts of weapon-grade material to the Chinese. In the current state of their relations with the Chinese, they would almost certainly not furnish fissionable materials to them.

6. Obviously, it is incongruous to bring a test site to a state of readiness described in paragraph 1 without having a device nearly ready for testing. It would be technically undesirable to install much of the instrumentation more than a few weeks before the actual test. We cannot tell from available photography whether the installations have yet reached this point -- it seems unlikely that they have, mainly because some heavy construction is still going on. However, it is possible that the basic work will soon be completed, and that final preparations could be made this fall.

7. On the other hand, in such a complex undertaking as advanced weapons development -- especially when it is almost certain that there is heavy political pressure for at least some results -- it would not be surprising if there were uneven progress among various phases of the program. In a number of instances in the past, Peiping has been unable to prevent -- and has seemed willing to tolerate -- uneven development in various important programs. Indeed, in other parts of their advanced weapons program we have already observed this. Some facilities seem to be behind schedule -- notably the incomplete gaseous diffusion plant at Lanchou; others are

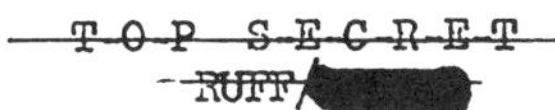

T-O-P S-E-C-R-E-T
RUFF/

~~T-O-P S-E-C-R-E-T~~
~~RUFF/~~

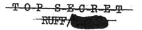

larger and more elaborate than present Chinese capabilities warrant -- for example, the possible nuclear weapons complex near Koko Nor.

8. As for the test site itself, Lop Nor is extremely remote, with poor transportation and communication facilities, and we might expect to see the Chinese taking a long leadtime in preparing this installation. They have relatively few men with the necessary scientific competence and and they cannot be fully confident that unexpected difficulties will not appear. We believe the Chinese would do everything in their power to prevent a last minute hitch on the testing facility from delaying, even briefly, China's advent as a nuclear "power."

9. The evidence and argument reviewed above do not permit a very confident estimate of the chances of a Chinese Communist nuclear detonation in the next few months. Clearly the possibility of such a detonation before the end of this year cannot be ruled out -- the test may occur during this period. On balance, however, we believe that it will not occur until sometime after the end of 1964.*

* 

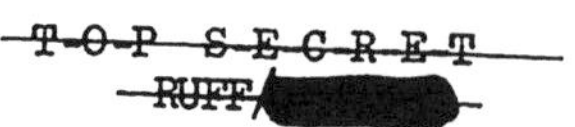

~~T-O-P S-E-C-R-E-T~~
~~RUFF/~~

CENTRAL INTELLIGENCE AGENCY

DISSEMINATION NOTICE

1. This document was disseminated by the Central Intelligence Agency. This copy is for the information and use of the recipient and of persons under his jurisdiction on a need to know basis. Additional essential dissemination may be authorized by the following officials within their respective departments:

a. Director of Intelligence and Research, for the Department of State
b. Director, Defense Intelligence Agency, for the Office of the Secretary of Defense and the organization of the Joint Chiefs of Staff
c. Assistant Chief of Staff for Intelligence, Department of the Army, for the Department of the Army
d. Assistant Chief of Naval Operations (Intelligence), for the Department of the Navy
e. Assistant Chief of Staff, Intelligence, USAF, for the Department of the Air Force
f. Director of Intelligence, AEC, for the Atomic Energy Commission
g. Assistant Director, FBI, for the Federal Bureau of Investigation
h. Director of NSA, for the National Security Agency
i. Director, National Photographic Interpretation Center, for the Central Intelligence Agency

2. This document may be retained, or destroyed by burning in accordance with applicable security regulations, or returned to the Central Intelligence Agency by arrangement with the National Photographic Interpretation Center.

3. When this document is disseminated overseas, the overseas recipients may retain it for a period not in excess of one year. At the end of this period, the document should either be destroyed, returned to the forwarding agency, or permission should be requested of the forwarding agency to retain it in accordance with IAC-D-69/2, 22 June 1953.

DISTRIBUTION:
White House
National Security Council
Department of State
Department of Defense
Atomic Energy Commission
Federal Bureau of Investigation

24. CIA/NPIC, Technical Publication, Photographic Evaluation Report, "Mission 1012-1, 17–20 October 1964/Mission 1012-2, 21–23 October 1964," April 1965 (Excerpt)

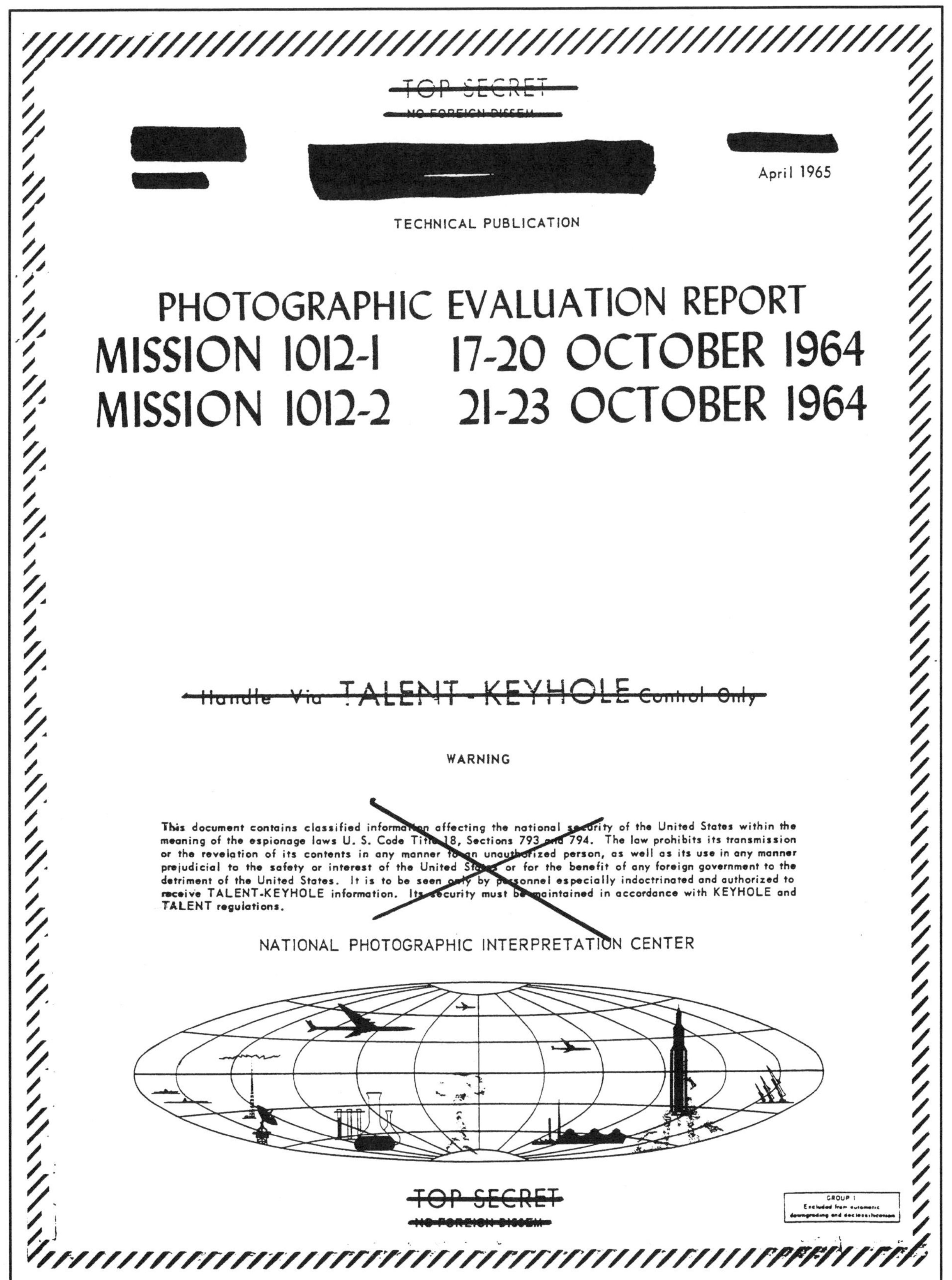

~~TOP SECRET~~

~~NO FOREIGN DISSEM~~

April 1965

TECHNICAL PUBLICATION

PHOTOGRAPHIC EVALUATION REPORT

MISSION 1012-1 17-20 OCTOBER 1964

MISSION 1012-2 21-23 OCTOBER 1964

~~Handle Via TALENT-KEYHOLE Control Only~~

WARNING

~~This document contains classified information affecting the national security of the United States within the meaning of the espionage laws U. S. Code Title 18, Sections 793 and 794. The law prohibits its transmission or the revelation of its contents in any manner to an unauthorized person, as well as its use in any manner prejudicial to the safety or interest of the United States or for the benefit of any foreign government to the detriment of the United States. It is to be seen only by personnel especially indoctrinated and authorized to receive TALENT-KEYHOLE information. Its security must be maintained in accordance with KEYHOLE and TALENT regulations.~~

NATIONAL PHOTOGRAPHIC INTERPRETATION CENTER

~~TOP SECRET~~

~~NO FOREIGN DISSEM~~

GROUP 1
Excluded from automatic downgrading and declassification

~~TOP SECRET RUFF~~
~~NO FOREIGN DISSEM~~

~~Handle Via~~
~~TALENT-KEYHOLE~~
~~Control System Only~~

SYNOPSIS

Mission 1012 (System No J-13), the twelfth of the "J" reconnaisance series, was launched 17 October 1964 and consisted of 2 operational phases, designated Missions 1012-1 and 1012-2, respectively. Mission 1012-1 accomplished 36 photographic revolutions, including 3 domestic and 3 engineering (dark side) passes. The first-phase payload was recovered by air catch on 20 October and second-phase operations were initiated on the following day. Mission 1012-2 accomplished 17 photographic revolutions, including 1 domestic and 1 engineering pass. Recovery of the second payload on 23 October terminated the mission. The capsule was retrieved from water but subsequent inspection of the contents revealed no immersion damage.

All cameras functioned satisfactorily except in Mission 1012-1, where the stellar/index unit was not operational due to a command system anomaly or program malfunction.

The quality of the panoramic photography is good and is considered comparable with the results achieved in Mission 1008. The next-to-last frames of most passes following 9AE contain light-struck areas. These traces resemble corona static discharges, but investigation has firmly established them to be light leak patterns. In any case, the resultant degradation is relatively slight. The horizon cameras associated with the panoramic instruments produced comparatively good images. Slight vignetting of the format corners does not hamper use of the horizon images for determination of vehicle attitude, which was normal until the terminal revolution, 73D, where an extreme departure from normal occurred.

The stellar/index unit operated satisfactorily in Mission 1012-2 and produced good-quality stellar and terrestrial photography. However, the vehicle attitude abnormality in the last photographic pass was responsible for gross overexposure of the last 5 stellar frames and distortion (off-axis photography) of the last 4 index frames, which contain images of the horizons.

Cloud cover obscured approximately 55 percent of the panoramic photography in Mission 1012-1 and 45 percent of Mission 1012-2. Solar elevations ranged from 3 degrees to 42 degrees.

~~Handle Via~~
~~TALENT-KEYHOLE~~
~~Control System Only~~

~~TOP SECRET RUFF~~
~~NO FOREIGN DISSEM~~

24. *(Continued)*

~~TOP SECRET RUFF~~
~~NO FOREIGN DISSEM~~

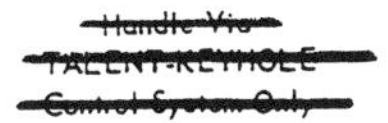

GENERAL FLIGHT DATA

Launch Date, Mission 1012-1 17 October 1964
Recovery Date, Mission 1012-1 20 October 1964

Activation Date, Mission 1012-2 21 October 1964
Recovery Date, Mission 1012-2 23 October 1964

Orbital Parameters

	Mission 1012-1		Mission 1012-2	
	Planned	Actual (Rev 10)	Planned	Actual (Rev 72)
Period	91.00 min	90.55 min	90.51 min	90.44 min
Perigee	96.28 nm	96.30 nm	98.78 nm	98.90 nm
Apogee	237.68 nm	237.60 nm	234.02 nm	235.79 nm
Eccentricity	0.0196	0.0196	0.0187	0.0188
Inclination Angle	75.05 deg	75.07 deg	74.99 deg	74.99 deg

Photographic Operations

	1012-1	1012-2
Operational Passes	30	15
Domestic Passes	3	1
Engineering Passes	3	1
Recovery Revolutions	49M	73D

~~Handle Via~~
~~TALENT-KEYHOLE~~
~~Control System Only~~

~~TOP SECRET RUFF~~
~~NO FOREIGN DISSEM~~

~~TOP SECRET RUFF~~
~~NO FOREIGN DISSEM~~

~~Handle Via~~
~~TALENT KEYHOLE~~
~~Control System Only~~

MISSION COVERAGE STATISTICS

1. Summary of Plottable Photographic Coverage

MISSION 1012-1

Country	Master (FWD) Camera		Slave (AFT) Camera		Combined Coverage	
	Linear nm	Square nm	Linear nm	Square nm	Linear nm	Square nm
USSR	13,032	1,965,626	12,525	1,916,272	25,557	3,881,898
China	1,982	280,662	2,608	348,510	4,590	629,172
Algeria	326	47,264	428	62,116	754	109,380
Cuba	335	33,596	335	33,004	670	66,600
Mongolia	218	33,900	124	18,500	342	52,400
Rumania	203	30,464	123	18,204	326	48,668
Greece	144	13,724	137	20,112	281	33,836
Poland	147	29,414	106	16,292	253	45,706
Bulgaria	106	15,688	95	13,994	201	29,682
Yugoslavia	76	11,150	113	16,724	189	27,874
Sweden	--	--	123	15,288	123	15,288
North Korea	123	1,752	--	--	123	1,752
Mexico	81	9,360	39	5,040	120	14,400
Turkey	62	7,300	49	7,154	111	14,454
Nigeria	68	9,928	27	3,942	95	13,870
South Korea	82	1,168	--	--	82	1,168
Afghanistan	57	8,240	21	3,066	78	11,306
Albania	33	3,358	37	5,476	70	8,834
Czechoslovakia	25	8,008	20	3,040	45	11,048
Jamaica	29	4,292	--	--	29	4,292
Bahama Islands	--	--	25	3,700	25	3,700
Norway	--	--	12	900	12	900
East Germany	8	8,008	4	608	12	8,616
Cayman Island	9	148	--	--	9	148
TOTAL	17,146	2,523,050	16,951	2,511,942	34,097	5,034,992
Continental United States	548	78,912	568	81,792	1,116	160,704
GRAND TOTAL	17,694	2,601,962	17,519	2,593,734	35,213	5,195,696

~~Handle Via~~
~~TALENT KEYHOLE~~
~~Control System Only~~

~~TOP SECRET RUFF~~
~~NO FOREIGN DISSEM~~

~~TOP SECRET RUFF~~
~~NO FOREIGN DISSEM~~

~~Handle Via~~
~~TALENT-KEYHOLE~~
~~Control System Only~~

MISSION 1012-2

Country	Master (FWD) Camera		Slave (AFT) Camera		Combined Coverage	
	Linear nm	Square nm	Linear nm	Square nm	Linear nm	Square nm
USSR	6,968	980,322	7,173	988,770	14,141	1,969,092
China	2,096	292,298	2,029	283,270	4,125	575,568
Mongolia	243	34,020	291	40,878	534	74,898
Congo	203	31,668	215	33,540	418	65,208
North Korea	115	6,348	148	6,160	263	12,508
Morocco	83	11,454	116	16,008	199	27,462
Rhodesia	87	13,572	54	8,424	141	21,996
Algeria	83	11,454	50	6,900	133	18,354
North Vietnam	57	7,980	33	4,620	90	12,600
Nepal	41	5,658	25	3,450	66	9,108
India	41	5,658	12	1,656	53	7,314
South Korea	49	2,760	--	--	49	2,760
Finland	--	--	41	5,550	41	5,550
Bhutan	10	1,380	4	552	14	1,932
Pakistan	10	1,380	--	--	10	1,380
TOTAL	10,086	1,405,952	10,191	1,399,778	20,277	2,805,730
Continental United States	400	44,160	410	50,922	810	95,082
GRAND TOTAL	10,486	1,450,112	10,601	1,450,700	21,087	2,900,812

~~Handle Via~~
~~TALENT-KEYHOLE~~
~~Control System Only~~

~~TOP SECRET RUFF~~
~~NO FOREIGN DISSEM~~

25. CIA/NPIC, Photographic Intelligence Report, "Severodvinsk Naval Base and Shipyard 402, Severodvinsk, USSR," November 1964

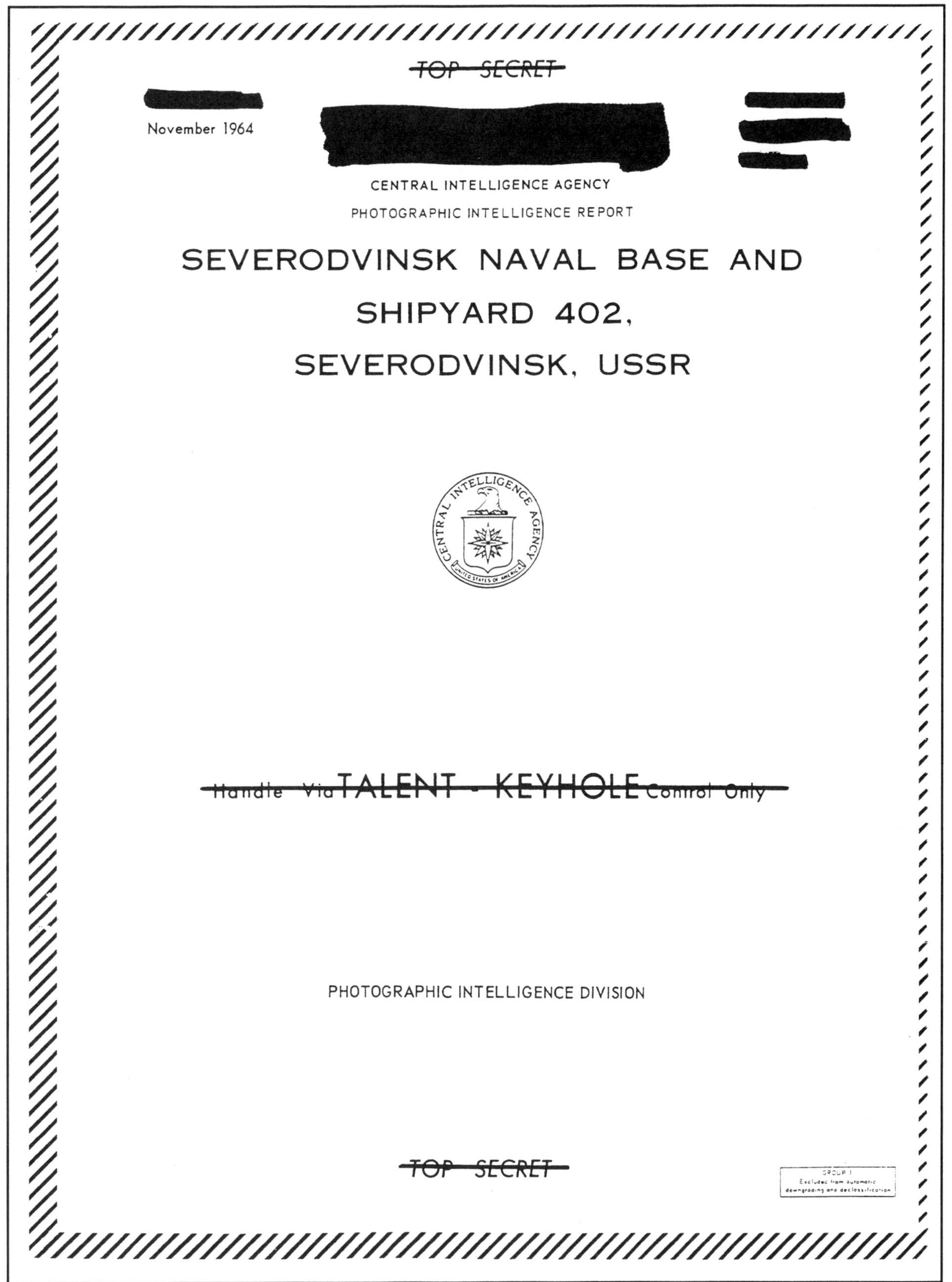

~~TOP SECRET~~

November 1964

CENTRAL INTELLIGENCE AGENCY

PHOTOGRAPHIC INTELLIGENCE REPORT

SEVERODVINSK NAVAL BASE AND SHIPYARD 402, SEVERODVINSK, USSR

~~Handle Via TALENT - KEYHOLE Control Only~~

PHOTOGRAPHIC INTELLIGENCE DIVISION

~~TOP SECRET~~

GROUP 1
Excluded from automatic downgrading and declassification

25. *(Continued)*

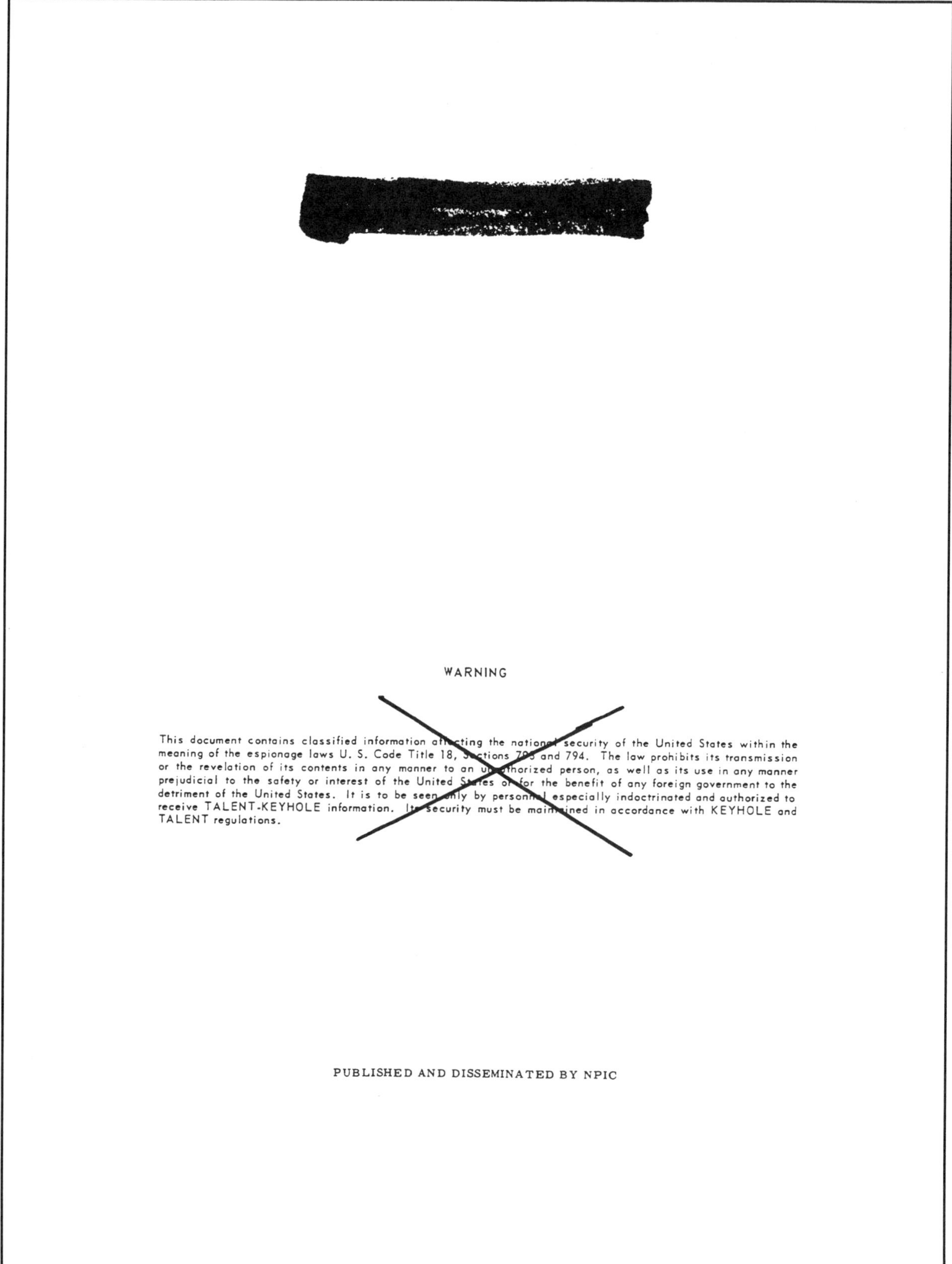

WARNING

This document contains classified information affecting the national security of the United States within the meaning of the espionage laws U. S. Code Title 18, Sections 793 and 794. The law prohibits its transmission or the revelation of its contents in any manner to an unauthorized person, as well as its use in any manner prejudicial to the safety or interest of the United States or for the benefit of any foreign government to the detriment of the United States. It is to be seen only by personnel especially indoctrinated and authorized to receive TALENT-KEYHOLE information. Its security must be maintained in accordance with KEYHOLE and TALENT regulations.

PUBLISHED AND DISSEMINATED BY NPIC

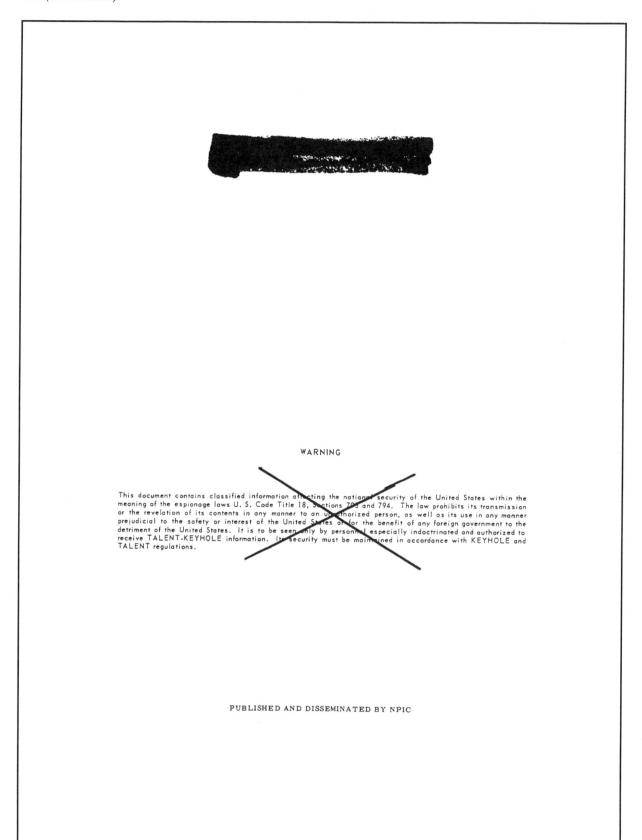

TOP SECRET [redacted] RUFF

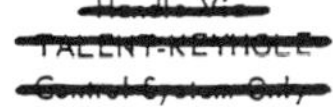

SEVERODVINSK NAVAL BASE AND SHIPYARD 402, SEVERODVINSK, USSR

Severodvinsk Naval Base and Shipyard 402 (65-35N 039-50E; [redacted]) is located in the city of Severodvinsk, USSR, on the White Sea, approximately 19 nautical miles west of Arkhangelsk (Figure 1). This installation, formerly known as the Molotovsk Shipyard, is situated around a small natural harbor formed by the mainland and Yagry Island. The harbor is dredged periodically to maintain a channel from the sea to the naval facility and city. Components of the naval base and shipyard are shown in Figure 2; item numbers are keyed to Figure 2 and Table 1. This installation is adjacent to Severodvinsk Naval Base Yagry Island [redacted]

Severodvinsk Naval Base and Shipyard 402 has been greatly expanded since World War II. Comparative photography of 1943 and 1964 is shown in Figure 3. Most of this expansion has taken place on Yagry Island. Shipyard 402 now includes a small shipyard on Yagry Island capable of handling vessels up to 500 feet long. When construction is completed it may be used for maintenance and repair, possibly including the recoring of nuclear submarines. This small shipyard is similar in size and facilities to the Petrovka Shipyard on the Pacific Ocean near Vladivostok, USSR.

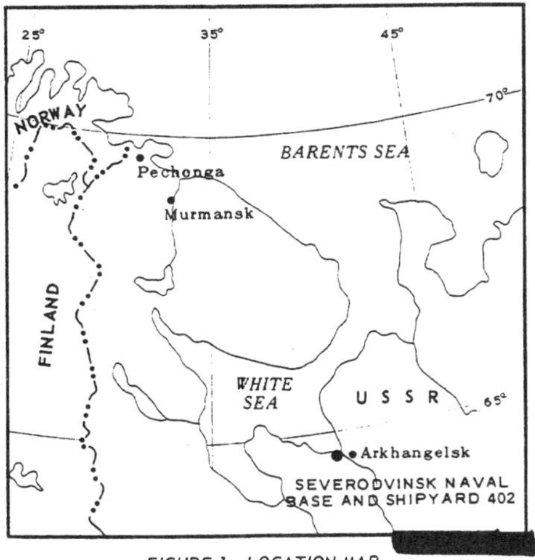

FIGURE 1. LOCATION MAP.

Expansion of Shipyard 402 on the mainland has been principally at the western end with the construction of a fabrication/construction hall (item 20) and adjacent launching way (item 18). Excavations for this building were visible on photography of 1943. The construction way of this building measures 1,040 by 105 feet and is 115 feet high. The launching way is capable of launching ships up to 500 feet long. The remainder of the shipyard on the mainland is relatively unchanged since 1943 except for the completion of the launching basin, removal of a wharf, and the addition of a few new buildings.

Severodvinsk Naval Base and Shipyard 402 is probably the largest producer of nuclear submarines in the Soviet Union. N-class SSN (nuclear-powered submarine) and E-class SSGN (nuclear-powered guided-missile submarine) have been observed here in recent months. With the exception of Komsomolsk Shipyard Amur 199, which is involved in the E-class SSGN program, 1/ Severodvinsk Shipyard 402 is the only shipyard known to be producing nuclear submarines.

Two of the fabrication buildings (items 6 and 47) are T-shaped, separately secured, and have two white objects on the roof. These objects may be ventilating, air-conditioning, or vacuum units used to create a "clean room" condition which is mandatory when working with the stainless steel piping employed in nuclear propulsion systems. A building identical to these has been identified at Komsomolsk Shipyard Amur 199.

TOP SECRET [redacted] RUFF

25. *(Continued)*

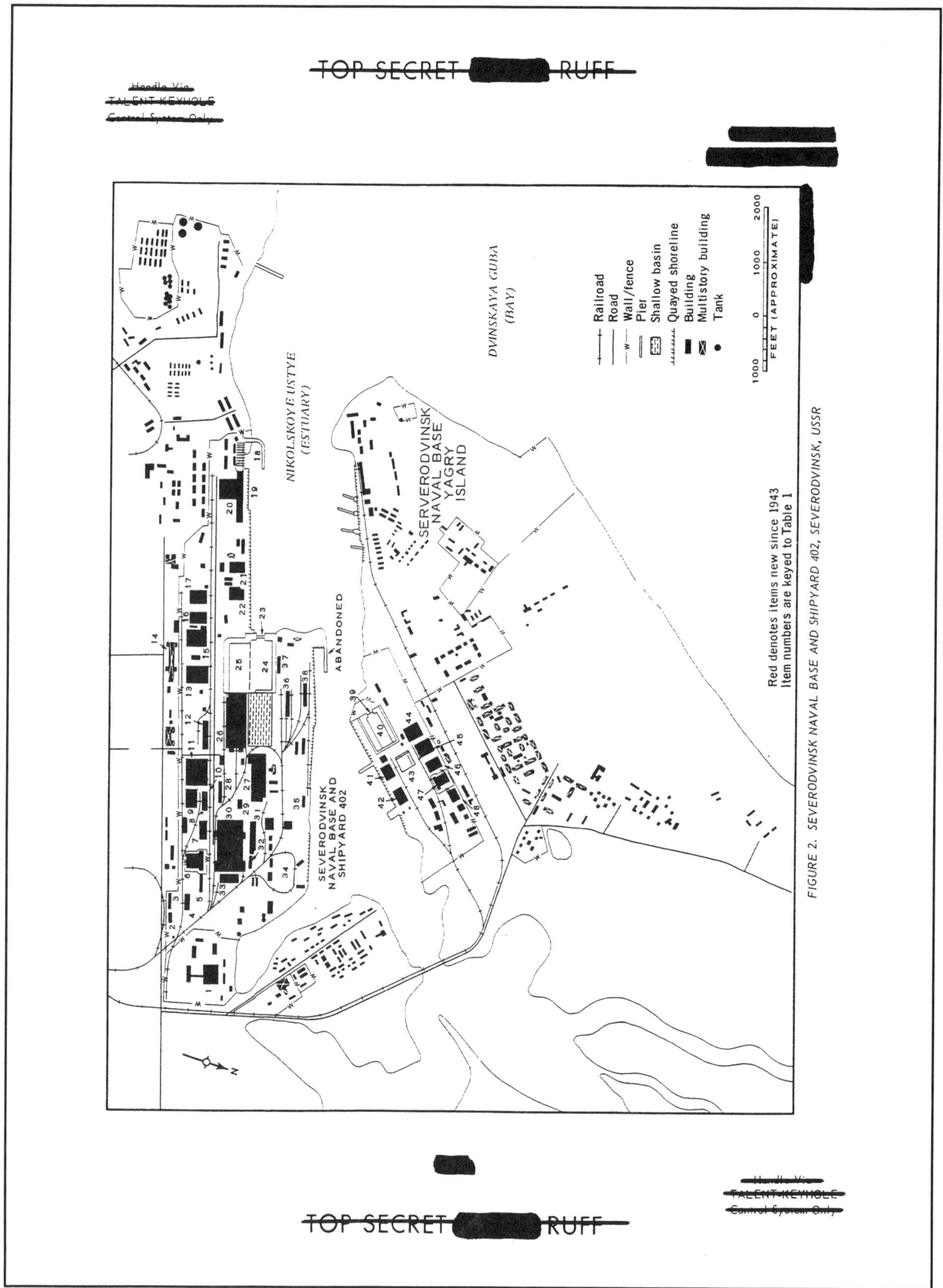

FIGURE 2. SEVERODVINSK NAVAL BASE AND SHIPYARD 402, SEVERODVINSK, USSR

25. *(Continued)*

~~TOP SECRET~~ [redacted] ~~RUFF~~

~~Handle Via~~
~~TALENT-KEYHOLE~~
~~Control System Only~~

Table 1. Components of Severodvinsk Naval Base and Shipyard 402
(Item numbers are keyed to Figure 2)

Item	Description	Dimensions (ft approx)	Item	Description	Dimensions (ft approx)
1	Powerplant	Irregular	24	Launching basin	1,075 x 325
2	Warehouse	190 x 115	25	Launching way	1,175 x 530
3	Warehouse	355 x 65	26	Construction hall	1,230 x 430 x 130h
4	Warehouse	325 x 90	27	Machine shop	775 x 210
5	Warehouse	470 x 65			(overall)
6	Fabrication building	bar: 260 x 75	28	Warehouse	500 x 60
		stem: 315 x 180	29	Machine shop	370 x 120
7	Machine shop	450 x 150	30	Fabrication building	
8	Machine shop	335 x 80		Fabrication section	845 x 565
9	Machine shop	430 x 180		Mold loft	655 x 275
10	Machine shop	330 x 100	31	Warehouse	590 x 95
11	Fabrication building	485 x 475	32	Warehouse	245 x 50
12	Machine shop	470 x 125	33	Probable pattern shop	400 x 170
13	Fabrication building	490 x 390 x 85h	34	Open storage area	Irregular
14	Administration building	Irregular	35	Warehouse	270 x 55
15	Fabrication building	base: 385 x 75	36	Warehouse	550 x 70
		leg: 350 x 320	37	Warehouse	305 x 45
16	Fabrication building	365 x 260	38	Warehouse	480 x 70
17	Fabrication building	365 x 260	39	Watertight gate	70 wide
18	Launching way	500 x 255	40	Launching basin under construction	550 x 195
19	Fitting-out quay	Irregular	41	Machine shop/fitting-out shop	355 x 130
20	Fabrication/construction hall		42	Machine shop/fitting-out shop	375 x 130
	Subassembly section	475 x 405	43	Transverser table	--
	Construction way	1,040 x 105 x 115h	44	Machine shop	375 x 260
	Machine shop	420 x 45	45	Machine shop	325 x 305
21	Fitting-out shop	340 x 195	46	Machine shop	315 x 200
22	Fitting-out shop	base: 310 x 185	47	Fabrication building	bar: 175 x 65
		leg: 230 x 60			stem: 330 x 190
23	Watertight gate	185 wide	48	Warehouse	220 x 125

~~Handle Via~~
~~TALENT-KEYHOLE~~
~~Control System Only~~

~~TOP SECRET~~ [redacted] ~~RUFF~~

25. *(Continued)*

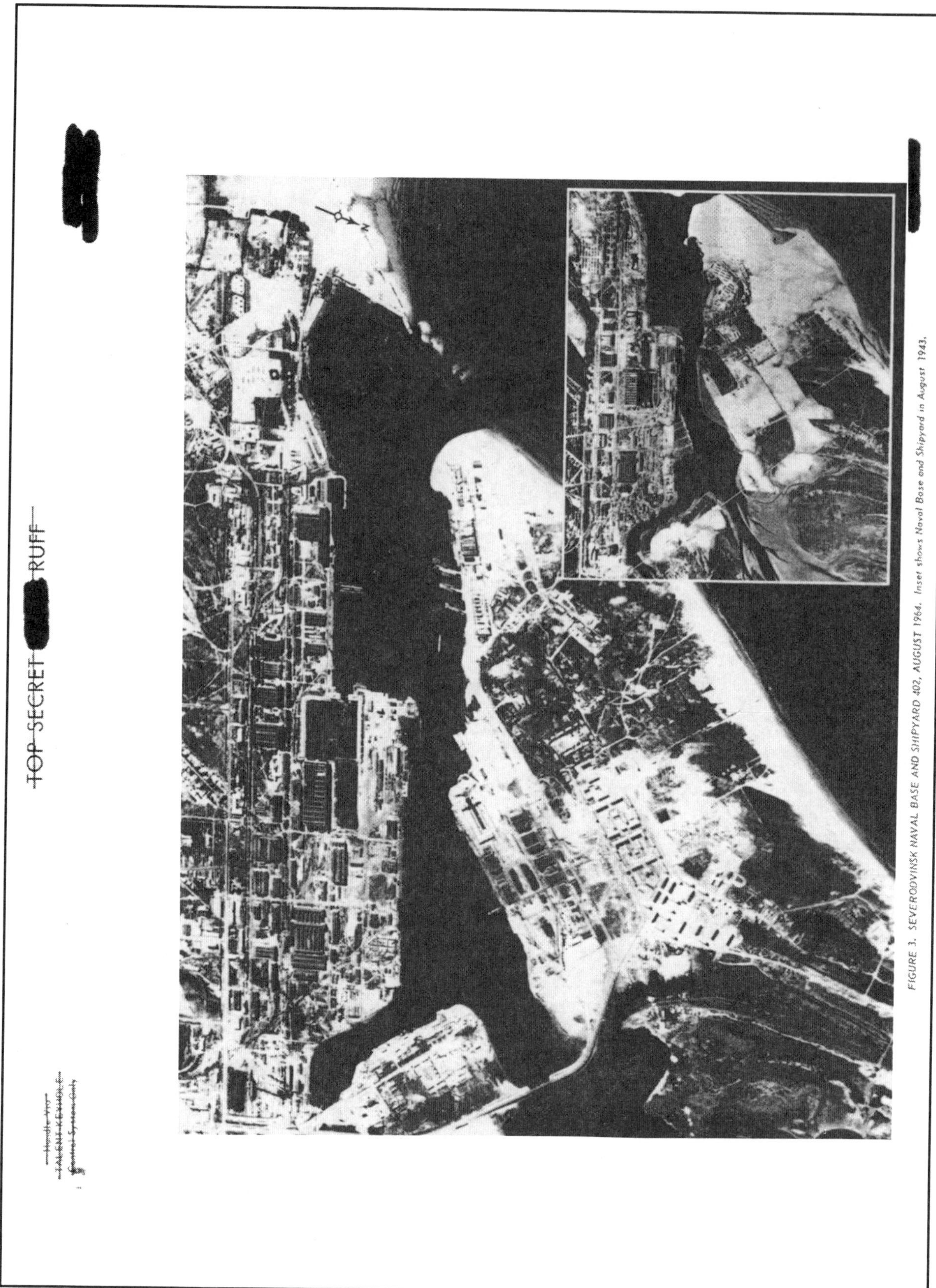

FIGURE 3. SEVERODVINSK NAVAL BASE AND SHIPYARD 402, AUGUST 1964. Inset shows Naval Base and Shipyard in August 1943.

25. *(Continued)*

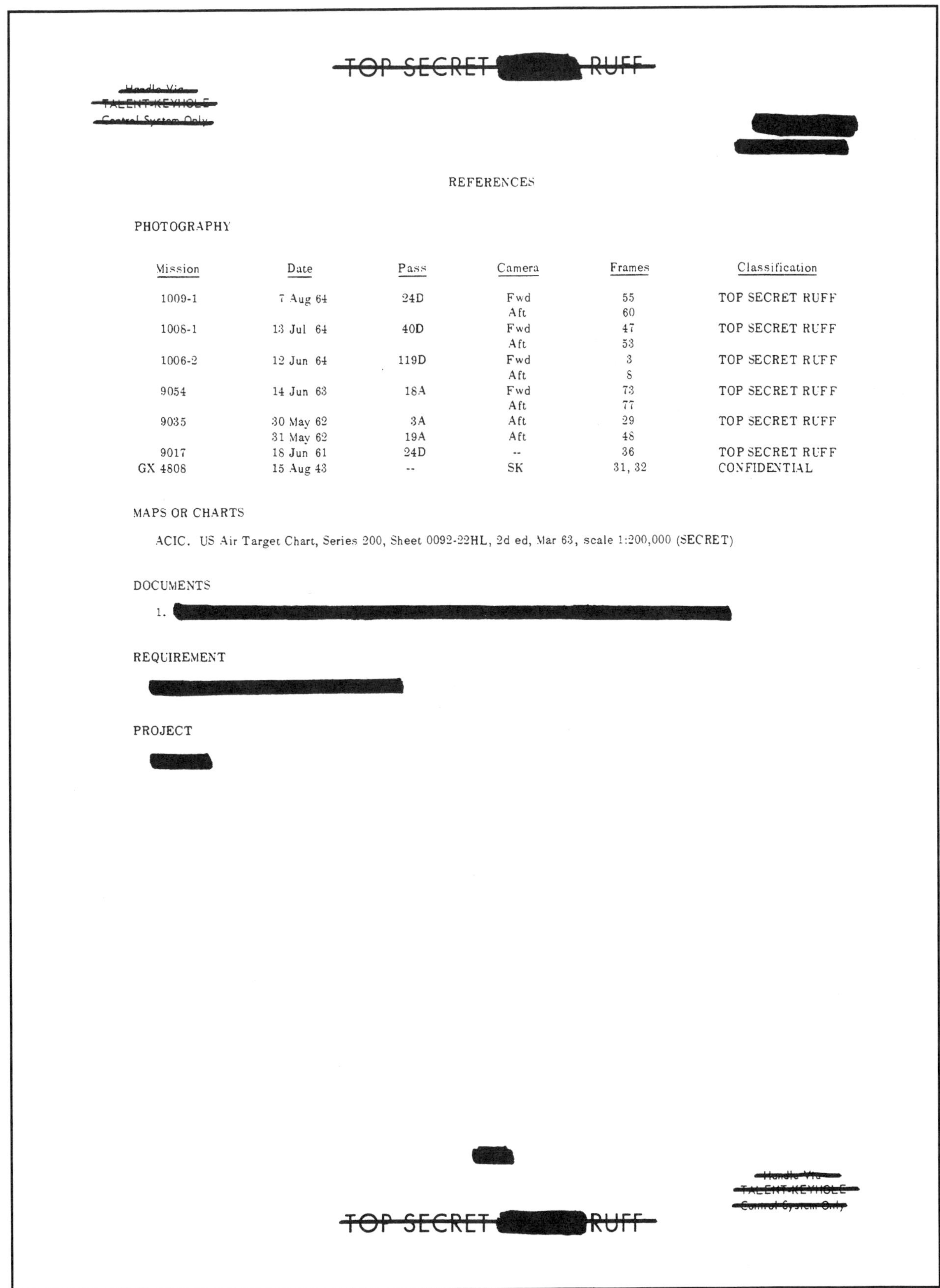

~~TOP SECRET~~ [redacted] ~~RUFF~~

~~Handle Via~~
~~TALENT-KEYHOLE~~
~~Control System Only~~

REFERENCES

PHOTOGRAPHY

Mission	Date	Pass	Camera	Frames	Classification
1009-1	7 Aug 64	24D	Fwd	55	TOP SECRET RUFF
			Aft	60	
1008-1	13 Jul 64	40D	Fwd	47	TOP SECRET RUFF
			Aft	53	
1006-2	12 Jun 64	119D	Fwd	3	TOP SECRET RUFF
			Aft	8	
9054	14 Jun 63	18A	Fwd	73	TOP SECRET RUFF
			Aft	77	
9035	30 May 62	3A	Aft	29	TOP SECRET RUFF
	31 May 62	19A	Aft	48	
9017	18 Jun 61	24D	--	36	TOP SECRET RUFF
GX 4808	15 Aug 43	--	SK	31, 32	CONFIDENTIAL

MAPS OR CHARTS

ACIC. US Air Target Chart, Series 200, Sheet 0092-22HL, 2d ed, Mar 63, scale 1:200,000 (SECRET)

DOCUMENTS

1. [redacted]

REQUIREMENT

PROJECT

~~Handle Via~~
~~TALENT-KEYHOLE~~
~~Control System Only~~

~~TOP SECRET~~ [redacted] ~~RUFF~~

26. CIA/NPIC, Photographic Interpretation Report, "KH-4 Mission 1042-1, 17–22 June 1967," June 1967 (Excerpt)

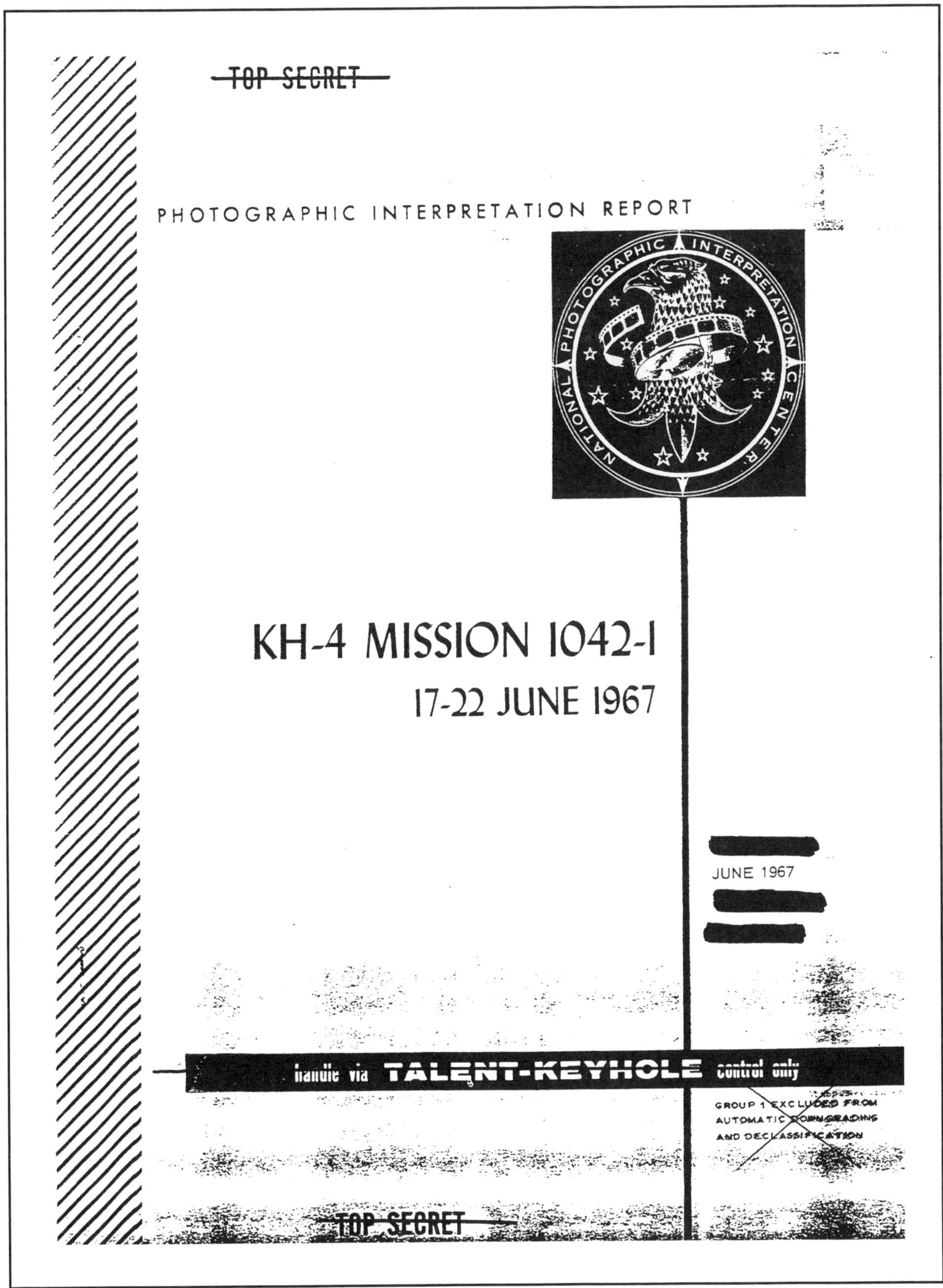

~~TOP SECRET~~

PHOTOGRAPHIC INTERPRETATION REPORT

NATIONAL PHOTOGRAPHIC INTERPRETATION CENTER

KH-4 MISSION 1042-1

17-22 JUNE 1967

[redacted]

JUNE 1967

[redacted]

handle via TALENT-KEYHOLE control only

GROUP 1 EXCLUDED FROM AUTOMATIC DOWNGRADING AND DECLASSIFICATION

~~TOP SECRET~~

26. *(Continued)*

WARNING

This document contains information affecting the national security of the United States within the meaning of the espionage laws U. S. Code Title 18, Sections 793 and 794. The law prohibits its transmission or the revelation of its contents in any manner to an unauthorized person, as well as its use in any manner prejudicial to the safety or interest of the United States or for the benefit of any foreign government to the detriment of the United States. It is to be seen only by personnel especially indoctrinated and authorized to receive information in the designated control channels. Its security must be maintained in accordance with regulations pertaining to TALENT-KEYHOLE Control System.

TARGET CATEGORIES

1. GUIDED MISSILES *
   - A. ICBM Deployment
   - B. IRBM and MRBM (including search areas)
   - C. Research and Development (including space)
   - D. Production Facilities (including test)
   - E. Suspect Missile (Search area or undetermined activity)
   - F.
   - G. Naval Launched Missiles
   - H. Anti-Missile Missile
   - I. SAM Sites
   - J. Short Range Surface-to-surface Missiles
   - K. Missile support/storage areas
   - L. SAM Training Complexes
2. AIRCRAFT
   - A. Long Range Bases
   - B. Production Facilities (including R and D)
   - C. Airfields
3. NUCLEAR ENERGY
   - A. Test Area
   - B. Production
   - C. Stockpiles
   - D. Research Institutes
   - E. Suspect Activity
4. NAVAL ACTIVITY
   - A. Operating Bases
   - B. Production Yards
   - C. Commercial Ports
   - D. Locks and Canals
5. BIOLOGICAL/CHEMICAL WARFARE
   - A. BW/CW Test Areas (including test)
   - B. Production
   - C. Storage
   - D. Research Institutes
   - E. Suspect Activity
6. ELECTRONICS
   - A. Missile Tracking Facilities
   - B. Electronics (general)
7. MILITARY
   - A. Military Installations
   - B. Special Areas
   - C.
   - D. Landing Beaches
   - E.
   - F.
   - G. Tactical SSM Support Facilities
8. URBAN/INDUSTRIAL
   - A. Complexes
   - B. Industrial Plants
   - C. Geodetic Control Points
9. OTHER
   - A. Unidentified Installations

* *Digits in these numbers indicate specific target complexes. A suffix identifies individual components within the complex.*

26. *(Continued)*

~~Handle Via~~
~~TALENT KEYHOLE~~
~~Control System Only~~

~~TOP SECRET [redacted] RUFF~~

KH-4 MISSION 1042-1, 17-22 JUN 67

TABLE OF CONTENTS

~~TOP SECRET [redacted] RUFF~~

~~Handle Via~~
~~TALENT KEYHOLE~~
~~Control System Only~~

~~Handle Via~~
~~TALENT-KEYHOLE~~
~~Control System Only~~

~~TOP SECRET [redacted] RUFF~~

KH-4 MISSION 1042-1, 17-22 JUN 67

PREFACE

THIS IS THE ONLY OAK REPORT ON MISSION 1042-1 COVERING THE HIGHEST PRIORITY TARGETS NOT REPORTED IN THE FORTHCOMING MIDDLE EAST EDITION ([redacted]) OR THE SOUTH CHINA/NORTH VIETNAM EDITION ([redacted]).

TARGETS ARE ARRANGED IN THE OAK BY SUBJECT AND WITHIN EACH SUBJECT BY COMOR TARGET NUMBER. TARGETS OF OPPORTUNITY OBSERVED DURING THE OAK REPORTING PERIOD ARE LISTED NUMERICALLY UNDER THE APPROPRIATE SUBJECT AFTER THE LISTING OF COMOR TARGETS. LOCATIONS OF TARGETS IN THIS REPORT ARE INDICATED BY THE COUNTRY CODE PRECEDING THE COORDINATES.

SELECTED PHOTOGRAPHIC REFERENCES FOLLOW THE DESCRIPTION OF THE TARGET. PASS NUMBERS ARE SUFFIXED WITH EITHER AN A FOR ASCENDING (SOUTH-TO-NORTH), D FOR DESCENDING (NORTH-TO-SOUTH) OR M FOR A MIXED ASCENDING AND DESCENDING PASS. THE LETTER F PRECEDING FRAME NUMBERS INDICATES THE FORWARD CAMERA, AND THE LETTER A INDICATES THE AFT CAMERA. UNIVERSAL REFERENCE GRID COORDINATES ARE GIVEN IN PARENTHESIS AFTER THE FRAME NUMBER TO WHICH THEY APPLY. THIS GRID IS UNIVERSAL GRID NO 1, FEB 64. SYMBOLS FOR CONDITIONS AFFECTING INTERPRETABILITY ARE -- C (CLEAR), SC (SCATTERED CLOUD COVER), HC (HEAVY CLOUD COVER), H (HAZE), CS (CLOUD SHADOW), S (SNOW), O (OBLIQUITY), SD (SEMIDARKNESS), D (DARKNESS), GC (GROUND COVER), CF (CAMOUFLAGE), GR (GROUND RESOLUTION), AND SS (SMALL SCALE).

INTERPRETABILITY OF THE PHOTOGRAPHY IS CATEGORIZED AS G (GOOD), F (FAIR), OR P (POOR). THE EXTENT OF PHOTOGRAPHIC COVERAGE IS DENOTED BY THE SYMBOLS T (TOTAL) OR PT (PARTIAL). THE MODE OF COVERAGE IS INDICATED BY ST (STEREO), NS (NON-STEREO), OR PS (PARTIAL STEREO).

RECIPIENTS ARE CAUTIONED THAT THE INITIAL SCAN OF THE PHOTOGRAPHY IS BEING ACCOMPLISHED IN A SHORT TIME AND PRIOR TO FINAL REFINEMENT OF EPHEMERIS DATA. CONSEQUENTLY, FUTURE DETAILED ANALYSIS MAY RESULT IN ADDITIONAL INFORMATION. MOREOVER, RECIPIENTS ARE CAUTIONED THAT THE IDENTIFICATION OF TARGETS IS BASED PRIMARILY ON PHOTOGRAPHY AND DOES NOT CONSTITUTE A FINISHED INTELLIGENCE JUDGMENT.

~~TOP SECRET [redacted] RUFF~~

~~Handle Via~~
~~TALENT-KEYHOLE~~
~~Control System Only~~

26. *(Continued)*

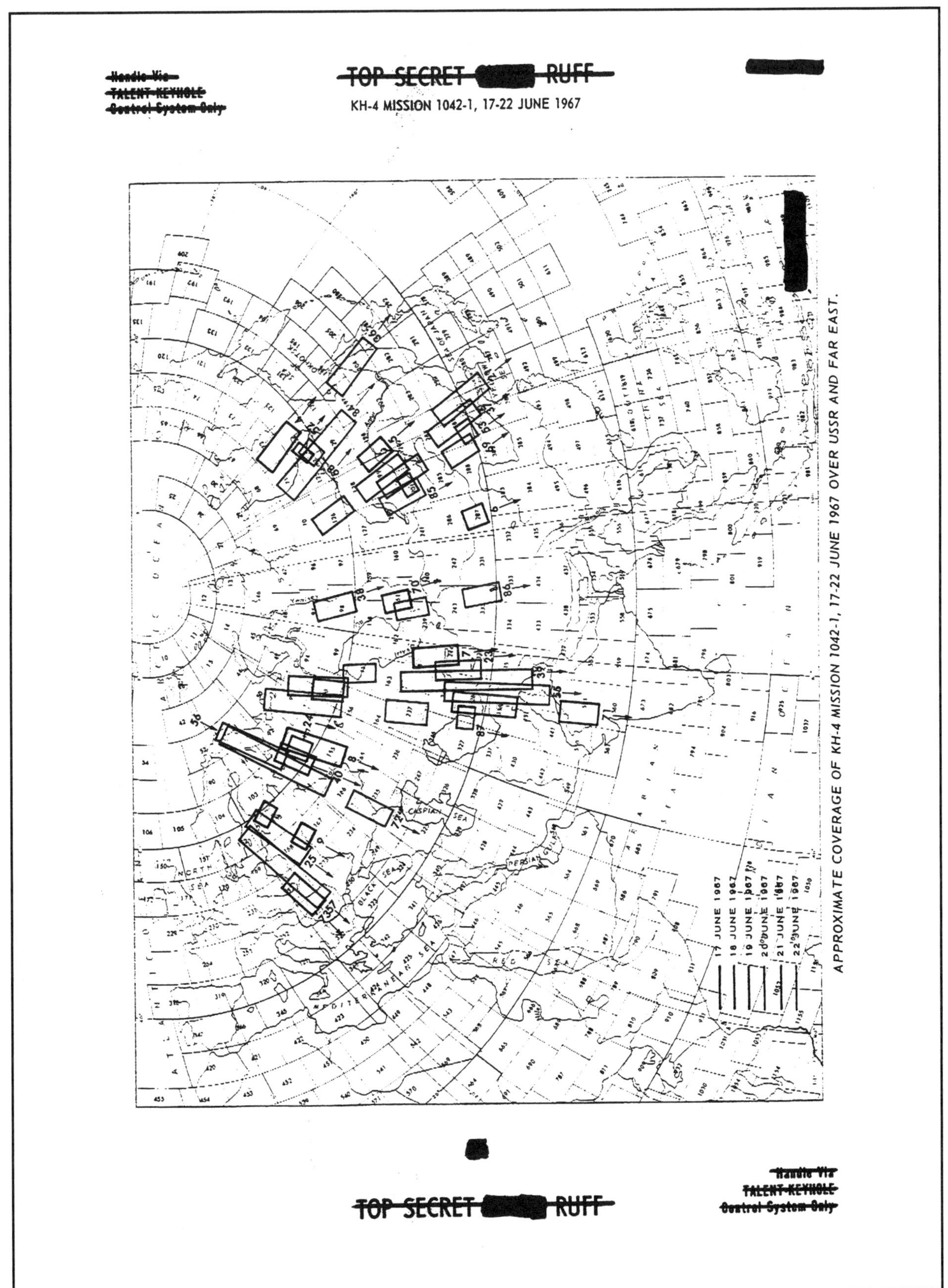

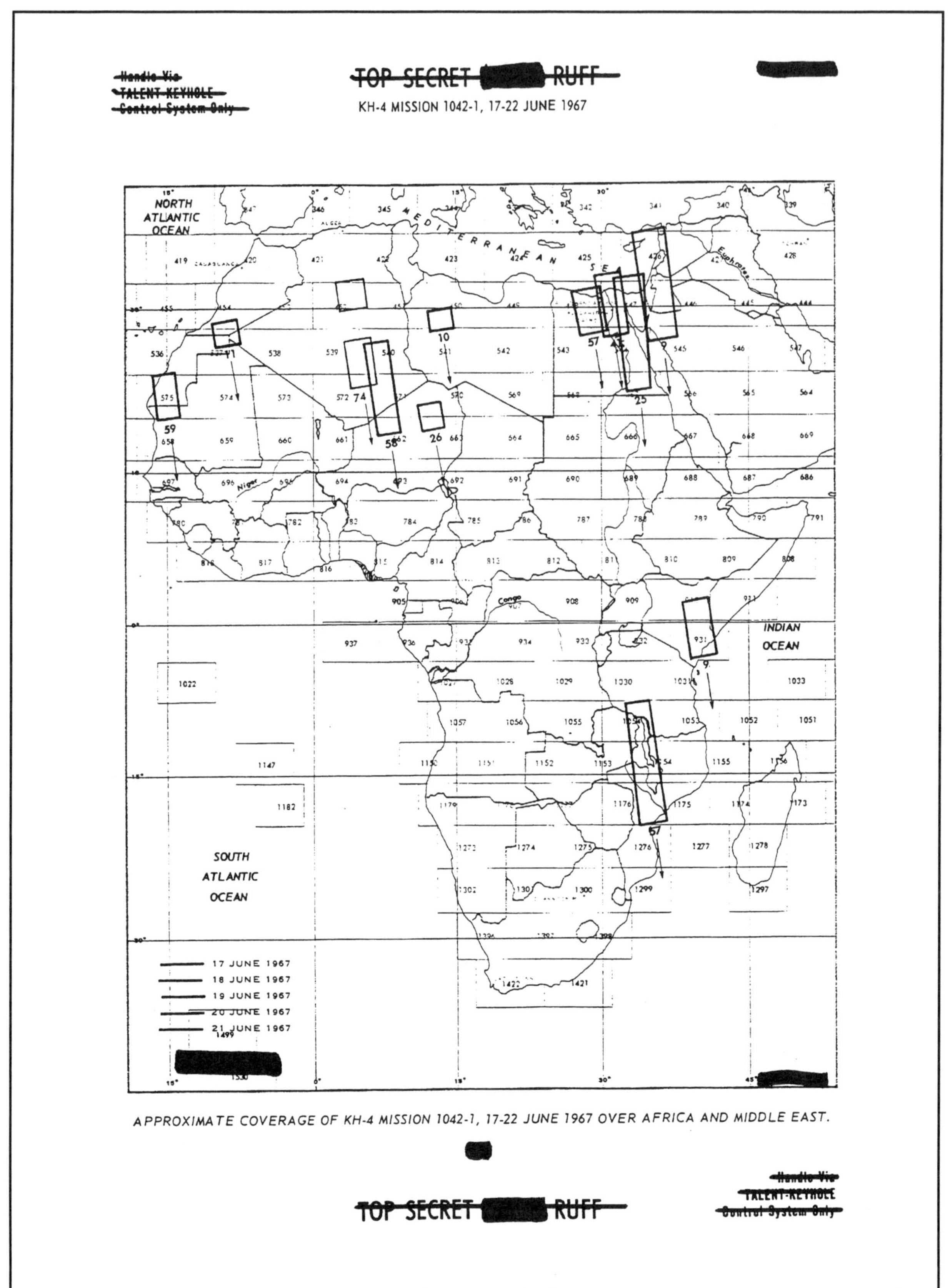

APPROXIMATE COVERAGE OF KH-4 MISSION 1042-1, 17-22 JUNE 1967 OVER AFRICA AND MIDDLE EAST.

26. *(Continued)*

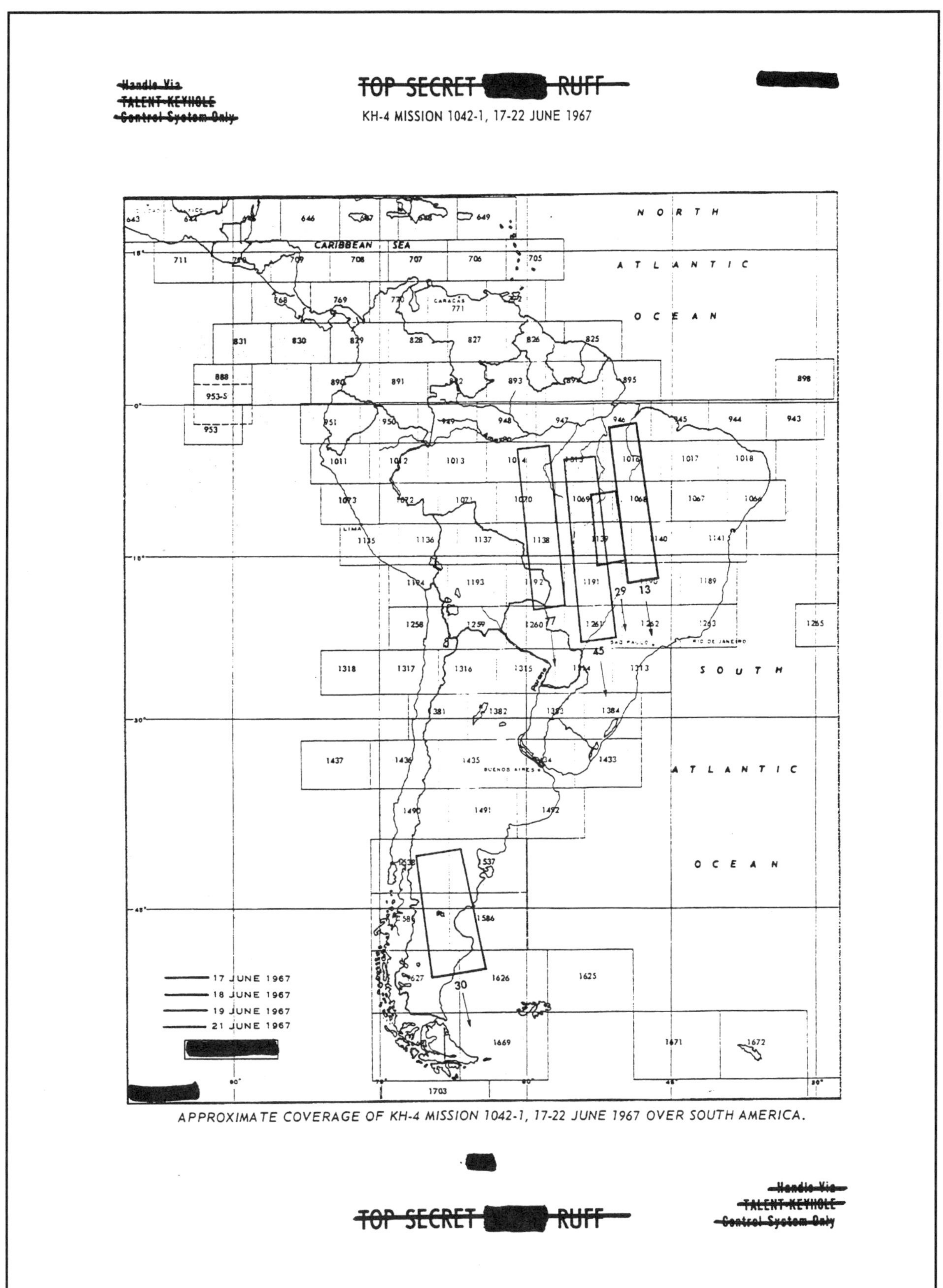

APPROXIMATE COVERAGE OF KH-4 MISSION 1042-1, 17-22 JUNE 1967 OVER SOUTH AMERICA.

26. *(Continued)*

Highlights

~~Handle Via TALENT KEYHOLE Control System Only~~

~~TOP SECRET [redacted] RUFF~~

KH-4 MISSION 1Ø42-1, 17-22 JUN 67

HIGHLIGHTS

1. NINE OF THE 24 DEPLOYED ICBM COMPLEXES ARE COVERED ON THIS MISSION. KOSTROMA AND YURYA ARE COMPLETELY COVERED ON CLEAR PHOTOGRAPHY. DROVYANAYA, IMENI GASTELLO, OLOVYANNAYA, TATISHCHEVO, TEYKOVO, UZHUR, AND YOSHKAR-OLA ARE OBSERVED THROUGH SCATTERED-TO-HEAVY CLOUDS. SIX TYPE IIID SITES ARE NEWLY IDENTIFIED. FIVE SITES, INCLUDING A CONTROL SITE, ARE IDENTIFIED AT OLOVYANNAYA. THESE SITES CONFIRM THE EXISTANCE OF THE NINTH GROUP AT THIS COMPLEX. [redacted] HOWEVER, THE FIRST EVIDENCE OF A SITE IS ON THIS MISSION. AT KOSTROMA, A POSSIBLE SITE IS IDENTIFIED AT THE TRANSFER POINT.

THE CURRENT NUMBER OF ICBM LAUNCHERS AT THE 24 KNOWN DEPLOYED COMPLEXES (THE FIGURES SHOWN IN PARENTHESIS INDICATE THE NUMBER OF POSSIBLE LAUNCHERS AND ARE NOT INCLUDED IN THE TOTALS) IS AS FOLLOWS --

TYPE IIIC	177 (1)
TYPE IIID	5Ø7 (4)
SUBTOTAL	684 (5)
OTHER LAUNCHERS	2Ø9
TOTAL	893 (5)

2. THE PLESETSK MISSILE AND SPACE CENTER IS IMAGED ON 3 PASSES OF PHOTOGRAPHY ON 19, 2Ø, AND 21 JUNE. NO NEW LAUNCH FACILITIES ARE IDENTIFIED.

3. PORTIONS OF THE KAPUSTIN YAR/VLADIMIROVKA MISSILE TEST CENTER INCLUDING THE SAM COMPLEX, LAUNCH COMPLEXES C, D, G, AND H, AND THE AHKTUBINSK/VLADIMIROVKA AIRFIELD ARE VISIBLE ON 21 JUNE THROUGH SCATTERED-TO-HEAVY CLOUDS. MISSILE EXERCISES ARE OBSERVED AT LAUNCH AREAS 1C, 2C, 3C, AND 4C AND THE 2 UNIDENTIFIED OBJECTS REMAIN ON THE PADS AT COMPLEX H. AN AERODYNAMIC VEHICLE IS OBSERVED ON A LAUNCHER AT LAUNCH AREA 3D.

4. THE SARY-SHAGAN ANTIMISSILE TEST CENTER IS COVERED ON CLEAR PHOTOGRAPHY. THE RADAR SCREEN UNDER CONSTRUCTION AT HEN ROOST ANTENNA NORTH, SARY-SHAGAN R&D RADAR FACILITY 2, HAS BEEN EXTENDED [redacted] TO 45Ø FEET AS OF 18 JUNE 67.

5. TWO PROBABLE LONG RANGE SAM LAUNCH COMPLEXES ARE NEWLY IDENTIFIED AT KALININGRAD (54-4ØN Ø21-Ø1E) AND VENTSPILS (57-16N Ø21-28E), USSR. KALININGRAD IS IN AN EARLY-TO-MID STAGE OF CONSTRUCTION WHILE VENTSPILS IS IN AN EARLY STAGE OF CONSTRUCTION.

~~Handle Via TALENT-KEYHOLE Control System Only~~

26. *(Continued)*

~~Handle Via~~
~~TALENT-KEYHOLE~~
~~Control System Only~~

~~TOP SECRET [redacted] RUFF~~

KH-4 MISSION 1042-1, 17-22 JUN 67

6. OF THE 32 PREVIOUSLY KNOWN PLRS COMPLEXES, 10 ARE OBSERVED ON THIS MISSION UNDER VARIOUS WEATHER CONDITIONS. THEY INCLUDE KURESSAARE, KAPUSTIN YAR, PLESETSK, TALLINN, CHEREPOVETS, LIEPAJA, MURASHI, KOSTROMA, AND SARY-SHAGAN 1 AND 2.

7. THE SHUANG-CHENG-TZU MISSILE TEST CENTER, CHINA, WITH THE EXCEPTION OF THE AIRFIELD IS COVERED ON 17 JUNE PHOTOGRAPHY WITH POOR-TO-FAIR INTERPRETABILITY. NO ACTIVITY IS OBSERVED AT THE SSM LAUNCH FACILITIES. UNIDENTIFIED CONSTRUCTION CONTINUES AT SUSPECT SSM TRAINING SITES 3 AND 5, LOCATED 56 NM NORTH AND 88 NM NNE, RESPECTIVELY, OF THE MAIN SUPPORT BASE.

8. AT THE SEMIPALATINSK NUCLEAR WEAPONS PROVING GROUND, USSR, THE UNDERGROUND TEST FACILITIES AND KONYSTAN TEST AREA ARE COVERED ON OBLIQUE PHOTOGRAPHY OF FAIR INTERPRETABILITY. THE POSSIBLE UNDERGROUND TEST AREA IS IDENTIFIABLE ONLY AT THE FRAME EDGE. APPROXIMATELY 95 PERCENT OF THE SHOT GROUND IS COVERED BY HEAVY CLOUDS AND HAZE. THE MAIN SUPPORT AREA AND SHAGAN RIVER TEST SITE ARE NOT COVERED. NO NEW CRATERS OR PREPARATIONS FOR TESTS NOT PREVIOUSLY REPORTED ARE VISIBLE IN THE AREAS COVERED.

9. THIS IS THE FIRST COVERAGE OF THE LOP NOR NUCLEAR TEST SITE, CHINA, 5 DAYS AFTER THE 17 JUNE 67 TEST (CHIC 6). SOME EFFECTS OF THE TEST ARE VISIBLE ON THE GROUND. ALL PARTS OF THE AIRDROP MARKER ARE LESS DISTINCT AND THE T-SIGNS NORTH AND SOUTH OF GZ-2 HAVE BEEN OBLITERATED. A PRONOUNCED LIGHT-TONED AREA CENTERED APPROXIMATELY 1,900 FEET SOUTH OF GZ-2 IS VISIBLE.

10. TWO TROPOSCATTER FACILITIES AND A TALL KING-AW RADAR FACILITY ARE NEWLY IDENTIFIED. THE FIRST TROPOSCATTER FACILITY IS LOCATED 21 NM NE OF ARANTUR AT 61-10N 064-00E AND THE OTHER IS LOCATED 9 NM SOUTH OF TATARSKOYE AT 64-36N 087-40E. THE TALL KING FACILITY IS LOCATED 19 NM SE OF KOTLAS AT 60-59N 046-57E.

11. THE SOVIET BUILDUP AT CHOYBALSAN, MONGOLIA, [redacted] CONTINUES TO SHOW A HIGH LEVEL OF ACTIVITY. A PROBABLE AIRFIELD IS UNDER CONSTRUCTION EAST OF THE COMPLEX AND A NEW TENT CAMP AND MOTOR PARK ARE OBSERVED JUST WEST OF CHOYBALSAN.

<u>STATUS OF TARGET READOUT</u>

THIS OAK REPORTS ON 53 COMOR TARGETS AND 6 BONUS TARGETS.

TABULAR SUMMARY

26. *(Continued)*

~~Handle Via~~
~~TALENT KEYHOLE~~
~~Control System Only~~

~~TOP SECRET~~ ~~RUFF~~

KH-4 MISSION 1Ø42-1, 17-22 JUN 67

A RECAPITULATION OF SIGNIFICANT ACTIVITY OBSERVED ON THIS MISSION AND NPIC CURRENT MISSILE LISTINGS IS PRESENTED IN THE FOLLOWING TABLES ON SUCCEEDING PAGES --

TABLE 1 -- SOVIET ICBM DEPLOYMENT
TABLE 2 -- SOVIET MRBM AND IRBM DEPLOYMENT
TABLE 3 -- SAM AND ABM DEPLOYMENT
TABLE 4 -- DEPLOYED PROBABLE LONG RANGE SAM COMPLEXES
TABLE 5 -- DATE/TIME OF PHOTOGRAPHIC PASSES

~~TOP SECRET~~ ~~RUFF~~

~~Handle Via TALENT-KEYHOLE Control System Only~~

~~TOP SECRET [redacted] RUFF~~

KH-4 MISSION 1042-1, 17-22 JUNE 1967

TABLE 1. DEPLOYED ICBM COMPLEXES, TYURATAM MTC AND PLESETSK MSC

COMPLEX	TYPE II A-D Sites	TYPE III A&B Sites	TYPE IIIC Sites (Launchers)	TYPE IIIC Complete	TYPE IIID Sites (Launchers)	TYPE IIID Complete	New Positions This Mission IIIC	New Positions This Mission IIID	Total Launch Positions	Launch Groups Identified IIIC	L-Electronic Facilities IIID	Estimated Start of Latest Group
ALEYSK			19	6					19	3		Jul 66
DOMBAROVSKIY			37	12					37	6		Dec 66
☆ DROVYANAYA	3	3			51	34			66		5	Jun 66
GLADKAYA	2	1			61	50			68		6	Mar 66
☆ IMENI GASTELLO			25	17					25	4		Jun 66
ITATKA	3								6			
KARTALY			29	12					29	5		Dec 66
☆ KOSTROMA	6	1			30(1#)			(1#)	45(1#)		3	Feb 67
KOZELSK	3	2			40	3			52		4	Feb 67
NOVOSIBIRSK	3	2							12			
☆ OLOVYANNAYA		3			86	38		5	95		9	Dec 66
OMSK		1							3			
PERM	5	1			41	40			54		4	Apr 66
SHADRINSK		3							9			
SVOBODNYY	7	1			49(1#)				66(1#)		5	Jan 67
☆ TATISHCHEVO					111(1#)	49			111(1#)		11	Jul 66
☆ TEYKOVO	6								12			
TYUMEN	2								4			
☆ UZHUR			42(1#)	18					42(1#)	7		Jan 67
VERKHNYAYA SALDA	7	2							20			
YEDROVO	6	2			38 (1#)	1			56(1#)		4	Feb 67
☆ YOSHKAR-OLA	6								12			
☆ YURYA	8	3							25			
ZHANGIZ-TOBE			25	12					25	4		Aug 66
TOTALS	67	25	177(1#)	77	507(4#)	215		6	893(5#)	29	51	

COMPLEX	TYPE II A-D Sites	TYPE III A&B Sites	TYPE IIIC Sites (Launchers)	TYPE IIIC Complete	TYPE IIID Sites (Launchers)	TYPE IIID Complete	Notes
TYURATAM MTC	2	3	9	6	11	11	60 total launchers includes 8 other single silos and 17 SSM R&D/Space Launch Positions.
☆ PLESETSK MSC	4	1					39 total SSM Launch Positions -- includes 6 Type I, 9 other Type II, 10 other Type III, and 1 U/I (site 12).

☆ Covered This Mission (#) Possible The Possible Sites Are Not Included In The Total Count

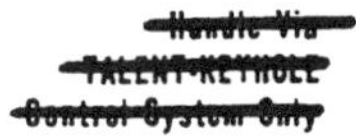

26. *(Continued)*

~~Handle Via TALENT-KEYHOLE Control System Only~~

~~TOP SECRET~~ ~~RUFF~~

KH-4 MISSION 1042-1, 17-22 JUNE 1967

TABLE 2. SOVIET MRBM AND IRBM DEPLOYMENT

Geographic Area	Soft MRBM Launch Areas	Soft IRBM Launch Areas	Hard MRBM Launch Areas	Hard IRBM Launch Areas	Total Launch Areas	Total Launch Positions
Western	118	9	18	10	155	608
Caucasus	3		2	3	8	29
Central Asia	2	2	1	3	8	29
Far Eastern	9	1		1	11	43
Totals	132	12	21	17	182	709

Totals do not include 100 fixed field-type sites with total of 374 launch positions and one launch facility at Sovetskaya Gavan

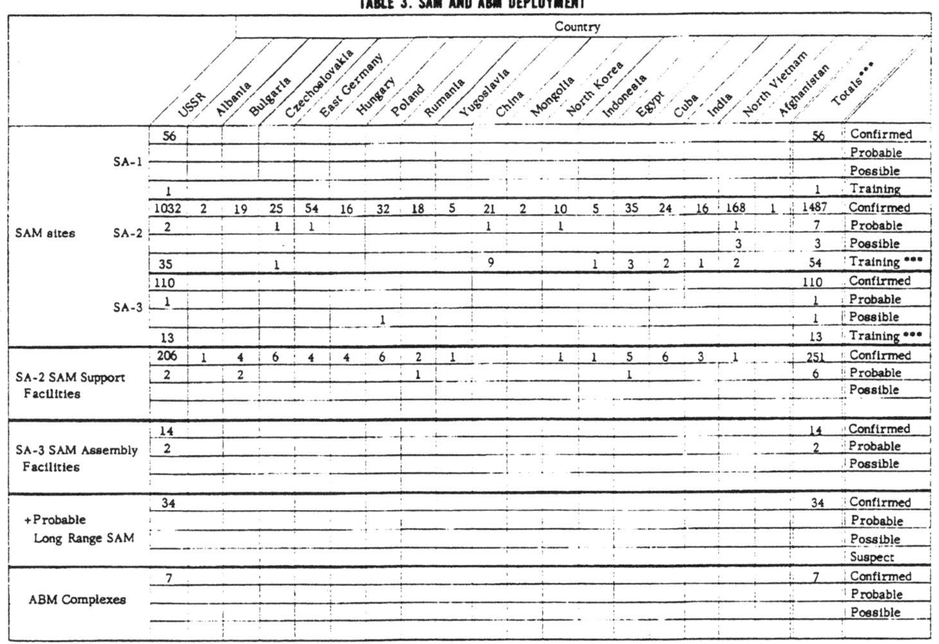

TABLE 3. SAM AND ABM DEPLOYMENT

		USSR	Albania	Bulgaria	Czechoslovakia	East Germany	Hungary	Poland	Rumania	Yugoslavia	China	Mongolia	North Korea	Indonesia	Egypt	Cuba	India	North Vietnam	Afghanistan	Totals***	
SAM sites	SA-1	56																		56	Confirmed
																					Probable
																					Possible
		1																		1	Training
	SA-2	1032	2	19	25	54	16	32	18	5	21	2	10	5	35	24	16	168	1	1487	Confirmed
		2			1	1					1		1					1		7	Probable
																		3		3	Possible
		35			1						9			1	3	2	1	2		54	Training ***
	SA-3	110																		110	Confirmed
		1																		1	Probable
								1												1	Possible
		13																		13	Training ***
SA-2 SAM Support Facilities		206	1	4	6	4	4	6	2	1			1	1	5	6	3	1		251	Confirmed
		2		2					1						1					6	Probable
																					Possible
SA-3 SAM Assembly Facilities		14																		14	Confirmed
		2																		2	Probable
																					Possible
+Probable Long Range SAM		34																		34	Confirmed
																					Probable
																					Possible
																					Suspect
ABM Complexes		7																		7	Confirmed
																					Probable
																					Possible

*Does not include alternate sites, USSR

**Does not include 31 abandoned sites

***Total reflects individual sites

+Includes 2 Probable R & D or Training sites at SSATC

~~TOP SECRET~~ ~~RUFF~~

~~Handle Via TALENT-KEYHOLE Control System Only~~

26. *(Continued)*

~~Handle Via TALENT-KEYHOLE Control System Only~~

~~TOP SECRET~~ ~~RUFF~~

KH-4 MISSION 1042-1, 17-22 JUNE 1967

TABLE 4. DEPLOYED PROBABLE LONG RANGE SAM COMPLEXES

Complex	Sites / Positions	Tracking/Guidance Facility No of Positions	Associated Air Warning Facility	Orientation	Complex	Sites / Positions	Tracking/Guidance Facility No of Positions	Associated Air Warning Facility	Orientation
ANGARSK	3 / 18	3	✓	355°	☆LIEPAJA	5 / 30	5	✓	244°
BABAYEVO	3 / 18	3		314°	MOZHAYSK	3 / 18	3		285°
BORSHCHEV	1 / 6				☆MURASHI	5 / 30	5	✓ Prob	341°
CHELYABINSK	3 / 18	3		321°	NEYA	3 / 18	3		298°
☆CHEREPOVETS	5 / 30	5	✓	360°	NIZHNIY TAGIL	3 / 18	3		360°
FEODOSIYA	3 / 18	3	✓	133°	NIZHNYAYA TURA	3 / 18	3		208°
KALININ	3 / 18	3		308°	PERESLAVL-ZALESSKIY	3 / 18	3	✓	323°
☆KALININGRAD	3 / 18	3		284°	☆PLESETSK	3 / 18	3	✓	356°
☆KAPUSTIN YAR	2 / 12	2		201°	☆SARY-SHAGAN 1*	3 / 18	3	✓	113°
KHABAROVSK	3 / 18	3		85°	☆SARY-SHAGAN 2*	2 / 12	2	✓	237°
KIMRY	3 / 18	3	✓	311°	SHARYA	2 / 12	2		295°
KIYEV	3 / 18	3	✓	230°	SVERDLOVSK	3 / 18	3	✓	325°
☆KOSTROMA	3 / 18	3		337°	☆TALLINN	5 / 30	5	✓	283°
KRASNOYARSK	3 / 18	3		36°	TOMSK	3 / 18	3		342°
☆KURESSAARE	3 / 18	3		280°	☆VENTSPILS	3 / 18	3		208°
LENINGRAD NE	5 / 30	5		18°	VOLGOGRAD	3 / 18	3	✓	150°
LENINGRAD NW	5 / 30	5		315°					
LENINGRAD SW	5 / 30	5		205°					
					TOTALS 34	111 / 666	110	13+1 Prob	

☆ Covered this mission * Probable R&D or training associated ✓ Identified at this complex

~~TOP SECRET~~ ~~RUFF~~

~~Handle Via TALENT-KEYHOLE Control System Only~~

26. *(Continued)*

~~Handle Via TALENT-KEYHOLE Control System Only~~

~~TOP SECRET~~ ~~RUFF~~

KH-4 MISSION 1042-1, 17-22 JUNE 1967

TABLE 5. DATE/TIME OF PHOTOGRAPHIC PASSES

Pass	Date	Camera ON		Camera OFF	
		Hr(Z)	Min(Z)	Hr(Z)	Min(Z)
4D	17 Jun 67	03	32	03	33
5D	17 Jun 67	05	05	05	06
6D (Part 1)	17 Jun 67	06	38	06	39
6D (Part 2)	17 Jun 67	06	43	06	46
7D	17 Jun 67	08	06	08	08
8D	17 Jun 67	09	33	09	35
9D (Part 1)	17 Jun 67	11	04	11	04
9D (Part 2)	17 Jun 67	11	05	11	06
9D (Part 3)	17 Jun 67	11	10	11	12
9D (Part 4)	17 Jun 67	11	18	11	19
10D	17 Jun 67	12	41	12	42
11D	17 Jun 67	14	12	14	12
13D	17 Jun 67	17	19	17	23
21D	18 Jun 67	05	08	05	10
22D	18 Jun 67	06	43	06	44
23D (Part 1)	18 Jun 67	08	03	08	05
23D (Part 2)	18 Jun 67	08	07	08	09
24D	18 Jun 67	09	33	09	34
25D (Part 1)	18 Jun 67	11	04	11	06
25D (Part 2)	18 Jun 67	11	11	11	13
26D	18 Jun 67	12	43	12	44
29D	18 Jun 67	17	21	17	23
30D	18 Jun 67	18	59	19	01
36D	19 Jun 67	03	35	03	37
37D (Part 1)	19 Jun 67	05	02	05	04
37D (Part 2)	19 Jun 67	05	04	05	06
37D (Part 3)	19 Jun 67	05	07	05	09
38D	19 Jun 67	06	32	06	33
39D	19 Jun 67	08	05	08	10
40D	19 Jun 67	09	32	09	33
41D	19 Jun 67	11	10	11	11
45D	19 Jun 67	17	19	17	24

Pass	Date	Camera ON		Camera OFF	
		Hr(Z)	Min(Z)	Hr(Z)	Min(Z)
52D	20 Jun 67	03	32	03	32
53D (Part 1)	20 Jun 67	05	04	05	06
53D (Part 2)	20 Jun 67	05	07	05	08
53D (Part 3)	20 Jun 67	05	23	05	25
55D (Part 1)	20 Jun 67	08	01	08	02
55D (Part 2)	20 Jun 67	08	06	08	10
56D	20 Jun 67	09	29	09	32
57D (Part 1)	20 Jun 67	11	02	11	05
57D (Part 2)	20 Jun 67	11	09	11	10
57D (Part 3)	20 Jun 67	11	19	11	22
58D	20 Jun 67	12	40	12	42
59D	20 Jun 67	14	11	14	12
68D	21 Jun 67	03	30	03	31
69D (Part 1)	21 Jun 67	05	03	05	04
69D (Part 2)	21 Jun 67	05	05	05	06
69D (Part 3)	21 Jun 67	05	22	05	24
70D	21 Jun 67	06	32	06	33
71D (Part 1)	21 Jun 67	08	00	08	01
71D (Part 2)	21 Jun 67	08	05	08	07
71D (Part 3)	21 Jun 67	08	08	08	10
72D (Part 1)	21 Jun 67	09	28	09	32
72D (Part 2)	21 Jun 67	09	33	09	34
73D	21 Jun 67	11	03	11	04
74D (Part 1)	21 Jun 67	12	37	12	38
74D (Part 2)	21 Jun 67	12	39	12	40
77D	21 Jun 67	17	16	17	20
84D	22 Jun 67	03	27	03	31
85D	22 Jun 67	05	01	05	02
86D (Part 1)	22 Jun 67	06	30	06	31
86D (Part 2)	22 Jun 67	06	33	06	34
87D (Part 1)	22 Jun 67	07	56	07	59
87D (Part 2)	22 Jun 67	08	01	08	02
87D (Part 3)	22 Jun 67	08	03	08	03

~~TOP SECRET~~ ~~RUFF~~

~~Handle Via TALENT-KEYHOLE Control System Only~~

26. *(Continued)*

Missiles

26. *(Continued)*

~~Handle Via TALENT KEYHOLE Control System Only~~

~~TOP SECRET [redacted] RUFF~~

KH-4 MISSION 1042-1, 17-22 JUN 67

87D F/72-89 & A/72-89 SC,H F PT PS

TATISHCHEVO ICBM COMPLEX UR 5140N 04534E

APPROXIMATELY 95 PERCENT OF THE SEARCH AREA IS CLOUD COVERED. NINE TYPE IIID LAUNCH SITES ARE INTERPRETABLE AND 8 ADDITIONAL SITES CAN BE IDENTIFIED ONLY. NO NEW MISSILE ACTIVITY OBSERVED.

72D F/100-109 & A/98-108 HC,H,O P PT PS

KAPUSTIN YAR/VLADIMIROVKA MSL TEST CTR UR 4835N 04545E

THE ENTIRE CENTER IS COVERED ON PHOTOGRAPHY BUT SCATTERED-TO-HEAVY CLOUD COVER AND CLOUD SHADOW SEVERELY LIMIT INTERPRETABILITY. NO NEW MISSILE FACILITIES OBSERVED IN THE VISIBLE PORTIONS OF THE 50-NM SEARCH AREA.

72D F/122-129 & A/121-128 SC,CS,H P T PS

KAPUSTIN YAR LAUNCH COMPLEX B UR 4841N 04616E

ONLY A PORTION OF THE LAUNCH AREA IS VISIBLE.

NO MISSILES, MISSILE-RELATED EQUIPMENT, NEW CONSTRUCTION, CHANGES IN FACILITIES, OR PREVIOUSLY UNREPORTED FEATURES OBSERVED.

72D F/125 (71.6-10.9) HC,CS,H P PT NS

KAPUSTIN YAR LAUNCH COMPLEX C UR 4836N 04616E

SCATTERED CLOUDS, CLOUD SHADOW, AND HAZE SEVERELY LIMIT INTERPRETATION OF LAUNCH COMPLEX C.

LAUNCH AREA 1C -- PAD 1C-1 IS CLOUD COVERED. THE APPROXIMATELY 80-FT-LONG RAIL-SERVED CYLINDER-ERECTOR REMAINS ON PAD 1C-2. ON PAD 1C-3, A PROBABLE MISSILE/ERECTOR IS ERECTED AT THE SAME POSITION. THE APPROXIMATELY 44-FT-LONG CYLINDRICAL OBJECT [redacted]

~~TOP SECRET [redacted] RUFF~~

~~Handle Via TALENT KEYHOLE Control System Only~~

~~Handle Via TALENT-KEYHOLE Control System Only~~

~~TOP SECRET [redacted] RUFF~~

KH-4 MISSION 1Ø42-1, 17-22 JUN 67

LAUNCH AREA 2C -- PAD 2C-1 IS CLOUD COVERED. PAD 2C-2 IS ACTIVE WITH A PROBABLE TRANSPORTER AND ERECTOR ON THE PAD.

LAUNCH AREA 3C -- A PROBABLE ERECTOR IS ON THE LAUNCH PAD AND A CHECKOUT TENT IS OBSERVED ON THE SW DUMBBELL PAD.

LAUNCH AREA 4C -- AT LAUNCH SITE 4C-1, THE SE SILO IS OPEN, ALL OTHER SILOS APPEAR CLOSED. AT LAUNCH SITE 4C-2, THE SW SILO IS OPEN, ALL OTHER SILOS APPEAR CLOSED.

LAUNCH AREA 5C -- THE AREA IS CLOUD COVERED.

LAUNCH AREA 6C -- CONSTRUCTION CONTINUES. DETAILS ARE NOT DISCERNIBLE.

72D F/126 (71.8-12.8) & A/124,125 SC,H,HC
P PT PS

KAPUSTIN YAR LAUNCH COMPLEX D UR 4828N Ø4618E

AN AERODYNAMIC VEHICLE, SIMILAR TO OTHERS OBSERVED PREVIOUSLY, IS ON A LAUNCHER AT LAUNCH AREA 3D. VEHICLES/PIECES OF EQUIPMENT ARE ON THE ROADS BEHIND THE CHECKOUT BUILDING.

72D F/126,127 (71.7-14.3) & A/125 SC,H,CS
P PT PS

KAPUSTIN YAR LAUNCH COMPLEX G UR 4822N Ø4617E

A ROW OF VEHICLES/PIECES OF EQUIPMENT IS JUST NORTH OF THE BARRACKS AREA.

NO NEW CONSTRUCTION, CHANGES IN FACILITIES, OR PREVIOUSLY UNREPORTED FEATURES OBSERVED.

72D F/127 (69.9-12.1) & A/125 SC,H,CS P
PT PS

KAPUSTIN YAR LAUNCH COMPLEX H UR 4848N Ø462ØE

VEHICLES/PIECES OF EQUIPMENT ARE ON BOTH LAUNCH PADS.

~~TOP SECRET [redacted] RUFF~~

~~Handle Via TALENT-KEYHOLE Control System Only~~

~~Handle Via TALENT KEYHOLE Control System Only~~

~~TOP SECRET [redacted] RUFF~~

KH-4 MISSION 1Ø42-1, 17-22 JUN 67

NO NEW CONSTRUCTION, CHANGES IN FACILITIES, OR PREVIOUSLY UNREPORTED FEATURES OBSERVED.

72D F/124,125 (73.8-14.6) & A/123 SC,H,CS
P T PS

KAPUSTIN YAR SAM LAUNCH COMPLEX UR 4848N Ø4545E

THE SAM LAUNCH COMPLEX IS COVERED ON PHOTOGRAPHY WITH FAIR-TO-POOR INTERPRETABILITY.

PRECISION TRACKING RADAR FACILITY -- UNIDENTIFIED GROUND SCARRING IS OBSERVED WITHIN THE SECURED FACILITY ALONG THE SE FENCE LINE. [redacted]

NO APPARENT CHANGES IN FACILITIES ARE OBSERVED AT THE PROBABLE LONG RANGE SAM LAUNCH FACILITY, THE R & D LAUNCH AREA, THE SA-1 LAUNCH SITE, OR THE SA-2 AND SA-3 LAUNCH AREAS. ALTHOUGH LIMITING CONDITIONS PRECLUDE AN ACCURATE COUNT AND IDENTIFICATION OF VEHICLES/PIECES OF EQUIPMENT, THERE DOES NOT APPEAR TO BE ANY INCREASE IN ACTIVITY.

OPERATIONAL SAM SITE B3Ø-2 -- OCCUPANCY IS UNDETERMINED.

NO APPARENT CHANGE IS OBSERVED AT THE R & D AND SA-1 YOYO GUIDANCE SITES, THE SAM HOUSING AND SUPPORT AREA, THE SAM WARHEAD AREA (5Ø PERCENT CLOUD COVERED), THE TROPOSCATTER AND MICROWAVE COMMUNICATIONS FACILITY (5Ø PERCENT CLOUD SHADOW), THE SAM CHECKOUT AND STORAGE AREA (95 PERCENT CLOUD COVERED), THE ELECTRONIC TEST FACILITY, KY AIRFIELD (5Ø PERCENT SCATTERED CLOUD COVER), AND THE BASE SUPPORT COMPLEX.

SAM MARSHALLING AREA -- MARSHALLING APRONS 1 THROUGH 6 ARE OCCUPIED BY VEHICLES/PIECES OF EQUIPMENT. A REDUCTION IN THE NUMBER OF VEHICLES/PIECES OF EQUIPMENT ON APRON 6 IS APPARENT [redacted]. APRONS 7, 8, AND 9 APPEAR UNOCCUPIED.

SAM TRAINING SITES A AND B -- NO ACTIVITY AND NO APPARENT CHANGE IN FACILITIES OBSERVED. SAM TRAINING SITE C IS CLOUD COVERED.

INSTRUMENTATION SITES 1 AND 4-9 CAN BE IDENTIFIED ONLY. SITE 2 IS CLOUD COVERED AND SITE 3 IS NEARLY OBSCURED

~~Handle Via TALENT KEYHOLE Control System Only~~

~~Handle Via TALENT KEYHOLE Control System Only~~

~~TOP SECRET [redacted] RUFF~~

KH-4 MISSION 1042-1, 17-22 JUN 67

BY CLOUD SHADOW.

72D F/123-125 & A/122,123 (30.3-10.3)
H,SC,CS P PT PS

KAPUSTIN YAR MISSILE STOR & HANDLING AREA UR 4822N 04612E

PROBABLE ASM CRATES REMAIN IN THE SAME POSITIONS AS ON PREVIOUS COVERAGE.

72D F/127 (69.9-12.1) & A/126 SC F T
PS

KAPUSTIN YAR MSL TEST & SUPPORT COMPLEX UR 4834N 04553E

NO MISSILES, MISSILE-RELATED EQUIPMENT, NEW CONSTRUCTION, CHANGES IN FACILITIES, OR PREVIOUSLY UNREPORTED FEATURES OBSERVED.

72D F/126 (63.9-13.9) & A/124 SC,H,CS F
T PS

SARY-SHAGAN ANTIMISSILE TEST CENTER UR 4602N 07334E

ALL FACILITIES AT THE TEST CENTER EAST OF AND INCLUDING TRACKING FACILITIES 6 AND 8 ARE COVERED ON CLEAR, STEREO PHOTOGRAPHY OF GENERALLY FAIR INTERPRETABILITY. NO NEW MISSILE FACILITIES OBSERVED.

23D F/36-45 & A/36-45 H,SS F PT PS
39D F/34-45 & A/33-41 C,SS F PT PS
55D F/44-46 & A/43-45 C,SS F PT PS

SARY-SHAGAN LAUNCH COMPLEX A (ZONE A) UR 4625N 07252E

LAUNCH COMPLEX A AND PROBABLE LONG RANGE SAM LAUNCH COMPLEX 2 ARE COVERED ON 19 JUN 67 ON ONE PASS OF CLEAR, STEREO PHOTOGRAPHY OF FAIR-TO-POOR INTERPRETABILITY. NO SIGNIFICANT CHANGES ARE OBSERVED [redacted] IN LAUNCH SITES 5 AND 6, ELECTRONIC SITES D AND G, THE SITE SUPPORT FACILITY, THE HEADQUARTERS AND ADMINISTRATION AREA, THE ON-SITE SAM SUPPORT FACILITY, AND THE AIRSTRIP.

AT LAUNCH SITE 1, LAUNCH POSITIONS 1/1 THROUGH 1/5 AND

~~TOP SECRET [redacted] RUFF~~

~~Handle Via TALENT KEYHOLE Control System Only~~

~~Handle Via TALENT-KEYHOLE Control System Only~~

~~TOP SECRET [redacted] RUFF~~

KH-4 MISSION 1Ø42-1, 17-22 JUN 67

POSITION 1/7 APPEAR OCCUPIED. POSITION 1/6 APPEARS UNOCCUPIED. IDENTIFICATION OF EQUIPMENT PARKED OFF THE LAUNCH POSITIONS AT LAUNCH SITES 1 AND 2 IS NOT POSSIBLE.

AT LAUNCH SITE 2, LAUNCH POSITIONS 2/2, 2/5, AND 2/6 APPEAR OCCUPIED. POSITIONS 2/1, 2/3, AND 2/4 APPEAR UNOCCUPIED.

AT LAUNCH SITE 3, ALL LAUNCH POSITIONS ARE OCCUPIED, PRESUMABLY BY LAUNCHERS, EXCEPT POSITION 3/3. MISSILE DOLLIES APPEAR IN THE SAME LOCATIONS AS ON 12 JUNE -- TWO AT POSITIONS 3/4, 3/5, AND 3/6, ONE AT POSITION 3/2, AND NONE AT POSITIONS 3/1 AND 3/3. ALTHOUGH THE RIGHT DOLLY AT POSITION 3/5 AND THE LAUNCHER AT POSITION 3/6 APPEAR BULKIER THAN A LAUNCHER OR DOLLY ALONE, POOR INTERPRETABILITY OF THE COVERAGE OVER THIS PARTICULAR AREA DOES NOT PERMIT DETERMINATION OF THE PRESENCE OF MISSILES. THE LAUNCHER OBSERVED ON THE TRANSPORTER ON THE ROAD CLOSE TO POSITION 3/3 ON 12 JUNE IS NO LONGER PRESENT. EQUIPMENT REMAINS NORTH OF POSITION 3/3, BUT CANNOT BE IDENTIFIED.

AT LAUNCH SITE 4, OCCUPANCY OF THE 6 LAUNCH POSITIONS APPEARS TO REMAIN AS SEEN ON 12 JUNE, WITH LAUNCHERS ON ALL POSITIONS EXCEPT POSITION 4/5. TWO DOLLIES ARE OBSERVED AT POSITIONS 4/1 AND 4/3, ONE AT POSITION 4/2 AND 4/4, AND NONE AT POSITIONS 4/5 AND 4/6. THE PROBABLE TRANSPORTER ON THE ROAD AT THE SITE CENTER ON 12 JUNE IS NO LONGER PRESENT. THE CONTINUED PRESENCE OF THE 3 MISSILES SEEN FROM 6 TO 12 JUNE ON THE 215-FT-LONG RAIL-EQUIPPED STRUCTURES IN THE MISSILE HANDLING AREA CANNOT BE DETERMINED.

AT ELECTRONIC SITE B, RADAR POSITIONS B-1 AND B-2 REMAIN OCCUPIED, AND POSITION B-3 IS UNOCCUPIED.

AT ELECTRONIC SITE C, RADAR POSITIONS C-1 AND C-2 ARE UNCHANGED. THE POSSIBLE ENGAGEMENT RADAR [redacted] UNDER ASSEMBLY ON THE NEW PAD AT RADAR POSITION C-3 WITH ASSOCIATED EQUIPMENT AND THE ENGAGEMENT RADAR OBSERVED WITH ASSOCIATED EQUIPMENT ON THE APRON AT THE NW CORNER OF THE CENTRAL CONTROL BUILDING APPEAR ESSENTIALLY AS [redacted] AT ELECTRONIC SITE E, THE 2 NORTHERN TOWERS ARE OCCUPIED BY RADAR. THE SOUTHERN TOWER APPEARS UNOCCUPIED.

AT ELECTRONIC SITE F, FACILITIES REMAIN UNCHANGED SINCE 12 JUNE. OF INTEREST IS THE FACT THAT FROM 27 MAY TO 8

~~TOP SECRET [redacted] RUFF~~

~~Handle Via TALENT-KEYHOLE Control System Only~~

~~Handle Via TALENT-KEYHOLE Control System Only~~

~~TOP SECRET [redacted] RUFF~~

KH-4 MISSION 1042-1, 17-22 JUN 67

JUNE, ALMOST ALL VANS/PIECES OF EQUIPMENT HAD LEFT THE SOUTHERN PORTION OF THE SITE AND RADARS AND VANS ARE NOW CONCENTRATED LARGELY IN THE CENTER OF THE SITE. COVERAGE OF THE MISSILE STORAGE AND CHECKOUT AREA IS NOT SUFFICIENTLY DISTINCT FOR DETERMINATION OF THE STATUS OF RECENT BUILDING EXTENSION AND TRENCHING.

AT PROBABLE LONG RANGE SAM LAUNCH COMPLEX 2, LAUNCHERS ARE PRESENT AT ALL LAUNCH POSITIONS EXCEPT POSITION B/2, WHICH REMAINS UNOCCUPIED. TWO DOLLIES ARE OBSERVED AT THE 6 POSITIONS AT LAUNCH SITE A AND APPEAR PRESENT AT THE 6 POSITIONS OF LAUNCH SITE B, ALTHOUGH AT TIMES THEY ARE BARELY DISCERNIBLE. THE DEGREE OF MISSILE ACTIVITY CANNOT BE ASCERTAINED BECAUSE OF POOR INTERPRETABILITY.

NO SIGNIFICANT CHANGES ARE OBSERVED IN THE ASSOCIATED TRACKING/GUIDANCE RADAR SITE, IN THE BACK NET/SIDE NET SITE TO THE NE, OR IN THE 2 NEARBY UNIDENTIFIED SITES.

39D F/39,40 & A/39 (43.3-12.0) C,SS P T ST

SARY-SHAGAN LAUNCH COMPLEX B (ZONE B) UR 4559N 07231E

TWO UNIDENTIFIED PIECES OF EQUIPMENT [redacted] REMAIN PARKED ON THE EAST EDGE OF LAUNCH PAD B-2. TWO PROBABLE ELECTRONIC VANS ARE PARKED AT INSTRUMENTATION SITE B-1. ALL 6 POSITIONS AT SAM SITE B21-2 ARE OCCUPIED.

NO NEW CONSTRUCTION, CHANGES IN FACILITIES, OR PREVIOUSLY UNREPORTED FEATURES OBSERVED.

39D F/42 (36.8-14.2) & A/41 C F T ST

SARY-SHAGAN LAUNCH COMPLEX C UR 4549N 07325E

NO SIGNIFICANT CHANGE OBSERVED IN THE LAUNCH AREA OR SUPPORT FACILITIES.

23D F/41 & A/41 C,SS P T ST
39D F/44 & A/43 (34.2-13.4) C F T ST

~~TOP SECRET [redacted] RUFF~~

~~Handle Via TALENT-KEYHOLE Control System Only~~

~~Handle Via~~
~~TALENT-KEYHOLE~~
~~Control System Only~~

~~TOP SECRET [redacted] RUFF~~

KH-4 MISSION 1Ø42-1, 17-22 JUN 67

SARY-SHAGAN PROB LR SAM LAUNCH CPLX 1 UR 46Ø5N Ø7327E

ALL LAUNCH POSITIONS AT THE 3 LAUNCH SITES APPEAR TO BE OCCUPIED BY LAUNCHERS. TWO DOLLIES ARE AT EACH OF THE 12 LAUNCH POSITIONS OF LAUNCH SITES B AND C. THE 3 RADAR POSITIONS IN THE TRACKING/GUIDANCE FACILITY ARE OCCUPIED. CONSTRUCTION HAS CONTINUED SINCE MISSION 1Ø4Ø, MAR-APR 67, ON THE BUILDINGS UNDER CONSTRUCTION CLOSE TO THE ENTRANCE TO THE LAUNCH COMPLEX. SINCE MISSION 1Ø4Ø, TRENCHING HAS BEGUN FROM THIS BUILDING SITE IN A GENERALLY SW DIRECTION.

IN THE ELECTRONIC AREA 2.5 NM NW OF THE LAUNCH COMPLEX, THE 2 BACK NET AND 2 SIDE NET RADARS REMAIN. NO SIGNIFICANT CHANGE IS APPARENT SINCE MISSION 1Ø4Ø IN THE 2 UNIDENTIFIED SITES JUST EAST OF THE BACK NET/SIDE NET AREA OR IN THE PROBABLE MISSILE-HANDLING FACILITY APPROXIMATELY 1Ø NM WNW OF THE LAUNCH COMPLEX.

23D F/39 & A/39 (81.8-11.9) C,SS F T
ST
39D F/42 & A/41,42 C,SS P T ST

SHUANG-CHENG-TZU MISSILE TEST CENTER CH 41Ø7N 1ØØ2ØE

THE MISSILE TEST CENTER, EXCEPT FOR THE AIRFIELD AND ADJACENT INSTALLATIONS, IS COVERED ON PHOTOGRAPHY OF GENERALLY FAIR-TO-POOR INTERPRETABILITY. THE LAUNCH COMPLEXES APPEAR INACTIVE AND NO SIGNIFICANT CHANGES IN EXISTING FACILITIES ARE OBSERVED.

THE UNIDENTIFIED CONSTRUCTION OBSERVED ON MISSION 1Ø41, MAY 67, CONTINUES AT THE 2 FORMERLY SUSPECT SSM TRAINING SITES 3 AND 5, 56 NM NORTH AND 88 NM NE, RESPECTIVELY, OF THE MAIN SUPPORT BASE. THE NATURE OF THE CONSTRUCTION IS NOT IDENTIFIABLE, BUT THERE IS A GENERAL SIMILARITY BETWEEN THE 2 SITES.

THE NORTHERNMOST SITE CONTAINS A GENERALLY OVAL CENTRAL AREA OF CONSTRUCTION IN WHICH A PLUS-SHAPED AND A T-SHAPED STRUCTURE ARE BEING ERECTED. NEW GROUND SCARRING IS VISIBLE AROUND THE AREA AND 45 OR MORE LARGE VEHICLES ARE NEARBY. A PROBABLE CAUSEWAY HAS BEEN CONSTRUCTED ACROSS THE STREAM WEST OF THE SITE, AND A SMALL DIVERSION DAM HAS BEEN PLACED PARTIALLY ACROSS THE STREAM TO THE SW. A PREPARED-SURFACE ROAD LEADS TO A MOTOR POOL AREA WITH A FEW TEMPORARY BUILDINGS/TENTS AND APPROXIMATELY 3Ø NEARBY TENTS. A

~~TOP SECRET [redacted] RUFF~~

~~Handle Via~~
~~TALENT-KEYHOLE~~
~~Control System Only~~

26. *(Continued)*

Attachments

26. *(Continued)*

~~Handle Via~~
~~TALENT-KEYHOLE~~
~~Control System Only~~

~~TOP SECRET~~ ~~RUFF~~

KH-4 MISSION 1042-1, 17-22 JUNE 1967

ATTACHMENT 1

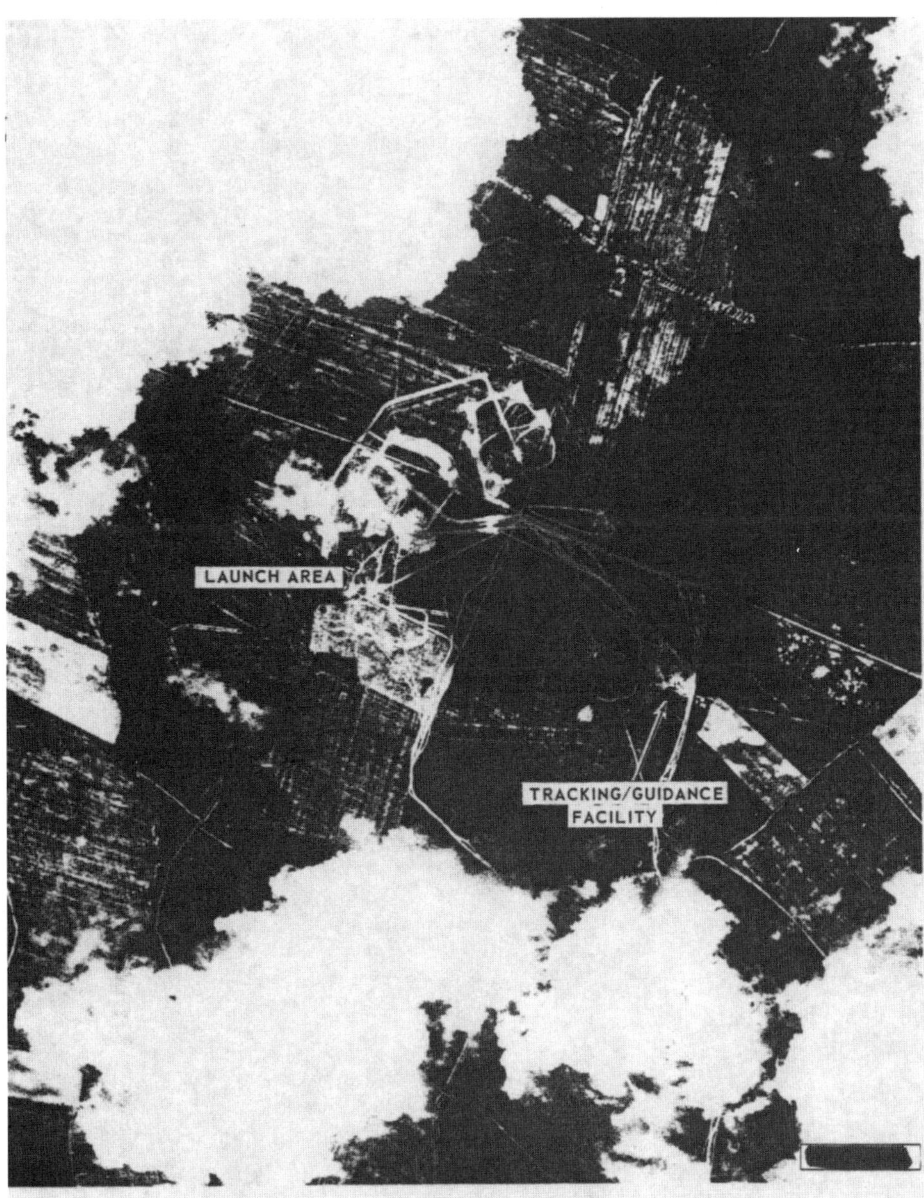

PROBABLE LONG RANGE SAM LAUNCH COMPLEX, KALININGRAD, USSR (MISSILE BONUS)

~~TOP SECRET~~ ~~RUFF~~

~~Handle Via~~
~~TALENT-KEYHOLE~~
~~Control System Only~~

26. *(Continued)*

~~Handle Via TALENT-KEYHOLE Control System Only~~

~~TOP SECRET~~ ~~RUFF~~

KH-4 MISSION 1042-1, 17-22 JUNE 1967

ATTACHMENT 2

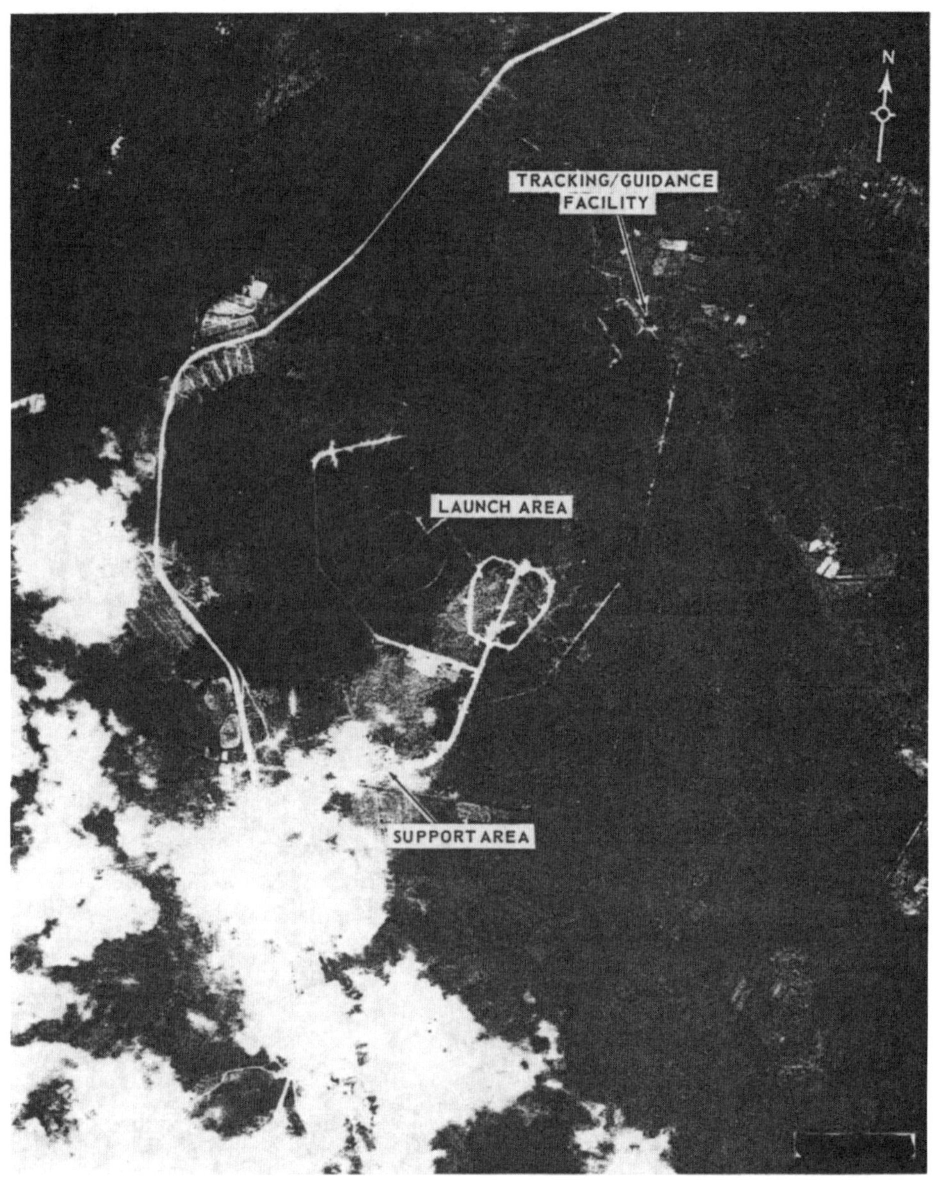

PROBABLE LONG RANGE SAM LAUNCH COMPLEX, VENTSPILS, USSR (MISSILE BONUS)

~~TOP SECRET~~ ~~RUFF~~

~~Handle Via TALENT-KEYHOLE Control System Only~~

26. *(Continued)*

~~Handle Via TALENT KEYHOLE Control System Only~~

~~TOP SECRET~~ ~~RUFF~~

KH-4 MISSION 1042-1, 17-22 JUNE 1967

ATTACHMENT 3

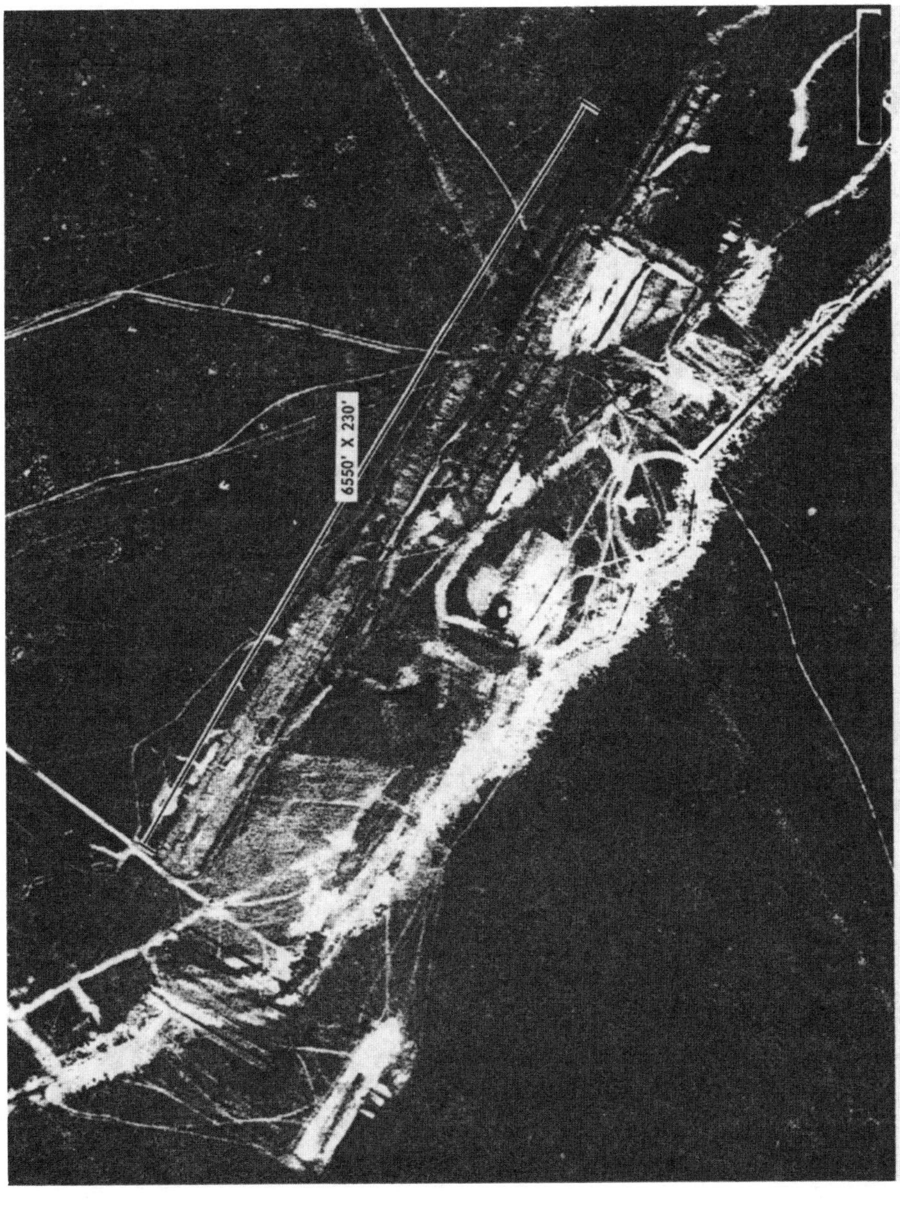

6550' X 230'

AIRFIELD UNDER CONSTRUCTION, CHOYBALSAN, MONGOLIA (AIRFIELD BONUS)

~~TOP SECRET~~ ~~RUFF~~

~~Handle Via TALENT KEYHOLE Control System Only~~

26. *(Continued)*

~~Handle Via~~
~~TALENT-KEYHOLE~~
~~Control System Only~~

~~TOP SECRET~~ ~~RUFF~~

KH-4 MISSION 1042-1, 17-22 JUNE 1967

ATTACHMENT 4

GROUND ZERO 2, NUCLEAR TEST SITE, LOP NOR, CHINA

~~TOP SECRET~~ ~~RUFF~~

~~Handle Via~~
~~TALENT-KEYHOLE~~
~~Control System Only~~

27. CIA/NPIC, Photographic Interpretation Report, "KH-4 Mission 1042-1, 17–22 June 1967, Middle East Edition," June 1967 (Excerpt)

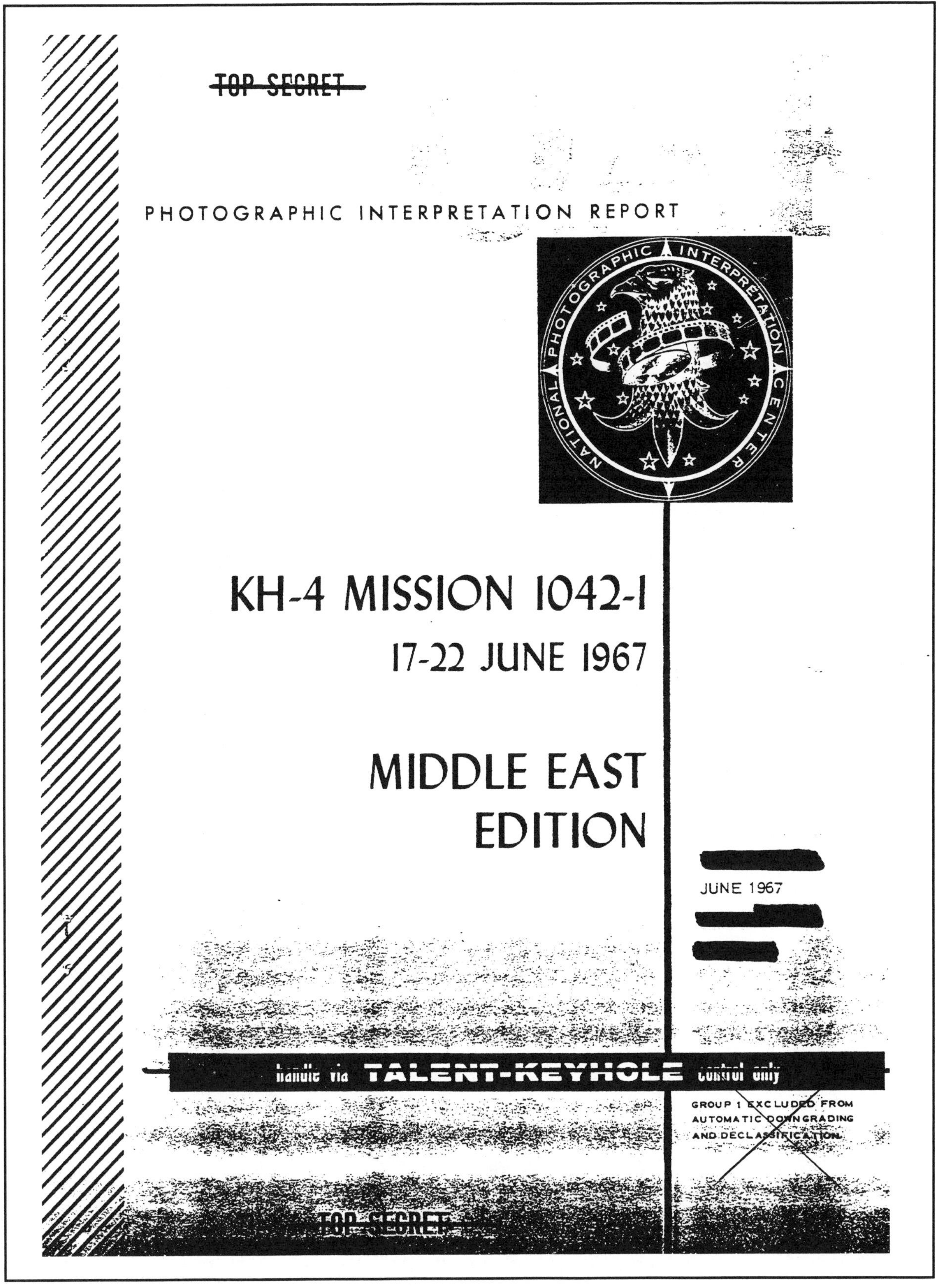

~~TOP SECRET~~

PHOTOGRAPHIC INTERPRETATION REPORT

KH-4 MISSION 1042-1

17-22 JUNE 1967

MIDDLE EAST EDITION

JUNE 1967

handle via TALENT-KEYHOLE control only

GROUP 1 EXCLUDED FROM AUTOMATIC DOWNGRADING AND DECLASSIFICATION

~~TOP SECRET~~

27. *(Continued)*

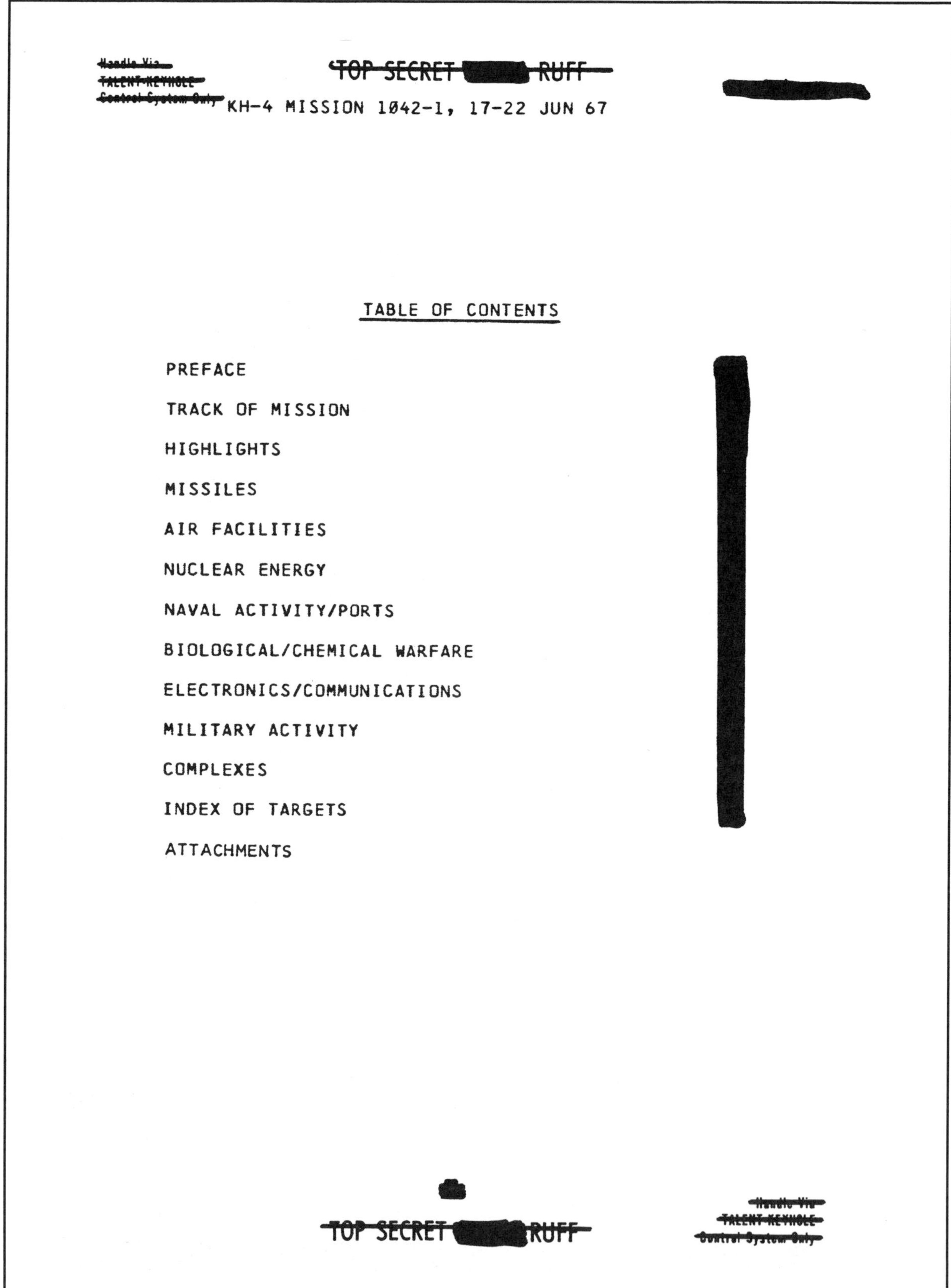

~~Handle Via TALENT-KEYHOLE Control System Only~~

~~TOP SECRET [redacted] RUFF~~

KH-4 MISSION 1042-1, 17-22 JUN 67

TABLE OF CONTENTS

~~TOP SECRET [redacted] RUFF~~

~~Handle Via TALENT-KEYHOLE Control System Only~~

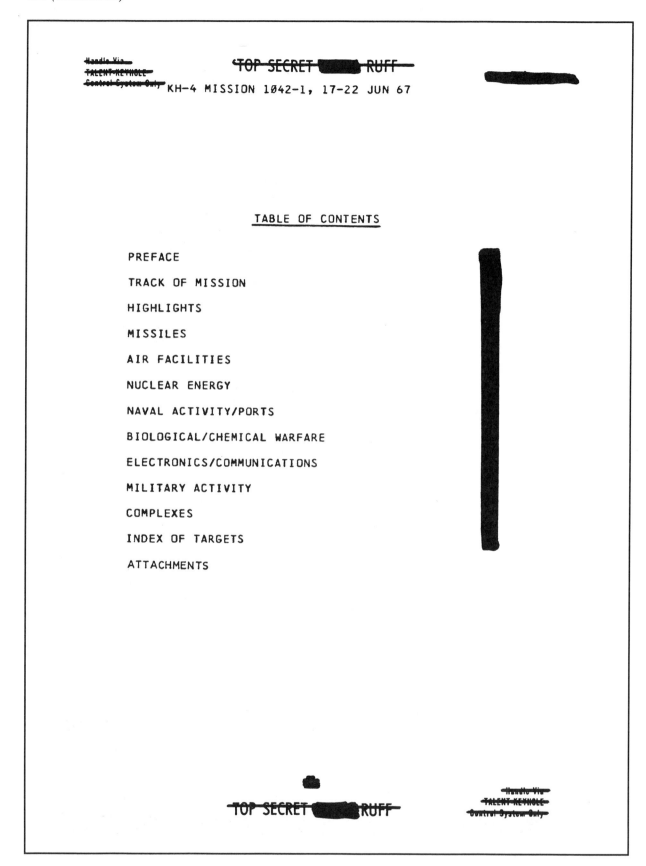

27. *(Continued)*

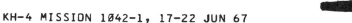

~~Handle Via TALENT-KEYHOLE Control System Only~~

~~TOP SECRET [redacted] RUFF~~

KH-4 MISSION 1Ø42-1, 17-22 JUN 67

PREFACE

THIS EDITION OF THE OAK ON MISSION 1Ø42-1 REPORTS ON TARGETS IN THE MIDDLE EAST. SOUTH CHINA AND NORTH VIETNAM TARGETS ARE REPORTED IN [redacted]. THE REMAINDER OF THE HIGHEST PRIORITY TARGETS, PRIMARILY IN THE USSR, ARE REPORTED IN [redacted].

TARGETS ARE ARRANGED IN THE OAK BY SUBJECT AND WITHIN EACH SUBJECT BY COMOR TARGET NUMBER. TARGETS OF OPPORTUNITY OBSERVED DURING THE OAK REPORTING PERIOD ARE LISTED NUMERICALLY UNDER THE APPROPRIATE SUBJECT AFTER THE LISTING OF COMOR TARGETS. LOCATIONS OF TARGETS IN THIS REPORT ARE INDICATED BY THE COUNTRY CODE PRECEDING THE COORDINATES.

SELECTED PHOTOGRAPHIC REFERENCES FOLLOW THE DESCRIPTION OF THE TARGET. PASS NUMBERS ARE SUFFIXED WITH EITHER AN A FOR ASCENDING (SOUTH-TO-NORTH), D FOR DESCENDING (NORTH-TO-SOUTH) OR M FOR A MIXED ASCENDING AND DESCENDING PASS. THE LETTER F PRECEDING FRAME NUMBERS INDICATES THE FORWARD CAMERA, AND THE LETTER A INDICATES THE AFT CAMERA. UNIVERSAL REFERENCE GRID COORDINATES ARE GIVEN IN PARENTHESIS AFTER THE FRAME NUMBER TO WHICH THEY APPLY. THIS GRID IS UNIVERSAL GRID NO 1, FEB 64. SYMBOLS FOR CONDITIONS AFFECTING INTERPRETABILITY ARE -- C (CLEAR), SC (SCATTERED CLOUD COVER), HC (HEAVY CLOUD COVER), H (HAZE), CS (CLOUD SHADOW), S (SNOW), O (OBLIQUITY), SD (SEMIDARKNESS), D (DARKNESS), GC (GROUND COVER), CF (CAMOUFLAGE), GR (GROUND RESOLUTION), AND SS (SMALL SCALE).

INTERPRETABILITY OF THE PHOTOGRAPHY IS CATEGORIZED AS G (GOOD), F (FAIR), OR P (POOR). THE EXTENT OF PHOTOGRAPHIC COVERAGE IS DENOTED BY THE SYMBOLS T (TOTAL) OR PT (PARTIAL). THE MODE OF COVERAGE IS INDICATED BY ST (STEREO), NS (NON-STEREO), OR PS (PARTIAL STEREO).

RECIPIENTS ARE CAUTIONED THAT THE INITIAL SCAN OF THE PHOTOGRAPHY IS BEING ACCOMPLISHED IN A SHORT TIME AND PRIOR TO FINAL REFINEMENT OF EPHEMERIS DATA. CONSEQUENTLY, FUTURE DETAILED ANALYSIS MAY RESULT IN ADDITIONAL INFORMATION. MOREOVER, RECIPIENTS ARE CAUTIONED THAT THE IDENTIFICATION OF TARGETS IS BASED PRIMARILY ON PHOTOGRAPHY AND DOES NOT CONSTITUTE A FINISHED INTELLIGENCE JUDGMENT.

~~Handle Via TALENT-KEYHOLE Control System Only~~

27. *(Continued)*

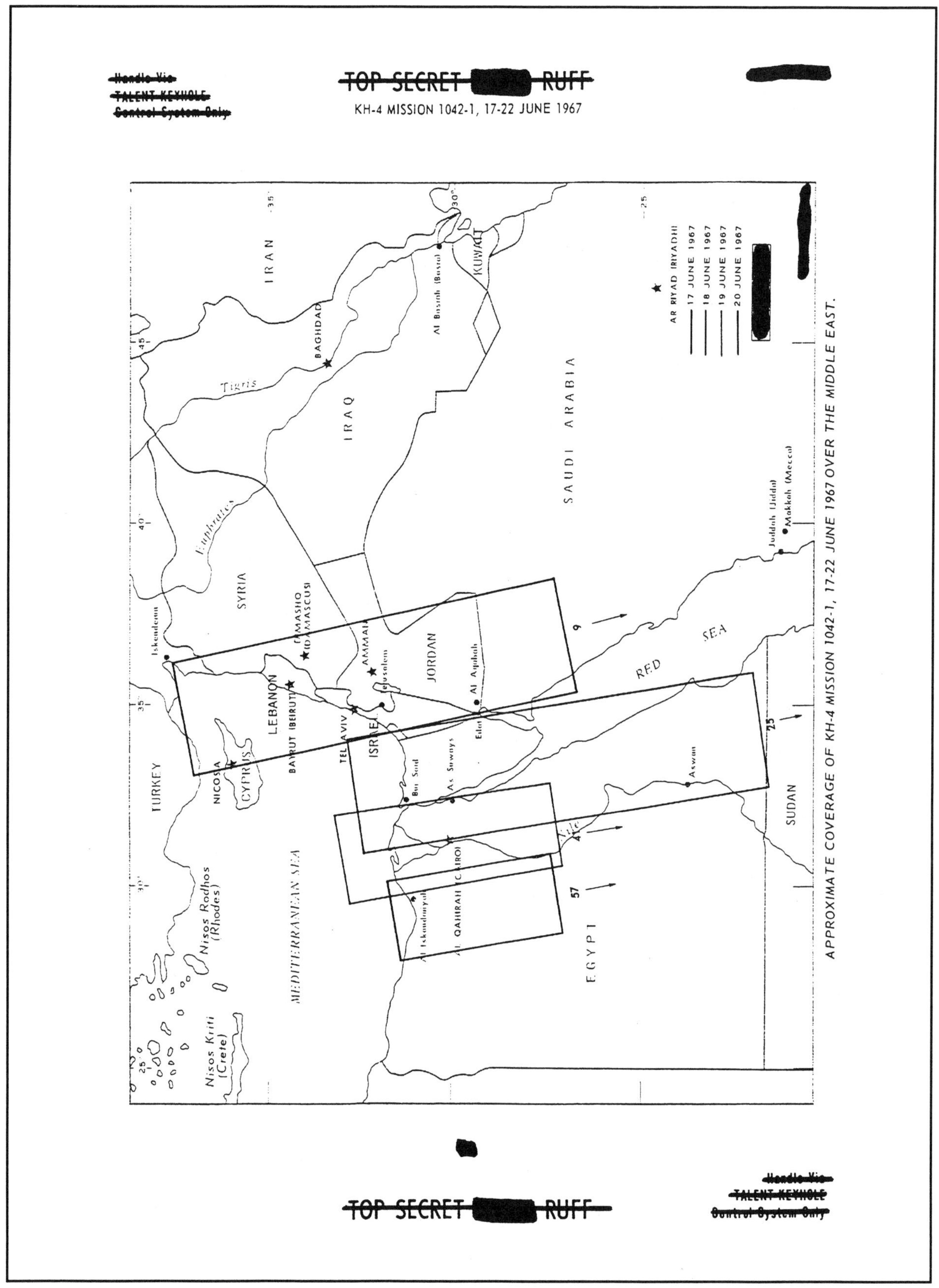

APPROXIMATE COVERAGE OF KH-4 MISSION 1042-1, 17-22 JUNE 1967 OVER THE MIDDLE EAST.

27. *(Continued)*

Highlights

27. *(Continued)*

~~Handle Via TALENT-KEYHOLE Control System Only~~

~~TOP SECRET [redacted] RUFF~~

KH-4 MISSION 1Ø42-1, 17-22 JUN 67

HIGHLIGHTS

1. ALL HIGH PRIORITY TARGETS PROGRAMMED FOR THIS MISSION IN EGYPT, ISRAEL, JORDAN, AND SOUTHWESTERN SYRIA ARE COVERED ON FOUR CLOUD-FREE PASSES IMAGED FROM 17-2Ø JUNE 1967.

2. A TOTAL OF 245 PROBABLY DESTROYED AIRCRAFT ARE OBSERVED AT AIRFIELDS IN 3 ARAB COUNTRIES -- 2Ø1 IN EGYPT, 26 IN JORDAN, AND 18 IN SYRIA. (SEE TABLE 1 FOR TABULAR LISTING)

THE RELATIVE ABSENCE OF COMBAT AIRCRAFT OBSERVED AT THE NORTHERN OR CENTRAL EGYPTIAN AIRFIELDS PROBABLY INDICATES THAT THE REMNANTS OF THE EGYPTIAN AIR FORCE HAVE BEEN DEPLOYED TO ASWAN, LUXOR, AND OTHER SOUTHERN AIRFIELDS WHICH WERE NOT IMAGED ON THIS MISSION.

SIXTEEN CAT/CUB TRANSPORT AIRCRAFT WERE IMAGED AT CAIRO AIRFIELD ON 19 JUNE 1967. USUALLY, NO MORE THAN 8 OF THIS TYPE OF AIRCRAFT ARE OBSERVED AT THIS FIELD.

3. A PASSENGER VESSEL HAS BEEN SUNK 3 NM SOUTH OF THE BUR SAID PORT FACILITIES CREATING A NAVIGATIONAL OBSTRUCTION IN THE SUEZ CANAL.

NO APPARENT DAMAGE IS DISCERNIBLE TO THE NAVAL VESSELS OR FACILITIES AT AL ISKANDARIYAH, EGYPT.

NO OTHER SIGNIFICANT NAVAL, AIR, OR GROUND MILITARY ACTIVITY IS OBSERVED ON THIS PHOTOGRAPHY.

4. OF THE 35 KNOWN SAM SITES IN EGYPT (INCLUDING 2 NEWLY IDENTIFIED SITES), 3Ø ARE OBSERVED ON THIS MISSION. TWELVE ARE OCCUPIED, ONE IS POSSIBLY OCCUPIED, FIVE ARE OF UNDETERMINED OCCUPANCY, NINE ARE UNOCCUPIED (INCLUDING THE TWO NEW SITES), AND THREE CAN BE IDENTIFIED ONLY. FIVE OF THE 6 KNOWN SAM SUPPORT FACILITIES ARE ALSO OBSERVED.

STATUS OF TARGET READOUT

THIS OAK REPORTS ON 129 COMOR TARGETS, 12 NON-COMOR TARGETS, AND 3 BONUS TARGETS.

~~Handle Via TALENT-KEYHOLE Control System Only~~

27. *(Continued)*

~~Handle Via~~
~~TALENT KEYHOLE~~
~~Control System Only~~

~~TOP SECRET~~ [redacted] ~~RUFF~~

KH-4 MISSION 1042-1, 17-22 JUNE 1967

TABLE 1. BOMB DAMAGE, ARAB AIRFIELDS

CN	AIRFIELD NAME	PHOTO DATE	PROBABLE DESTROYED AIRCRAFT	RUNWAY DAMAGE	OTHER DAMAGE	AOB	
						FIGHTERS	BOMBERS
EGYPT							
	ABU SUWEIR	18 & 19 JUNE 67	37	24 CRATERS	NONE	NONE	NONE
	AL ARISH	18 JUNE 67	7	6 CRATERS	NONE	7 POSS A/C	NONE
	BENI SUEF	19 JUNE 67	12	AT LEAST 13 CRATERS	NONE	NONE	NONE
	BIR HASANAH NEW	18 JUNE 67	13	NONE	NONE	NONE	NONE
	BIR JIFJAFAH	18 JUNE 67	22	NONE	NONE	NONE	NONE
*	CAIRO	19 JUNE 67	NONE	3 CRATERS	NONE	1 SMALL SWEPT	NONE
*	CAIRO WEST	19 JUNE 67	25	24 CRATERS	NONE	16 SMALL SWEPT	1 BADGER
	EL MANSURA	19 JUNE 67	NONE	9 CRATERS (REPAIRED)	NONE	NONE	NONE
	FAYID	18 & 19 JUNE 67	14	17 CRATERS/ CHARRED AREAS	NONE	NONE	NONE
	GAZA	18 JUNE 67	22	NONE	NONE	NONE	NONE
	HURGHADA NEW	18 JUNE 67	7	NONE	NONE	NONE	NONE
	INCHAS	19 JUNE 67	20	9 POSS CRATERS	PARTIALLY DE-STROYED HANGAR	NONE	NONE
	KABRIT	18 & 19 JUNE 67	22	8 CRATERS	NONE	NONE	NONE
	GEBEL LIBNI	18 JUNE 67	NONE	NONE	CHARRED AREA IN SUPPORT FAC	NONE	NONE
*	ALMAZA	19 JUNE 67		NO DAMAGE OBSERVED		7 SMALL SWEPT	NONE
	ISMAILIA	18 & 19 JUNE 67		NO DAMAGE OBSERVED		NONE	NONE
	PORT SAID	18 & 19 JUNE 67		NO DAMAGE OBSERVED		NONE	NONE
	RAS BANAS	18 JUNE 67		NO DAMAGE OBSERVED		NONE	NONE
			TOTAL 201				
SYRIA							
	DAMASCUS	17 JUNE 67	9	PROBABLE, EXTENT UNK	NONE	NONE	NONE
	DUMAYR	17 JUNE 67	9	NONE	NONE	NONE	NONE
	MARJ RHAYAL	17 JUNE 67		NO DAMAGE OBSERVED		NONE	NONE
	DAMASCUS NEW	17 JUNE 67		NO DAMAGE OBSERVED		NONE	NONE
			TOTAL 18				
JORDAN							
	AMMAN	17 JUNE 67	12	17 PROB CRATERS	NONE	NONE	NONE
*	KING HUSSEIN	17 JUNE 67	14	NONE	NONE	NONE	NONE
			TOTAL 26				
SAUDI ARABIA							
*	TABUK	17 JUNE 67		NO DAMAGE OBSERVED		NONE	NONE

GRAND TOTAL OF ARAB AIRCRAFT PROBABLY DESTROYED: 245

* CARGO AND TRANSPORT AIRCRAFT ARE EXCLUDED FROM AOB COUNT

~~TOP SECRET~~ [redacted] ~~RUFF~~

~~Handle Via~~
~~TALENT KEYHOLE~~
~~Control System Only~~

27. *(Continued)*

Air Facilities

27. *(Continued)*

Handle Via
TALENT KEYHOLE
Control System Only

TOP SECRET [redacted] RUFF

KH-4 MISSION 1Ø42-1, 17-22 JUN 67

DAMASCUS AIRFIELD SY 3328N Ø3613E

OB -- NONE DISCERNIBLE.

NINE CHARRED AREAS (PROBABLY DESTROYED AIRCRAFT) ON RUNWAYS AND PARKING AREAS ARE OBSERVED. THE MAIN RUNWAY APPEARS TO HAVE PROBABLE BOMB DAMAGE.

9D F/38 & A/38 (22.8-1Ø.8) H,D,SS P T ST

RAS BANAS AIRFIELD EG 2358N Ø3527E

OB -- NONE OBSERVED.

NO NEW CONSTRUCTION, CHANGES IN FACILITIES, OR PREVIOUSLY UNREPORTED FEATURES OBSERVED.

25D F/1Ø9 (74.2-1Ø.6) & A/1Ø9 C F T ST

GEBEL LIBNI AIRFIELD EG 3Ø45N Ø3345E

OB -- NONE OBSERVED.

ONE CHARRED AREA IS OBSERVED IN THE SUPPORT FACILITIES.

25D F/63 (56.7-13.1) & A/62 C,SS P T ST

BIR JIFJAFAH AIRFIELD EG 3Ø24N Ø33Ø8E

OB -- NONE OBSERVED.

TWENTY-TWO CHARRED AREAS (PROBABLY 22 DESTROYED AIRCRAFT) ARE OBSERVED.

25D F/65 (35.4-13.Ø) & A/64 C F T ST

AL ARISH AIRFIELD EG 31Ø4N Ø335ØE

OB -- 7 POSSIBLE AIRCRAFT.

SIX PROBABLE BOMB CRATERS ON NW/SE RUNWAY, 3 CHARRED AREAS (PROBABLY DESTROYED AIRCRAFT) ON THE EAST END OF THE E/W RUNWAY, AND 4 CHARRED AREAS (PROBABLY DESTROYED AIRCRAFT) ON TAXIWAY ARE OBSERVED.

Handle Via
TALENT KEYHOLE
Control System Only

27. *(Continued)*

Attachments

27. *(Continued)*

27. *(Continued)*

~~Handle Via~~
~~TALENT-KEYHOLE~~
~~Control System Only~~

~~TOP SECRET~~ ~~RUFF~~

KH-4 MISSION 1042-1, 17-22 JUNE 1967

ATTACHMENT 2

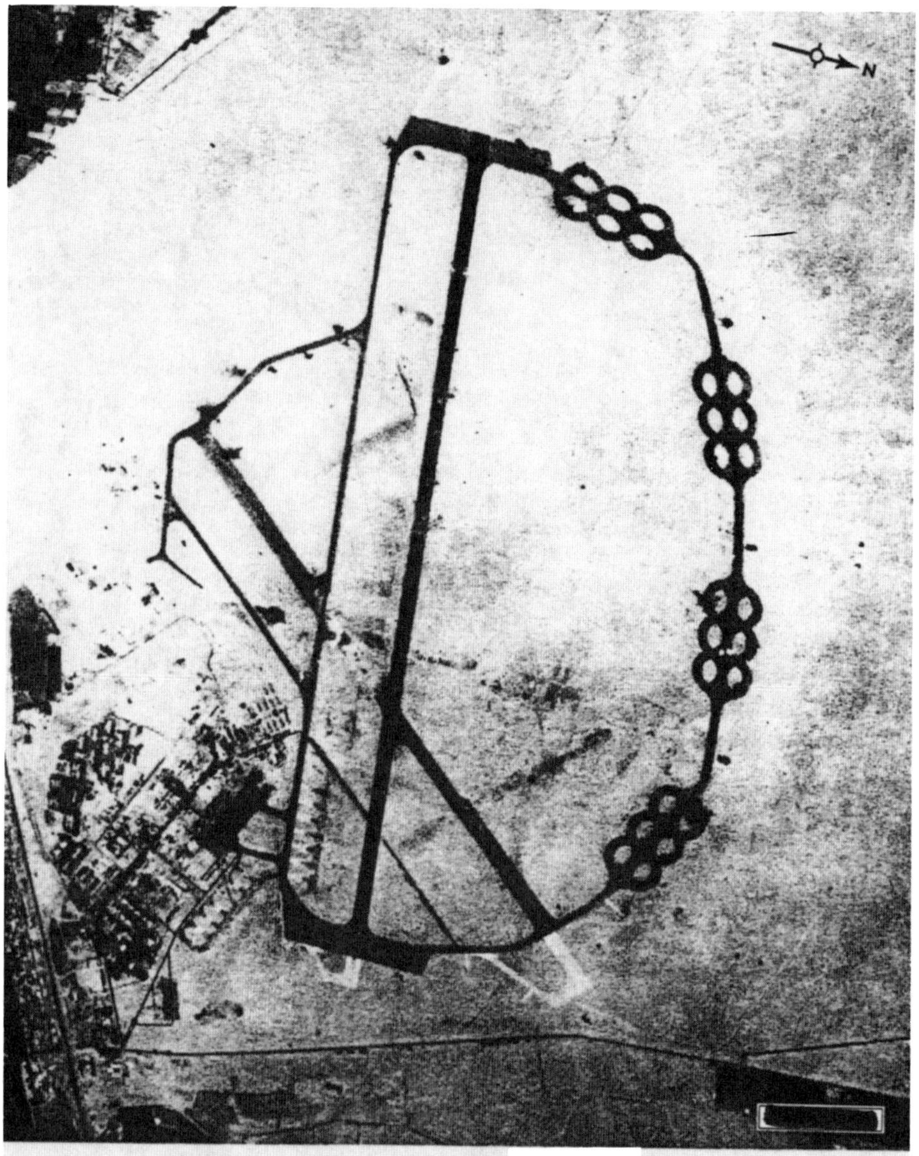

ABU SUWEIR AIRFIELD, EGYPT

~~TOP SECRET~~ ~~RUFF~~

~~Handle Via~~
~~TALENT-KEYHOLE~~
~~Control System Only~~

27. *(Continued)*

~~Handle Via~~
~~TALENT-KEYHOLE~~
~~Control System Only~~

~~TOP SECRET~~ ~~RUFF~~

KH-4 MISSION 1042-1, 17-22 JUNE 1967

ATTACHMENT 3

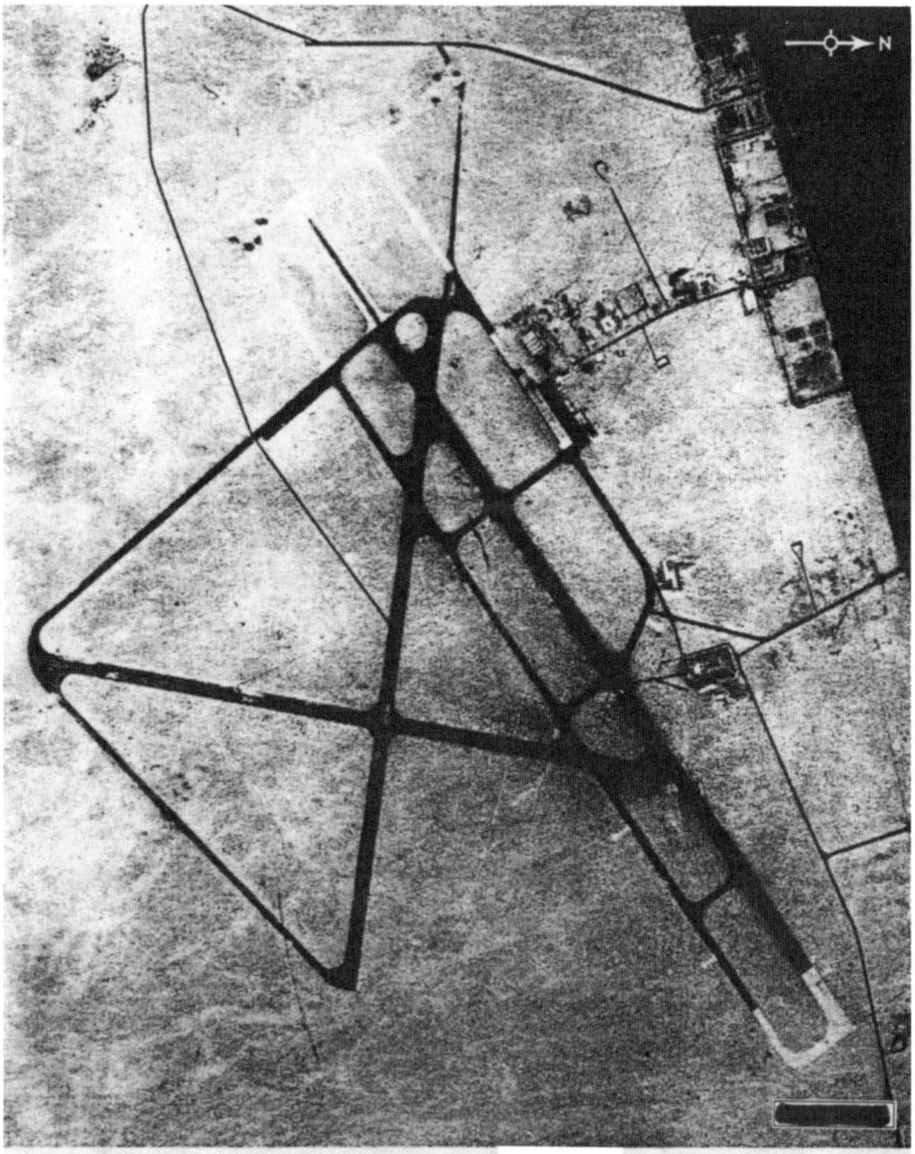

N

INCHAS AIRFIELD, EGYPT

~~TOP SECRET~~ ~~RUFF~~

~~Handle Via~~
~~TALENT-KEYHOLE~~
~~Control System Only~~

27. *(Continued)*

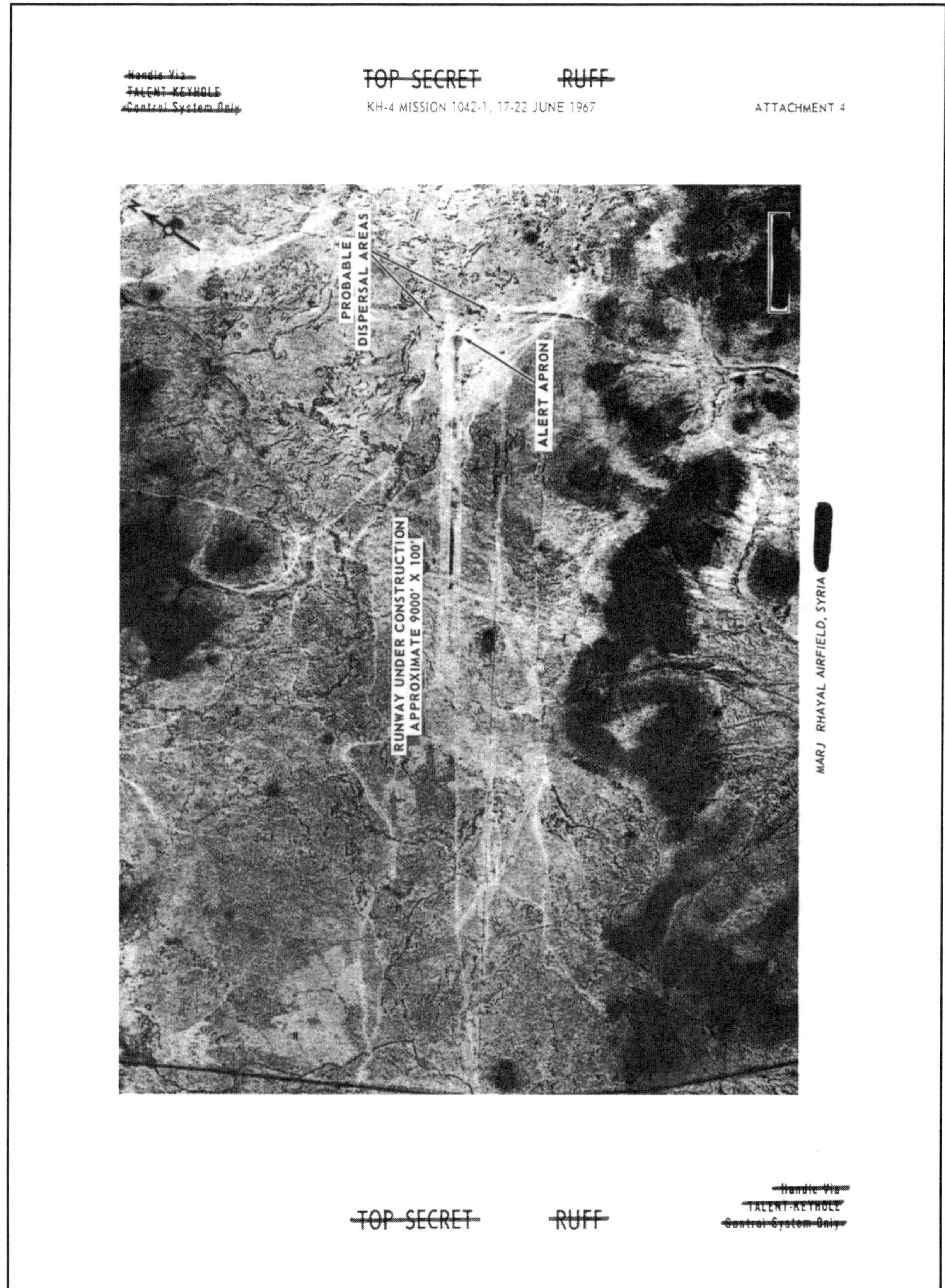

Handle Via
TALENT-KEYHOLE
Control System Only

TOP SECRET RUFF

KH-4 MISSION 1042-1, 17-22 JUNE 1967

ATTACHMENT 4

TOP SECRET RUFF

Handle Via
TALENT-KEYHOLE
Control System Only

27. *(Continued)*

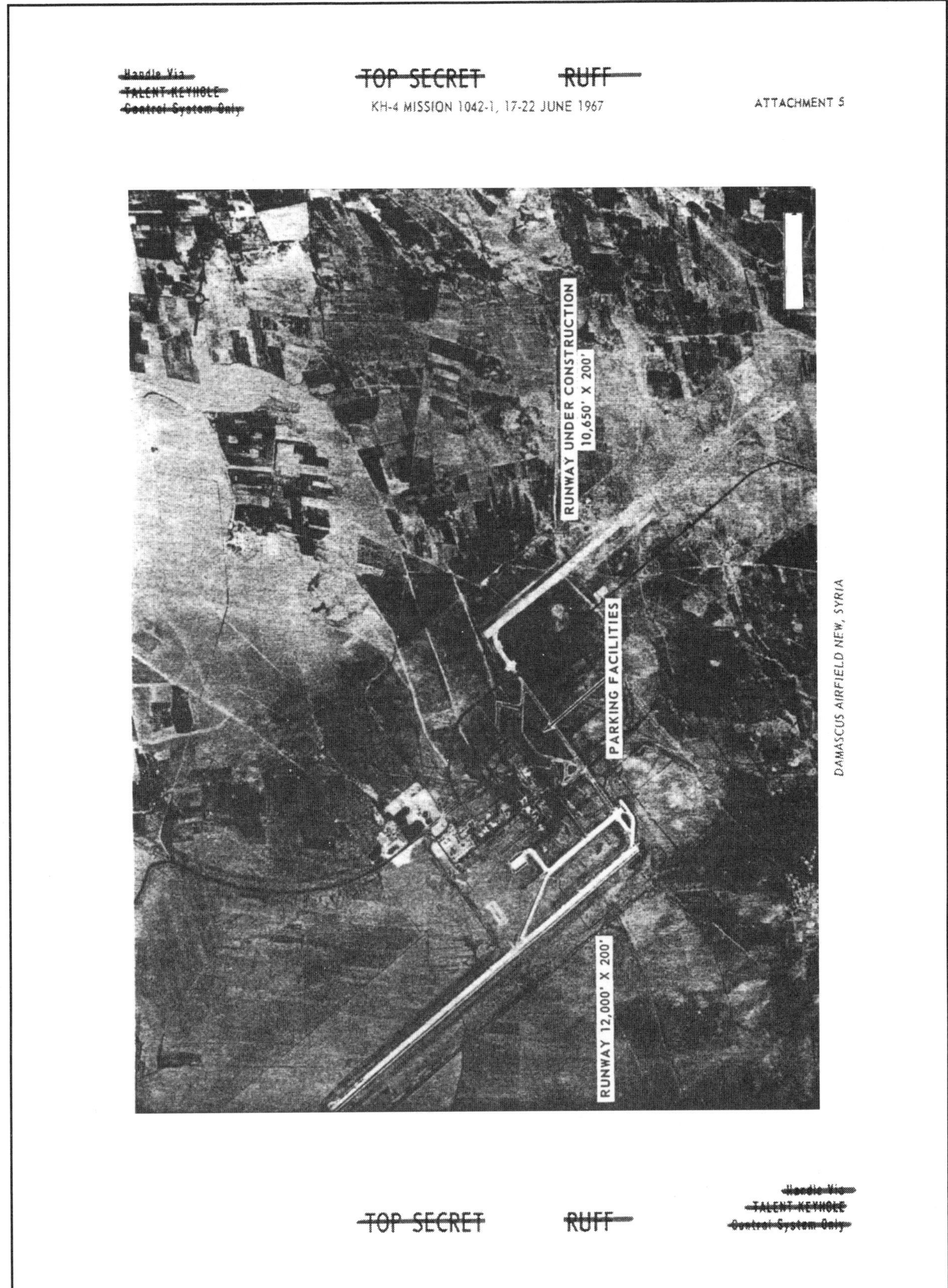
~~Handle Via~~
~~TALENT-KEYHOLE~~
~~Control System Only~~

~~TOP SECRET~~ ~~RUFF~~

KH-4 MISSION 1042-1, 17-22 JUNE 1967

ATTACHMENT 5

~~TOP SECRET~~ ~~RUFF~~

~~Handle Via~~
~~TALENT-KEYHOLE~~
~~Control System Only~~

27. *(Continued)*

27. *(Continued)*

~~Handle Via~~
~~TALENT-KEYHOLE~~
~~Control System Only~~

~~TOP SECRET~~ ~~RUFF~~

KH-4 MISSION 1042-1, 17-22 JUNE 1967

ATTACHMENT 7

CAPSIZED VESSEL, QANAT AS SUWAYS, EGYPT (NAVAL BONUS)

~~TOP SECRET~~ ~~RUFF~~

~~Handle Via~~
~~TALENT-KEYHOLE~~
~~Control System Only~~

28. CIA/NPIC, Photographic Interpretation Report, "KH-4 Mission 1042-1, 17–22 June 1967, South China and North Vietnam Edition," June 1967 (Excerpt)

TOP SECRET

OAK

PHOTOGRAPHIC INTERPRETATION REPORT

NATIONAL PHOTOGRAPHIC INTERPRETATION CENTER

KH-4 MISSION 1042-1

17-22 JUNE 1967

SOUTH CHINA AND NORTH VIETNAM EDITION

JUNE 1967

handle via TALENT-KEYHOLE control only

GROUP 1 EXCLUDED FROM

28. *(Continued)*

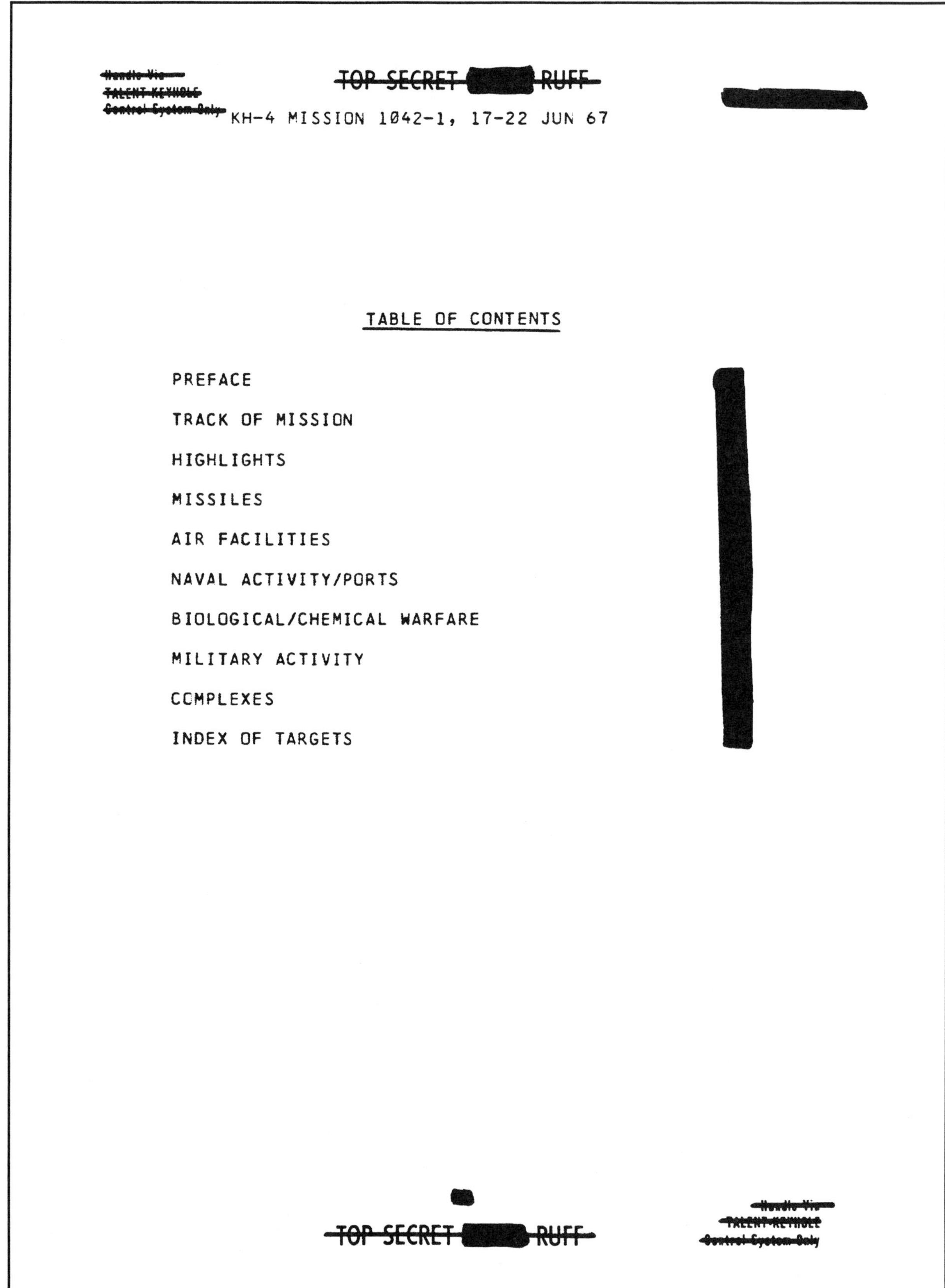

~~Handle Via~~
~~TALENT-KEYHOLE~~
~~Control System Only~~

~~TOP SECRET [redacted] RUFF~~

KH-4 MISSION 1042-1, 17-22 JUN 67

TABLE OF CONTENTS

~~TOP SECRET [redacted] RUFF~~

~~Handle Via~~
~~TALENT-KEYHOLE~~
~~Control System Only~~

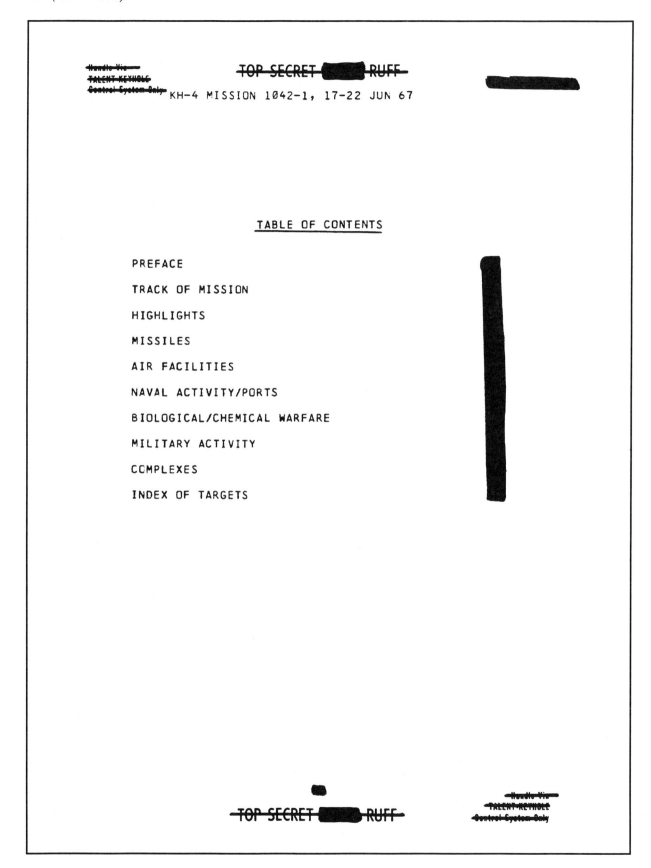

28. *(Continued)*

~~Handle Via~~
~~TALENT-KEYHOLE~~
~~Control System Only~~

~~TOP SECRET [redacted] RUFF~~

KH-4 MISSION 1Ø42-1, 17-22 JUN 67

PREFACE

THIS EDITION OF THE OAK ON MISSION 1Ø42-1 REPORTS ON THE HIGHEST PRIORITY TARGETS IN SOUTH CHINA AND NORTH VIETNAM. MIDDLE EAST TARGETS ARE REPORTED IN [redacted]. THE REST OF THE HIGHEST PRIORITY TARGETS, PRIMARILY IN THE USSR, ARE REPORTED IN [redacted].

TARGETS ARE ARRANGED IN THE OAK BY SUBJECT AND WITHIN EACH SUBJECT BY COMOR TARGET NUMBER. TARGETS OF OPPORTUNITY OBSERVED DURING THE OAK REPORTING PERIOD ARE LISTED NUMERICALLY UNDER THE APPROPRIATE SUBJECT AFTER THE LISTING OF COMOR TARGETS. LOCATIONS OF TARGETS IN THIS REPORT ARE INDICATED BY THE COUNTRY CODE PRECEDING THE COORDINATES.

SELECTED PHOTOGRAPHIC REFERENCES FOLLOW THE DESCRIPTION OF THE TARGET. PASS NUMBERS ARE SUFFIXED WITH EITHER AN A FOR ASCENDING (SOUTH-TO-NORTH), D FOR DESCENDING (NORTH-TO-SOUTH) OR M FOR A MIXED ASCENDING AND DESCENDING PASS. THE LETTER F PRECEDING FRAME NUMBERS INDICATES THE FORWARD CAMERA, AND THE LETTER A INDICATES THE AFT CAMERA. UNIVERSAL REFERENCE GRID COORDINATES ARE GIVEN IN PARENTHESIS AFTER THE FRAME NUMBER TO WHICH THEY APPLY. THIS GRID IS UNIVERSAL GRID NO 1, FEB 64. SYMBOLS FOR CONDITIONS AFFECTING INTERPRETABILITY ARE -- C (CLEAR), SC (SCATTERED CLOUD COVER), HC (HEAVY CLOUD COVER), H (HAZE), CS (CLOUD SHADOW), S (SNOW), O (OBLIQUITY), SD (SEMIDARKNESS), D (DARKNESS), GC (GROUND COVER), CF (CAMOUFLAGE), GR (GROUND RESOLUTION), AND SS (SMALL SCALE).

INTERPRETABILITY OF THE PHOTOGRAPHY IS CATEGORIZED AS G (GOOD), F (FAIR), OR P (POOR). THE EXTENT OF PHOTOGRAPHIC COVERAGE IS DENOTED BY THE SYMBOLS T (TOTAL) OR PT (PARTIAL). THE MODE OF COVERAGE IS INDICATED BY ST (STEREO), NS (NON-STEREO), OR PS (PARTIAL STEREO).

RECIPIENTS ARE CAUTIONED THAT THE INITIAL SCAN OF THE PHOTOGRAPHY IS BEING ACCOMPLISHED IN A SHORT TIME AND PRIOR TO FINAL REFINEMENT OF EPHEMERIS DATA. CONSEQUENTLY, FUTURE DETAILED ANALYSIS MAY RESULT IN ADDITIONAL INFORMATION. MOREOVER, RECIPIENTS ARE CAUTIONED THAT THE IDENTIFICATION OF TARGETS IS BASED PRIMARILY ON PHOTOGRAPHY AND DOES NOT CONSTITUTE A FINISHED INTELLIGENCE JUDGMENT.

~~Handle Via~~
~~TALENT-KEYHOLE~~
~~Control System Only~~

28. *(Continued)*

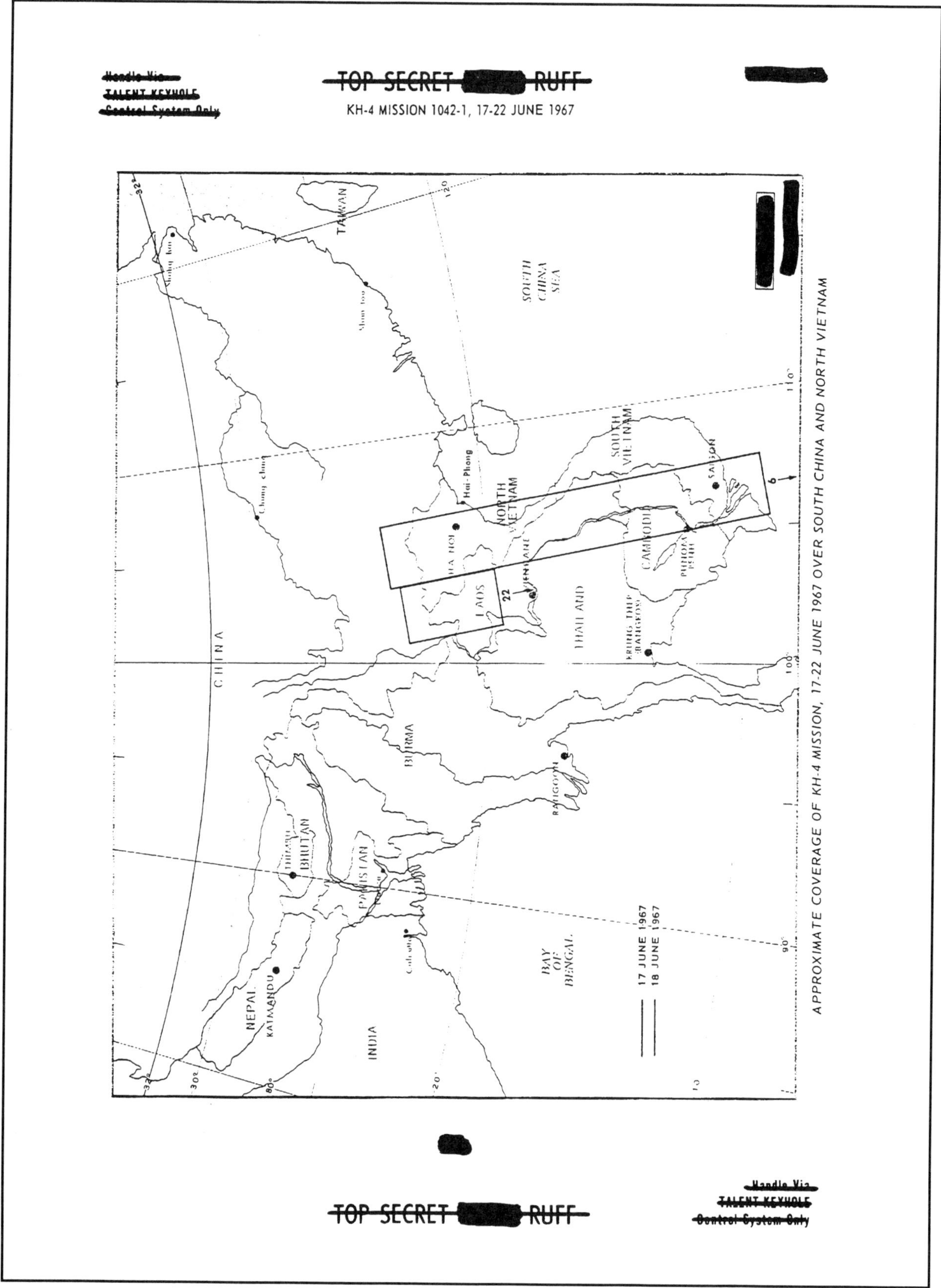

28. *(Continued)*

Highlights

28. *(Continued)*

~~Handle Via TALENT KEYHOLE Control System Only~~ ~~TOP SECRET~~ [redacted] ~~RUFF~~ [redacted]

KH-4 MISSION 1Ø42-1, 17-22 JUN 67

HIGHLIGHTS

1. NORTH VIETNAM

ALMOST ALL OF NORTH VIETNAM IS COVERED ON TWO PASSES OF PHOTOGRAPHY WHICH ARE ESTIMATED TO BE APPROXIMATELY 45 PERCENT CLOUD COVERED. A PRELIMINARY SEARCH OF ALL CLOUD-FREE AREAS REVEALS NO EVIDENCE OR INDICATION OF OFFENSIVE SURFACE-TO-SURFACE MISSILES. NO SIGNIFICANT NAVAL, AIR, OR GROUND MILITARY ACTIVITY IS OBSERVED.

OF THE 172 KNOWN SA-2 SAM SITES, 144 ARE OBSERVED IN THE CLOUD-FREE AREAS. THE FOLLOWING 14 SITES ARE OCCUPIED -- HA NOI A1Ø-2, A17A-2, A29-2, A33-2, BØ4-2, B12-2, B21-2, B29-2, B33-2, AND CØ2-2, HAI PHONG C28-2, HOA BINH B2Ø-2, VINH A36-2, AND YEN BAI C1Ø-2. EIGHTY-NINE SITES ARE UNOCCUPIED AND AT 4Ø SITES OCCUPANCY IS UNDETERMINED DUE TO LIMITING CONDITIONS OF THE PHOTOGRAPHY. SAM SITE VINH LINH B29-2 LOCATED AT 17-Ø9-2ØN 1Ø6-44-21E HAS BEEN BOMBED. ALL LAUNCH POSITIONS ARE DESTROYED AND THE GUIDANCE AREA IS PARTIALLY DESTROYED. THEREFORE, THIS SITE IS DROPPED FROM NPIC LISTINGS. NO NEW SAM INSTALLATIONS ARE IDENTIFIED.

2. SOUTH CHINA

PHOTOGRAPHY OF SOUTH CHINA IS LIMITED TO THE IMMEDIATE AREAS ADJACENT TO THE CHINA-NORTH VIETNAM BORDER. HEAVY CLOUD COVER PRECLUDES THE IDENTIFICATION OF ANY SIGNIFICANT ACTIVITY.

STATUS OF TARGET READOUT

THIS OAK REPORTS ON 138 COMOR TARGETS AND 3 NON-COMOR TARGETS.

~~TOP SECRET~~ [redacted] ~~RUFF~~ ~~Handle Via TALENT-KEYHOLE Control System Only~~

28. *(Continued)*

Missiles

Handle Via TALENT-KEYHOLE Control System Only

TOP SECRET [redacted] RUFF

KH-4 MISSION 1042-1, 17-22 JUN 67

THANH HOA SAM SITE A19-2 (87) VN 1941N 10546E

NO MISSILES, MISSILE-RELATED EQUIPMENT, NEW CONSTRUCTION, CHANGES IN FACILITIES, OR PREVIOUSLY UNREPORTED FEATURES OBSERVED.

6D F/45 & A/45 (38.8-10.6) C F T ST

THANH HOA SAM SITE B28A-2 (100) VN 1950N 10529E

NO MISSILES, MISSILE-RELATED EQUIPMENT, NEW CONSTRUCTION, CHANGES IN FACILITIES, OR PREVIOUSLY UNREPORTED FEATURES OBSERVED.

6D F/45 & A/44 (47.7-9.5) C F T ST

THANH HOA SAM SITE B27-2 (109) VN 1948N 10532E

NO MISSILES, MISSILE-RELATED EQUIPMENT, NEW CONSTRUCTION, CHANGES IN FACILITIES, OR PREVIOUSLY UNREPORTED FEATURES OBSERVED.

6D F/45 & A/44 (46.3-11.1) C F T ST

THANH HOA SAM SITE A14-2 (111) VN 1945N 10550E

NO MISSILES, MISSILE-RELATED EQUIPMENT, NEW CONSTRUCTION, CHANGES IN FACILITIES, OR PREVIOUSLY UNREPORTED FEATURES OBSERVED.

6D F/45 & A/44 (36.2-13.8) C F T ST

THANH HOA SAM SITE D04A-2 (112) VN 2020N 10611E

THIS SITE HAS BEEN RETURNED TO CULTIVATION AND WAS PREVIOUSLY DROPPED FROM NPIC LISTINGS.

6D F/42 & A/42 H F T ST

THANH HOA SAM SITE D20-2 (123) VN 1916N 10535E

SITE IS PROBABLY UNOCCUPIED AND APPEARS TO BE DETERIORATING.

TOP SECRET [redacted] RUFF

Handle Via TALENT-KEYHOLE Control System Only

Part IV

Nonmilitary Uses for Satellite Imagery

Part IV: Nonmilitary Uses for Satellite Imagery

While CORONA was born out of the need for military intelligence, it has proved to have an extraordinary range of other applications. As the system became more reliable and its users became more experienced, CORONA offered unprecedented possibilities. In 1970, when CIA's experimentation with color film in a KH-4B mission proved less than useful for military targets, NPIC reasoned that color film might nonetheless expand coverage for other forms of intelligence. This led CIA to offer a subcontract to a geology firm to assess the use of color imagery for mineral resources exploration. The subcontract resulted in Document No. 29, "Appraisal of Geologic Value for Mineral Resources Exploration."

Vice President Albert Gore announced the declassification of CORONA imagery and its transfer to the National Archives and Records Administration in February 1995. A key figure in the formation of CIA's Environmental Task Force, Gore recognized that satellite imagery "recorded, however, much more than the landscape of the cold war. In the process of acquiring this priceless data, we recorded for future generations, the environmental history of the earth."

The forthcoming declassification of CORONA's enormous files of global imagery may yet help turn swords into plowshares.

29. Report No. 9, KH-4B System Capability, "Appraisal of Geologic Value for Mineral Resources Exploration," March 1971

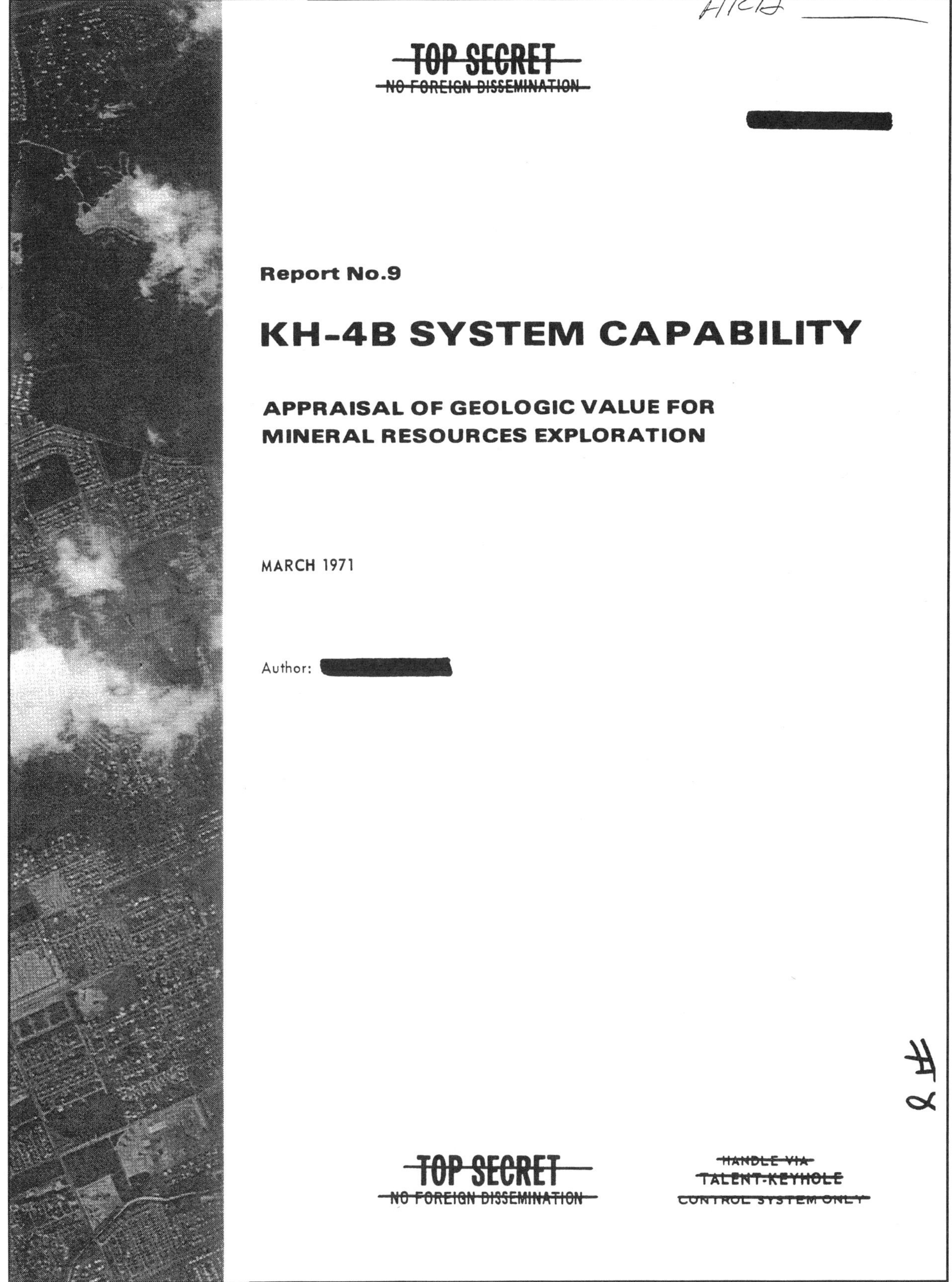

~~TOP SECRET~~

~~NO FOREIGN DISSEMINATION~~

Report No.9

KH-4B SYSTEM CAPABILITY

APPRAISAL OF GEOLOGIC VALUE FOR MINERAL RESOURCES EXPLORATION

MARCH 1971

Author: [redacted]

~~TOP SECRET~~

~~NO FOREIGN DISSEMINATION~~

~~HANDLE VIA~~
~~TALENT-KEYHOLE~~
~~CONTROL SYSTEM ONLY~~

29. *(Continued)*

~~TOP SECRET~~

~~NO FOREIGN DISSEMINATION~~

Report No.9

KH-4B SYSTEM CAPABILITY

APPRAISAL OF GEOLOGIC VALUE FOR MINERAL RESOURCES EXPLORATION

MARCH 1971

Author: [redacted]

~~TOP SECRET~~

~~NO FOREIGN DISSEMINATION~~

~~HANDLE VIA~~
~~TALENT-KEYHOLE~~
~~CONTROL SYSTEM ONLY~~

SEP 01 1971

29. *(Continued)*

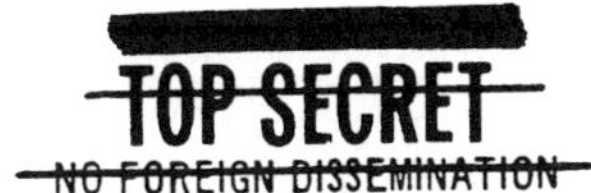

CONTENTS

HANDLE VIA
TALENT-KEYHOLE
CONTROL SYSTEM ONLY

FIGURES

~~TOP SECRET~~
~~NO FOREIGN DISSEMINATION~~

~~HANDLE VIA~~
~~TALENT-KEYHOLE~~
~~CONTROL SYSTEM ONLY~~

~~TOP SECRET~~

~~NO FOREIGN DISSEMINATION~~

1. INTRODUCTION

As part of a continuing investigation of the value of satellite photography, 800 feet of SO-242 was flown on mission 1108-2. SO-242 is a relatively new "high resolution" color film especially built for use in satellite and high flying aircraft systems. This film was flown at the end of mission 1108. Further, the two stereo cameras on the KH-4B System allowed us to acquire high resolution black and white coverage concurrent with the color. Unfortunately, the resolution of the color photography was not as good as hoped, due primarily to the poor color correction of the KH-4B lenses. NPIC did an evaluation of this photography and concluded,* in part, that:

1. The primary constraint on using SO-242 in this system is the incompatibility between the intelligence community requirements and the spatial resolution afforded by SO-242 at this scale.
2. The difference in ground resolution between the color and black and white material is approximately 2:1 in favor of the black and white.
3. Except for its stereo contribution, the color material was of no apparent value to first or second phase analysis.
4. Color photography as provided by this system is expected to contribute most to regional, agricultural, and geological studies.

The first three conclusions relate, of course, primarily to the military oriented intelligence community. That is, order of battle, missile readiness, etc. The last conclusion was, in our view, of significance to the nonmilitary intelligence community, such as the Offices of Basic and Geographic Intelligence and Economic Research in the CIA. It was, therefore, with these offices in mind that this study was undertaken. The purposes of this somewhat limited study were, therefore, to:

1. Evaluate the information content of color versus black and white for one type of natural resources task
2. Evaluate the potential of the KH-4B System for resources exploration
3. Perform sufficient analysis to indicate the potential advantages of color photography for geographic/economic intelligence purposes.

Review of the photography indicated that excellent cloud free coverage of the Tsaidam Basin in China and of Southern Russia had been obtained, and that these areas possessed interesting geological features. It was, therefore, decided to concentrate on the value of this photography for

---

*

~~TOP SECRET~~ ~~HANDLE VIA TALENT-KEYHOLE CONTROL SYSTEM ONLY~~ 

~~NO FOREIGN DISSEMINATION~~

29. *(Continued)*

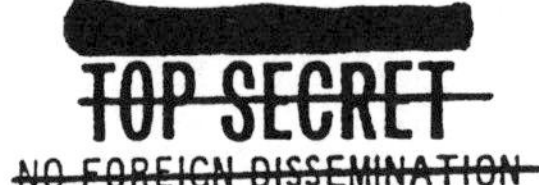

TOP SECRET

NO FOREIGN DISSEMINATION

mineral resources exploration. With this goal in mind, a contract was arranged with a commercial petroleum and mineral exploration firm of geologists [redacted] Contract support was also provided by Itek Corporation, Lexington, Massachusetts.

[redacted] were instructed to review the photography, prepare detailed geological maps and prepare an analysis on the mineral resources similar to that which would be done for another commercial customer (such as an oil company). This they have done, and this is the final report of that study.

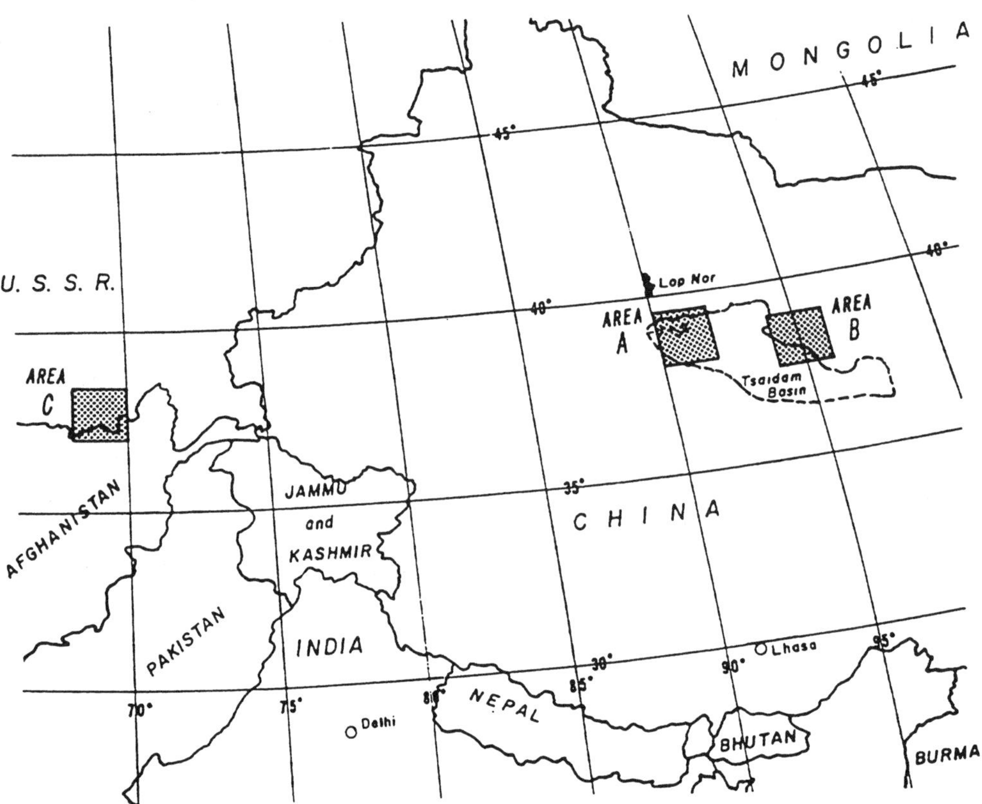

Fig. 1-1 — Index map showing three study areas

TOP SECRET

NO FOREIGN DISSEMINATION

HANDLE VIA TALENT-KEYHOLE CONTROL SYSTEM ONLY

2. GEOLOGIC EXPLORATION PRINCIPLES

Mineral deposits and fossil fuel resources occur at or beneath the earth's surface in irregular deposits under varying geologic conditions. Frequently these accumulations occur under conditions that can be relatively accurately predicted by an experienced exploration geologist.

Geologic investigation differs from conventional photo intelligence in one important respect. Instead of an expert searching for a specific feature, recognizing and pinpointing it, then describing it . . . the geologist must analyze the entire area and record all his findings on a map. Then, and only then, is he able to fully analyze geologic conditions and make accurate interpretations.

Thus, the exploration geologist, in searching for "hidden" mineral deposits, must first prepare a basic geologic map to guide him. The conventional geologic map contains substantive geologic information of two basic classes: (1) rock type, and (2) structure. When interpreted correctly, this map provides clues to the most likely areas for accumulation of mineral deposits.

Conventional geologic maps are prepared by ground parties traversing an area and taking rock samples at numerous localities. The rock type (lithologic description, geologic age and formation name, if possible) and structural conditions at each locale are plotted onto a base map. These data are then extrapolated across the areas not traversed by the ground parties and the map is thus completed.

2.1 STANDARD PHOTOGEOLOGIC TECHNIQUES

Photogeologic mapping has been used for the past 30 years to augment ground geologic mapping. It is now an accepted and fundamental tool for mineral and petroleum exploration and has been found to greatly reduce the time and expense of ground parties. Its main advantage is the enlarged vertical perspective where all areas, not just those that can be readily reached on foot, are analyzed in all their geologic detail.

Photogeologic mapping utilizing space photography enlarges this perspective even more. Using the KH-4B System, this is achieved without unduly sacrificing ground resolution necessary for reliable photogeologic mapping.

According to Gilluly, Waters, and Woodford*:

> "A geologic map is a valuable economic tool, useful in locating supplies of oil, water, coal, iron ore, and other substances buried beneath a cover of soil and rocks. Though such valuable prizes are completely hidden beneath the

---

* Gilluly, J., Waters, A., and Woodford, A., "Principles of Geology," Freeman, San Francisco, California, 1968, pp 2 and 88.

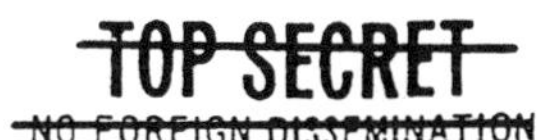

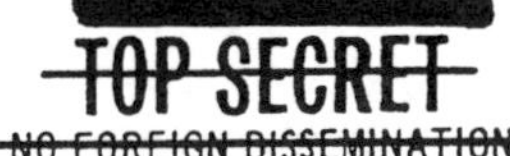

> surface, a geologic map often reveals where tunneling or drilling will be successful. The accuracy of such predictions has been proved again and again by discoveries of valuable ores, coal, and petroleum. Geologic maps are indeed the indispensable foundation of all geology—basic to our understanding of all subsurface processes, . . ." "On the international scene, the power and wealth of a nation is largely determined by its endowment of useful minerals, its skill in finding and utilizing them, or in obtaining needed supplies from other lands. In this age of political unrest and readjustment among nations, the vast accumulation of petroleum in such little-industrialized nations as Iran, Saudi Arabia, Iraq, and Kuwait is a potent force in world politics. We shall be wiser in world affairs if we know where and why petroleum occurs, how it is discovered, and how its quantity underground may be estimated."

Photogeologic mapping involves two basic functions: (1) differentiation of rock type, and (2) structural mapping.

2.1.1 Differentiation of Rock Type (lithologic/stratigraphic mapping)

This involves differentiating the various rock units exposed at the surface. In a virtually unmapped area of the world, this will involve distinguishing only between the basic rock types, as follows: (1) igneous (intrusive-granite, extrusive-basalt, etc.); (2) sedimentary (sandstone, shale, limestone, etc.); and (3) metamorphic (schist, gneiss, slate, marble, etc.). Distinguishing between these gross rock units is generally not too difficult for an experienced photogeologist. This is because each basic type usually exhibits an identifying "signature" such as color, texture, land form pattern, etc.

A more useful map will be prepared however, when some ground truth is available (see Section 3.2). Information such as lithologic descriptions of various land specimens from the various formations will be most useful, as will any information regarding the geologic age of individual units.

2.1.2 Structural Mapping

This involves mapping the structural relationships of the various rock units. Structural features, such as folds (anticlines, synclines, monoclines), faults (normal, reverse, thrust, etc.), fractures, joints, etc., are often better observed from the vertical stereoscopic perspective than from the ground. Comprehensive mapping of the structural features permits a proper understanding of the chronology of events affecting the subject area.

2.2 ECONOMIC ASSESSMENT

Interpretation of the geologic map is the next important step. What does all this information mean economically? The exploration geologist looks for certain clues to guide him to hidden mineral or petroleum deposits. For instance, the petroleum exploration geologist knows that oil is found in sedimentary basin areas. He restricts his study to these areas and does not search the mountainous hard-rock (igneous and metamorphic) regions. He knows that for the sedimentary basin area to hold economic petroleum deposits, it must contain: (1) source beds (generally marine shales), (2) reservoir beds (usually porous sandstones or limestones), and (3) traps (many types—the most common are anticlinal folds or faulted anticlines). After he has ascertained that

HANDLE VIA
TALENT KEYHOLE
CONTROL SYSTEM ONLY

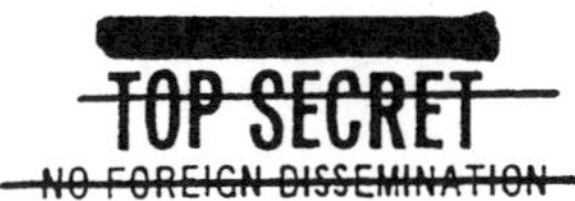

conditions (1) and (2) above are met for an area, he focuses his attention toward looking for traps. Surface geological maps are extremely useful for this purpose because many deep-seated structures are reflected in the structural conditions revealed at the surface.

For mineral (other than petroleum) exploration, however, he searches not only the sedimentary basin areas, but more particularly the mountainous "hard-rock" regions, depending on the types of minerals desired. He knows, for instance, that certain metallic deposits are often found in the vicinity of igneous intrusive activity, strong metamorphism, and faulting. Therefore, he searches for significant granitic intrusions within metamorphic rock regions and major fault zones. He does not always, however, restrict his search to the hard-rock regions because many nonmetallic mineral deposits (potash, gypsum, etc.) occur in sedimentary environments.

~~TOP SECRET~~
~~NO FOREIGN DISSEMINATION~~

~~HANDLE VIA~~
~~TALENT-KEYHOLE~~
~~CONTROL SYSTEM ONLY~~

29. *(Continued)*

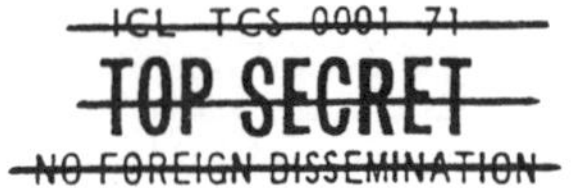

3. METHODS OF INTERPRETATION

3.1 COMPILATION AND INTERPRETATION PROCEDURES

Mapping of the three areas under consideration was achieved in the following sequence: (1) Area A—NW Tsaidam Basin, (2) Area C—Kafirnigan-Pyandzh Area, and (3) Area B—NE Tsaidam Basin. This was a matter of convenience rather than design, since that is the order in which the necessary reference and mapping materials became available.

The compilation and interpretation procedures varied slightly from area to area, because of differences in the available base map control, geologic references, and nature of the KH-4B materials. As a result of following these varied procedures, which are discussed in more detail below, it was possible to develop a preferred (optimum) set of procedures for future mapping projects.

For each area mapped, it was necessary to use photography from both the: (1) convergent panoramic (PAN) camera, and (2) the DISIC framing camera. The PAN photography was used for making the geologic interpretations. The DISIC photography was used for construction of the base maps and compilation of the geologic annotations.

For the purposes of these studies, the optimum mapping scale was determined to be 1:250,000. Each area embraces slightly less than 12,000 miles and includes 2 degrees of longitude and 1½ degrees of latitude. This is slightly larger than the conventional "2° × 1°" format of most 1:250,000 scale maps.

3.1.1 Procedures Used for Area A

Area A is bounded by Latitudes 38°00'N and 39°30'N and Longitudes 90°00'E and 92°00'E. A preliminary (pencil compiled) planimetric map of the subject area was constructed on 0.003-inch Herculene film to serve as a base for the photogeologic mapping. This was prepared, due to the lack of available up-to-date 1:250,000 scale topographic maps, in a make-shift manner as follows. A geographic coordinate network of 15-minute intervals was laid out using the Universal Transverse Mercator Projection. A 1:1,000,000 scale topographic map (ONC Chart) was photographically enlarged to the mapping scale of 1:250,000. By overlaying the control grid film onto this enlargement, a preliminary planimetric base map was generated by lightly tracing the major drainage network, roads, railroads, town, and additional cultural data. This provided the gross "horizontal control" for subsequent plotting of geologic detail from the space photography.

Because of the DISIC failure on mission 1108-2 before the SO-242 film was exposed in the panoramic camera, this imagery was not available. The void was filled, however, by bringing together comparable DISIC coverage from previous missions. This DISIC photography (missions 1102 and 1106) was enlarged (approximately 8 ×) into working prints at the 1:250,000 scale. The adjacent DISIC frames were similarly enlarged and cut into strips roughly equivalent to the width

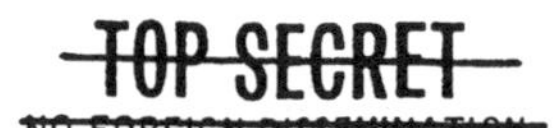

HANDLE VIA
TALENT-KEYHOLE
CONTROL SYSTEM ONLY

of two panoramic frames. This provided a crude but effective way to obtain stereoscopy. In this stereoscopic mode, detailed drainage patterns and other topographic and cultural data were plotted in colored pencil onto the enlarged DISIC print. Once completed, these drainage patterns and related cultural information were transferred to the planimetric base film overlay which was held in correct position by the gross drainage patterns plotted in the base film. This completed the planimetric base map preparation phase of the project.

Photogeologic interpretation of the panoramic photography was accomplished utilizing a Richards GFL-940 MCE Light Table mounted with a Bausch & Lomb zoom 70 Microscope modified with a Richards Stereodapter. In this way, the black and white and color records were transported in parallel across the light table and the imagery studied in stereoscopic perspective, one eye viewing the color record, the other the black and white. In this process, it soon became evident that color and stereo are essential requirements to extract the maximum amount of geologic information. Loss of one or the other results in a significant reduction in information content.

Plotting and transferring the geologic interpretations to the base map was a somewhat difficult and cumbersome process. The geologic information observed on the 3404 and SO-242 records had to be visually plotted onto the DISIC print. The adjacent DISIC strips were used to obtain the correlatable image in stereo. To say the least, this was not the most effective way to interpret and plot the observed geologic information.

3.1.2 Procedures Used for Area B

Area B is bounded by Latitudes 37°30' N and 39°N and Longitudes 94°E and 96°E. Preparation of the planimetric base map was achieved utilizing essentially the same procedures as for Area A.

The geologic interpretation and compilation procedures for Area B, however, were considerably improved over those for Areas A and C. The Area B study was begun last, and by this time it was possible to obtain transformed (rectified) and enlarged (2×) records of the PAN photography. This was accomplished by the Aeronautical Chart and Information Center (ACIC) using the Itek Gamma I Rectifier. These materials were placed in parallel on a standard light table and interpretations were made using an Old Delft scanning mirror stereoscope. Geologic annotations were made directly to acetate overlays, which were later transferred to the preliminary planimetric film by use of a scale-changing Kail Reflecting Projector. These procedures precluded the laborious, inefficient and often inaccurate process of transferring mental images to the DISIC print. Moreover, they allowed for discernment and annotation of considerably more geologic detail than on the earlier studies. It is an understatement to say the the use of the enlarged and transformed PAN imagery is the more desired procedure.

3.1.3 Procedures Used for Area C

Area C is bounded by Latitudes 37° 00' N and 38° 30' N and Longitudes 68° 00' E and 70° 00' E. The mapping and compilation procedures utilized were essentially the same as for Area A with one noteable exception: horizontal control was good. It was not necessary to blow up a small scale 1:1,000,000 map for this project. Classified 1:250,000 scale AMS topographic map sheets were obtained for this area through the assistance of CIA personnel. In this instance, the topographic maps were overlayed by the geographic control grid and the gross planimetric control was lifted off directly. The DISIC photography was enlarged to scale and the detail matched perfectly, confirming that the AMS topographic sheets were of very recent vintage and most accurate.

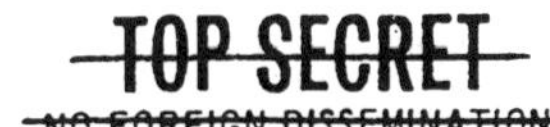

HANDLE VIA
TALENT-KEYHOLE
CONTROL SYSTEM ONLY

3.2 BACKGROUND RESEARCH

An effective photogeologic evaluation is always significantly enhanced when some basic geologic "ground truth" is available. The type of information most helpful is rock-type (lithologic) descriptions of the various rock units present and their geologic age determinations. From this basic data, a more detailed geologic map can be prepared.

In many areas (as is the case in most of the continental United States) published geologic maps provide an excellent foundation upon which to build a detailed and comprehensive photogeologic study. This fundamental data can be extrapolated to an extensive degree using the space photographs. Thus, the ground-derived information taken from several localities can be used to trace the geologic phenomena across the entire area in question including many locales never visited on foot by man.

In many regions, geologic maps are either nonexistent or are nothing more than small scale compilations of wide spread observations and are sometimes of questionable accuracy. In areas such as these, the effectiveness of the evaluation will be greatly dependent on the interpreter's degree of experience in photogeologic mapping.

A moderately extensive geologic research effort was undertaken for the three study areas under consideration. Since these areas lie within Iron Curtain countries, the search could not be conducted by the geologists working on the project using standard scientific research procedures without unduly risking a breach of security. Therefore, the research effort was conducted with the assistance of the Office of Basic Geographic Intelligence of the CIA.

A limited amount of useful reference material was found to be available for the subject areas. It is possible that additional published and unpublished reference materials exist; however, the necessity to adhere to strict security procedures, as well as lack of time, precluded a thorough and comprehensive research effort.

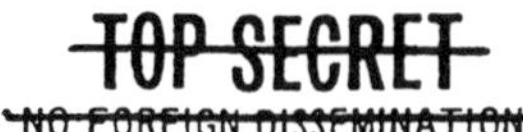

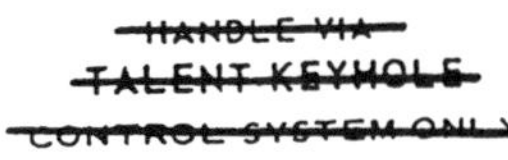

4. AREA A—NW TSAIDAM BASIN

4.1 GENERAL GEOLOGY

The primary reference found for Areas A and B. the Tsaidam Basin, was "The Geology of China" by Ch'ang Ta.* This publication contained the following: (1) very small scale sketch map of the geologic outcrop pattern (see Fig. 4-1), (2) generalized descriptions of the rock types, and (3) a brief discussion on the regional structure in the Tsaidam Basin Region. Although the information contained herein is meager, it provided the foundation upon which the information revealed by the space photography was applied.

Fig. 4-2 is a photographic reduction of the complete photogeologic map of Area A.

According to Ch'ang, the Tsaidam Basin belongs to the vast "NW Hercynian Zone of Fold" of northern China. The basin is shaped like a rhomb with a broad west end and a narrow eastern end. It is surrounded by moutain ranges. Ch'ang says:

> "The Tsaidam Basin has very rich underground treasures. It also has fertile soil and suitable agriculture and pastery climate. The area is populated mostly by our national minorities. Today, the Socialist sunlight has already shown on the good earth of the Tsaidam. We have the responsibility as well as the confidence to have this piece of our land built into a greater center of the future."

4.2 ROCKS EXPOSED

Five basic rock units are recognized by Ch'ang in the Tsaidam Basin region. Ch'ang's descriptions are summarized in the following paragraphs. The letter symbols in parentheses refer to the symbols used on the photogeologic map (Fig. 4-2).

4.2.1 Pre-Sinian Metamorphic System (PC or PCg)

The photogeologic interpretation indicates that this might include Precambrian metamorphic and/or granitic rocks. The metamorphic sequence occurs in the northwest part of the mapped area, along the northern edge of the Altin-Tagh Mountains. Several large intrusive areas in the northwest and the western part of the project are designated as Precambrian (?) granite. It is probable, however, that these intrusive areas are much younger than Precambrian in age.

---

* Ch'ang Ta, "The Geology of China," Joint Publication Research Service, JPRS no. 19,209, 16 May 1963.

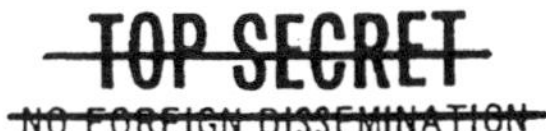

29. *(Continued)*

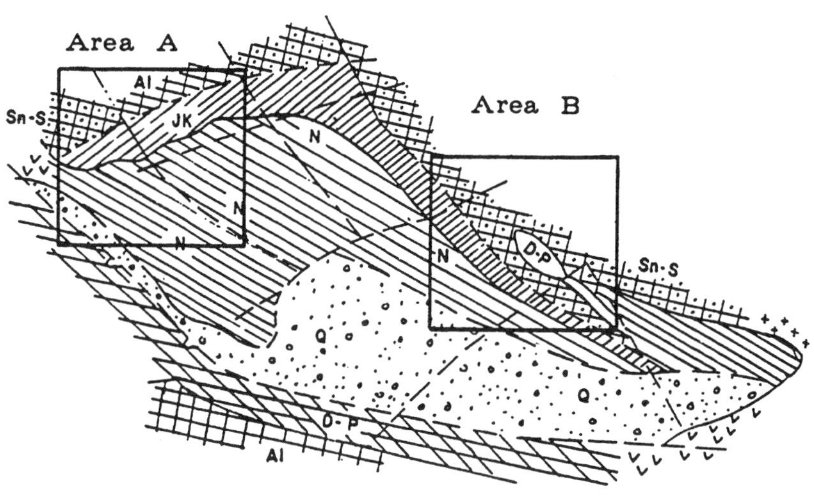

Fig. 4-1 — Outcrop map of the Tsaidam Basin (after Ch'ang Ta)

TOP SECRET
NO FOREIGN DISSEMINATION

HANDLE VIA
TALENT KEYHOLE
CONTROL SYSTEM ONLY

29. *(Continued)*

4.2.2 Lower Paleozoic System (Nan-Shan metamorphic rock series—LPm)

This is the "ancient metamorphic rock series" consisting mainly of slates, phyllites, schist, and various types of gneiss. It forms the main rock unit of the northern Altin-Tagh Mountains and occurs in the Kunlun Shan and Ch'i-lien Shan areas. In the latter area, this is called the Nan-Shan system.

4.2.3 Marine Devonian Through Permian Systems (D-P)

This sequence includes more than 2,150 meters of calcareous shale, shale, sandstone. argillaceous limestone, black schist, and light and dark gray limestone. Similar to the Sc.. China marine sequence, it occurs along the northern edge of the Kunlun Shan Mountains forming the southern flank of the basin. Within the project area, this sequence is also interpreted to be present in the foothills of the southern Altin-Tagh Mountains in the west central part of the project. A large inferred granite intrusion is mapped here, designated as "PCg" on the map. It is, however, more likely to be post-Permian in age.

4.2.4 Jurassic-Cretaceous System (JK)

This is a continental facies lake basin sequence. It consists of a lower interval, fro~ 900 to 2,600 meters thick of grayish-green conglomerate, sandstone, black shale, and some coal be... The Cretaceous system consists of as much as 1,800 meters of conglomerate, green sandstone, and purple shale. The Jurassic-Cretaceous beds crop out in the southern foothills of the Altin-Tagh Mountains and form the outer sedimentary rim of the Tsaidam Basin.

4.2.5 Tertiary Kansu System (Tk)

This is a continental facies sequence that represents the most widely distributed and thickest rock unit within the Tsaidam Basin proper. It consists of from 3,000 to 6,000 meters of relatively thin-bedded conglomerate, sandstone, shale, and gypsum strata. This sequenc.. greatly in thickness in different parts of the basin, being thickest in the southeast part.

Above the Kansu System are the Quaternary fluvio-lake accumulations. Quaternary deposits mapped include: lake beds (Q1), sand dunes (Qsd), terrace deposits (Qt), and undifferentiated materials (Q), including alluvium, colluvium, fans, bolson, and aeolian deposits.

In addition to the five basic rock units described above, Ch'ang reports the presence of various igneous bodies, including Caledonian and Hercynian age granites. On the photogeologic map, all apparent intrusive igneous rocks are labeled "PCg." Most of these are probably youn.. than Precambrian age. Along the northern edge of the Altin-Tagh Mountains a series of ...... toned, resistant beds appear in the stream cuts. These appear to be relatively young (Tertiary ?) volcanic rocks.

4.3 STRUCTURE

A majority of the structural features in the vicinity of the project are aligned toward the west-northwest. The Tertiary beds are considerably deformed into elongated, faulted anticlines and synclines. The Jurassic-Cretaceous rocks exhibit long-axial box and comb folds. The older rock sequences adjacent to the basin exhibit relatively complex folding and faulting with th.. dominant trends oriented toward the west-northwest.

Ch'ang reports that rift faults are the Tsaidam's most characteristic feature. These are reportedly of the high-angle reverse type, where the older rocks are thrust upon the younger.

TOP SECRET
NO FOREIGN DISSEMINATION

HANDLE VIA
TALENT-KEYHOLE
CONTROL SYSTEM ONLY

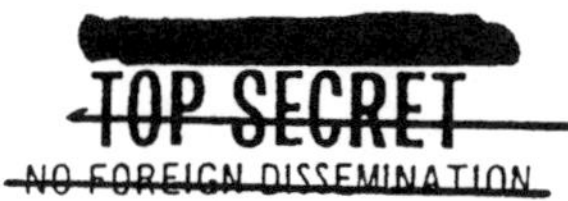

Most major faults and fault systems are aligned toward the west-northwest, generally parallel with the Kunlun Shan and Ch'i-lien Shan Mountains.

A notable exception to this is the outlining structures of the Altin-Tagh Mountains. This range is anomalously aligned toward the northeast and is bounded by major faults. This is probably a relatively young fault system since the strike of the interior folds within the range and schistocity of the older metamorphic rocks are generally aligned toward the older, west-northwest direction.

The geologic history and chronology of geologic events is postulated by Ch'ang. It is beyond the scope of this report to presume to question his findings, at least not until the adjacent areas are studied in more detail.

4.4 ILLUSTRATIONS

The following examples depict some of the more important geologic features revealed by the photogeologic evaluation in the NW Tsaidam Basin Area. They illustrate the value of the KH-4B imagery as well as the effectiveness of the techniques used. Their location is depicted on Fig. 4-2.

Fig. 4-3 is a dual illustration (3404 and SO-242 records) including an interpretation overlay showing the "contact" (interface) between two major Tsaidam rock units, i.e., the lower Paleozoic metamorphics (more resistant, darker colored) and the younger Jurassic-Cretaceous continental sediments (red and reddish-brown banding). Note how the contact is virtually indistinguishable on the black and white photograph, yet easily depicted on the color film. The dark-toned areas might be indicative of basic intrusive igneous rocks. Mineralization might occur along these interfaces, and along the traces of the numerous faults and fractures within the area. The fault zone on the right represents the eastern-most end of the northeastward-trending "major fault zone" rimming the Altin-Tagh Mountains.

Fig. 4-4 is a 3404 record showing a classic anticlinal structure within a central part of the basin proper . . . the "oil patch." Fig. 4-5 is the SO-242 companion photo of the same area. This well developed structural feature is mapped in thin-bedded Tertiary Kansu strata. The individual beds within this unit are essentially the same color, and hence the SO-242 color photography does not materially enhance the interpretation. This classic anticlinal fold is the type of "trap" that oil geologists continually seek. The circular, arcuate patterns that appear like rings around a tub are, in reality, individual rock layers that have been arched into an anticlinal upwarp and beveled off by erosion. The black dots in the crestal part of the fold are oil wells of the Yuchuantze Oil Field. These, found as a pleasant surprise during the interpretation, indicate that: (1) this is a petroliferous providence (petroleum source rocks and reservoir beds are present in the basin), and (2) that the photo resolution is more than adequate for geologic mapping purposes.

Fig. 4-6 is a schematic cross-sectional drawing of the anticlinal structure at the Yuchuantze Oil Field. No specific oil production data is available for this field, but no doubt the oil comes from porous sandstone reservoir beds within the Kansu sequence.

In the typical oil producing region, oil is believed to be formed in the basin deeps from marine shales and is squeezed by pressure into more porous rocks such as sandstone or limestone beds. Water is also often present and, being heavier than oil, pushes the oil up the dip of the porous reservoir bed. If the layer above the reservoir bed is impermeable (the cap rock), the oil continues to move within the reservoir bed up dip until it is trapped in the crest of an anticlinal upwarp, or similar trap.

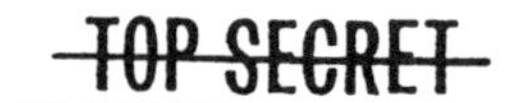

Fig. 4-7 is a stereo pair (3404 and SO-242 records) including an interpretation overlay showing a large uplifted anticlinal fold of the box type mapped in the Jurassic-Cretaceous beds along the outer margin of the Tsaidam Basin. The reddish-brown color of the strata is typical of this continental sequence and provides clues regarding the lithologic character of the various strata. Note that these color signatures are lacking in the 3404 record. The stereo pair here gives good evidence of the need for three-dimensional depth perception for accurate photogeologic structural mapping. Note the deep river canyon cut by erosion across the crest of the fold.

4.5 MINERAL RESOURCES

According to Ch'ang, the Tsaidam region offers considerable mineral resources potential. The Tsaidam Basin proper contains thick Mesozoic-Cenozoic oil bearing deposits. Numerous oil seepages have been reported. Other potential mineral resources indicated by Ch'ang are: various metallic mineral deposits, as indicated by the presence of various acidic to basic igneous rock bodies in the mountainous regions adjacent to the basin; coal in the Mesozoic and Cenozoic strata; and salt, soda, and gypsum in the basin interior.

The following general statements are made with respect to the possible mineral and petroleum potential of the subject area in light of the photogeologic study.

4.5.1 Petroleum

Several producing oil fields are known in the Tsaidam Basin, including the Yuchuantze Oil Field discussed above. No production information was available for the Yuchuantze field. However, the following data for the Leng-Hu Oil Field was found in an [redacted] the Leng-Hu Oil Field, located at about Latitude 38°50' N and Longitude 93°00' E (mid-way between Study Areas A and B) was discovered in 1958. It produces from a "swell" of about three east-west trending elongated anticlines from numerous, very thin (1 to 3 meters) sandstone layers of Tertiary Oligocene age (Kansu) at depths of about 1,000 meters. Each anticline is about 5 kilometers long and is complicated by numerous faults. The quality of the oil is good. Numerous wells have been drilled but few produce commercially. This is due to a high water/oil production ratio. In 1959/1960, the total output of the Tsaidam Basin was 700 tons (approximately 5,200 barrels) of crude oil per day.

The petroleum potential of Area A is restricted to the sedimentary basin in the southeast part of the map sheet, i.e., that part of the area covered by Tertiary Kansu and Jurassic-Cretaceous rocks. The Yuchuantze Oil Field produces from but one of about nine closed anticlinal folds mapped in the Tertiary rocks within Area A. All of these similar folds can be expected to be prolific in relation to the present production. One of these folds, positioned at approximately Latitude 38°20′ N and Longitude 91°30' E, is much better developed than the producing structure. This fold is about 35-kilometers long and 10-kilometers wide, many times larger than the producing Leng-Hu Field east of this area.

It is possible that some of the box-type anticlinal folds mapped in the Jurassic-Cretaceous beds along the margin of the basin might also prove productive, although the sedimentary section will be thin.

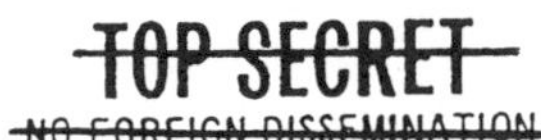

~~HANDLE VIA TALENT-KEYHOLE CONTROL SYSTEM ONLY~~

29. *(Continued)*

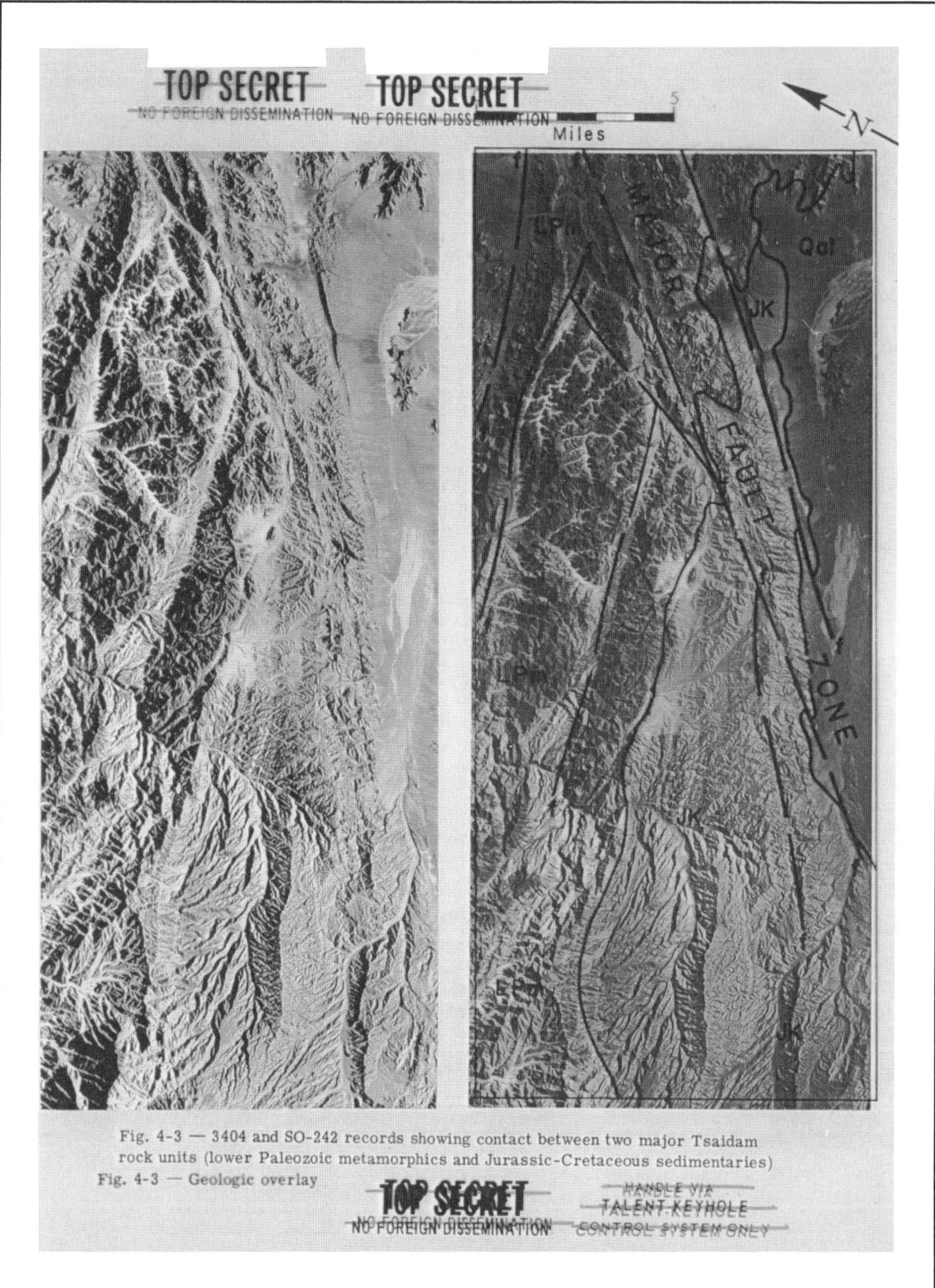

Fig. 4-3 — 3404 and SO-242 records showing contact between two major Tsaidam rock units (lower Paleozoic metamorphics and Jurassic-Cretaceous sedimentaries)

Fig. 4-3 — Geologic overlay

29. *(Continued)*

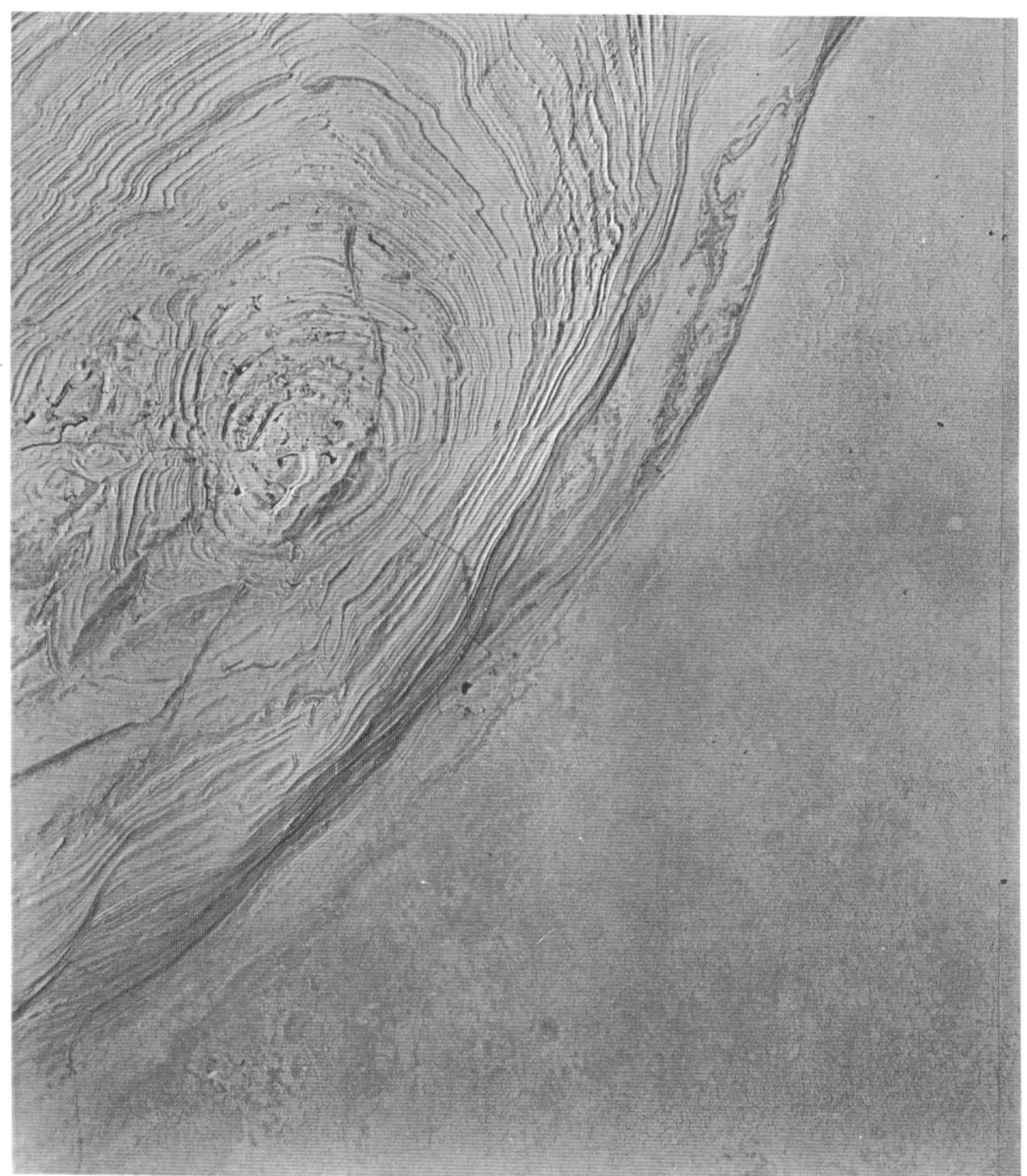

Fig. 4-4 — 3404 record showing classic anticlinal structure

29. *(Continued)*

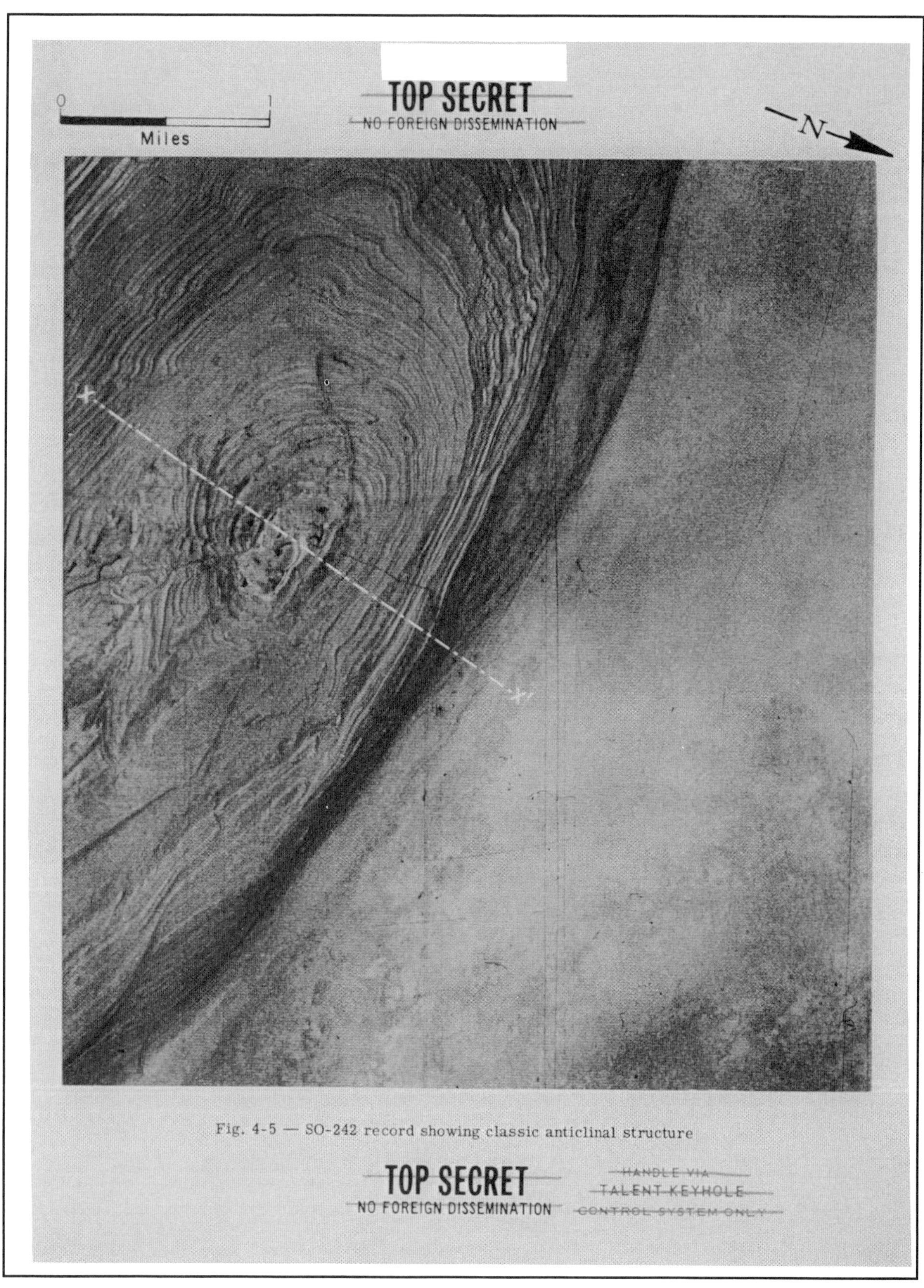

Fig. 4-5 — SO-242 record showing classic anticlinal structure

29. *(Continued)*

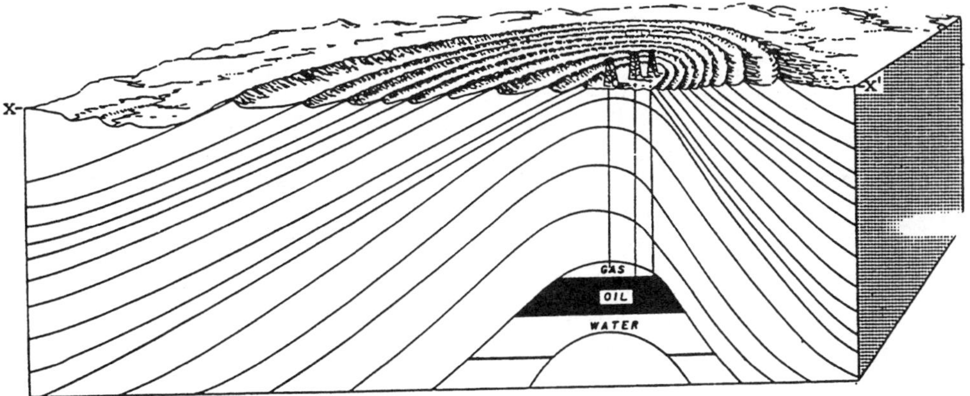

Fig. 4-6 — Schematic drawing—cross section of anticlinal structure, Yuchuantze oil field

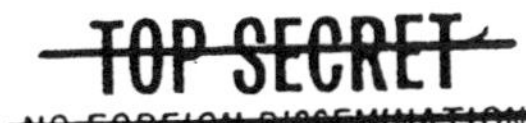

29. *(Continued)*

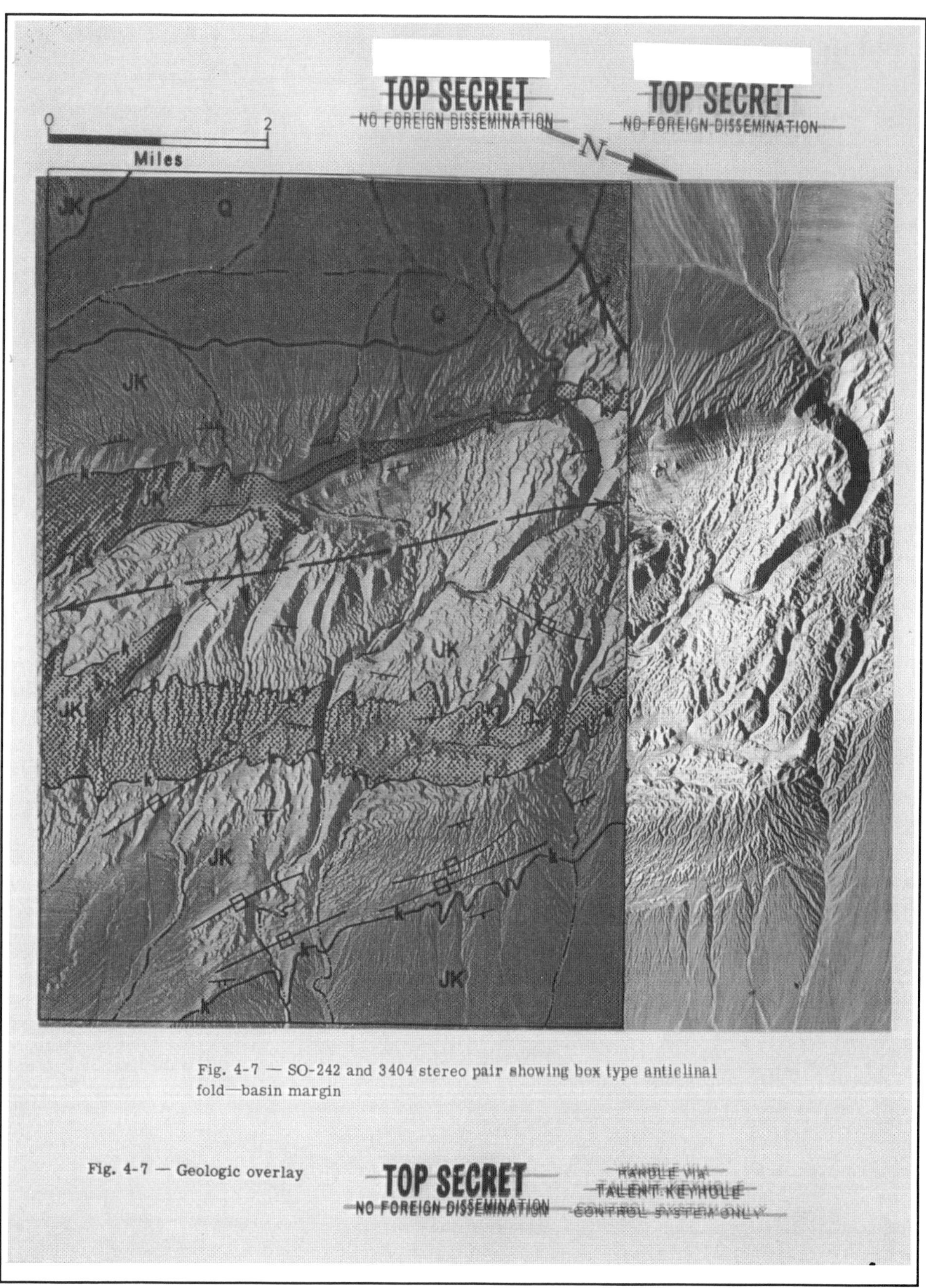

Fig. 4-7 — SO-242 and 3404 stereo pair showing box type anticlinal fold—basin margin

Fig. 4-7 — Geologic overlay

4.5.2 Metallic Minerals

Gold and silver deposits of unknown economic value have been reported* to the west and east of the project. Ch'ang recognizes metallic mineral potential in the vicinity of various igneous rock bodies within the metamorphic rock sequence. From this study, the most favorable areas appear to be along the major fault zones, particularly the northeastward-trending major fault zone crossing the central part of the project, and along the outer edges of the granitic or "PC" zones. dark-toned areas within the Lower Paleozoic sequence might also prove to be favorable areas for metallic mineral concentrations.

4.5.3 Nonmetallic Minerals

Commercial coal bearing beds are reported to occur along the outer edges of the basin within the Jurassic-Cretaceous sequence. Salt, potash, and gypsum in commercial quantities are likely to be found in the vicinity of the modern interior lake basins.

4.5.4 Other

No doubt other mineral possibilities exist in the subject area. The full potential can be thoroughly evaluated by more detailed photogeologic analysis in conjunction with additional ground truth.

*United Nations, "Mineral Distribution Map of Asia and the Far East," 1:5,000,000, 19[illegible].

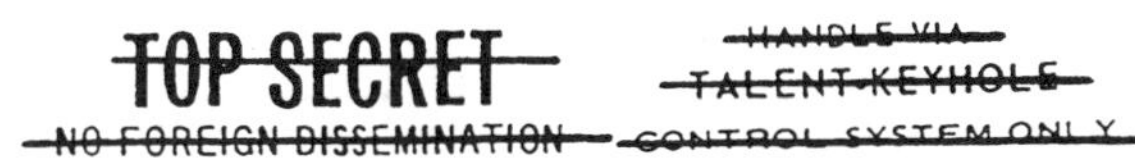

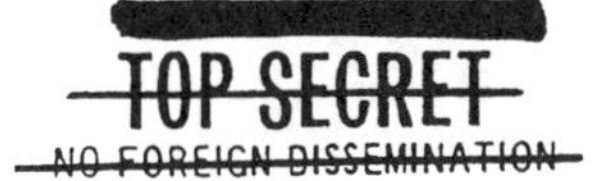

5. AREA B—NE TSAIDAM BASIN

5.1 GENERAL GEOLOGY

The general geologic conditions of the NE Tsaidam Area are similar to those of Area A, the NW Tsaidam Basin Area. As was the case for Area A, "The Geology of China" by Ch'ang Ta* provided the primary reference used for the mapping of Area B.

Fig. 5-1 is a photographic reduction of the completed photogeologic map of Area B.

5.2 ROCKS EXPOSED

The five basic rock units recognized by Ch'ang, described in Section 4.2, occur in Area B as well as Area A. Ch'ang's sketch map, Fig. 4-1, shows the generalized distribution of these rock units over the subject area.

For the photogeologic mapping of Area B, all apparent intrusive igneous rocks were labeled "gr." This undoubtedly includes basic intrusive igneous rocks as well as granites. No attempt has been made at age determination of these rocks.

5.3 STRUCTURE

The area under consideration includes an appreciable segment of the northeast edge of the Tsaidam Basin. The southwestern one-third of the map sheet falls within the basin proper. The remainder of the area is characterized by strongly deformed mountain ranges composed principally of the Lower Paleozoic metamorphic series.

Within the basin, the Tertiary Kansu beds are deformed into elongated faulted anticlines and synclines. Most of these structural features are characteristically aligned, like those of Area A, toward the northwest. In several places, however, east-west trending fault or fracture zones appear to intersect these dominant trends at an angle. The Tertiary folds appear to be more open to the west, becoming tighter and more faulted as the margin of the basin is approached.

In many places the Tertiary rocks are covered by a light veneer of sand dunes, obscuring the underlying structural details. It is possible therefore that more structural folds are present than mapped.

The Jurassic and Cretaceous beds outlining the basin edge are more strongly deformed than in Area A to the west. The folds are tighter and often pass laterally into faults or fault zones.

The older rock sequences adjacent to the basin exhibit complex folding and strong faulting with the dominant trends oriented toward the northwest. A secondary east-west set of fault trends is in evidence in several places within the area.

---

*Op cit.

HANDLE VIA
TALENT-KEYHOLE
CONTROL SYSTEM ONLY

29. *(Continued)*

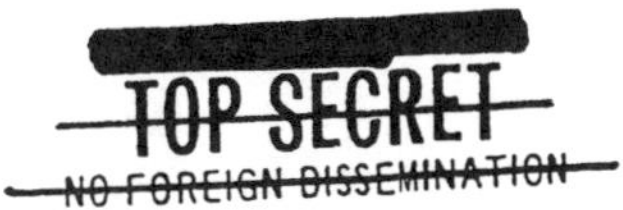

Major fault zones appear to outline the margins of most of the dominant mountain ranges within the area. These are probably of the high-angle reverse type, with the older rocks thrust upon the younger. The contact relationship between the Lower Paleozoic metamorphic rocks and the Jurassic-Cretaceous basin sediments is probably of this type.

A number of igneous intrusive bodies are mapped within the mountain ranges. These are postulated on the basis of their land form, texture, and color characteristics and appear to interrupt the characteristic metamorphic terrain. These probably include basic igneous rocks as well as granite; their relative ages cannot be determined on the basis of the evidence revealed by the photogeologic study.

The northwestern corner of the project area is a topographically low region covered by various Quaternary deposits. Although no Tertiary or Mesozoic rocks were observed here, it is possible that they are present beneath the superficial deposits and that this area represents an isolated, structural re-entrant of the Tsaidam Basin proper.

5.4 ILLUSTRATIONS

The following examples depict some of the more interesting geologic features revealed by the photogeologic evaluation in the subject area. Their location is depicted in Fig. 5-1.

Fig. 5-2 is a dual illustration (3404 and SO-242 records) including an interpretation overlay showing the contact zone between the Lower Paleozoic metamorphic rocks (dark-toned, mottled, and highly fractured) and the younger Jurassic-Cretaceous sedimentary strata (red, gray, and reddish-brown banding). Note how the distinct color banding highlights the tight folds in the younger beds. Within the older rock sequence, mineralization might occur along the interfaces between the light and dark banding and along the major fault traces.

Fig. 5-3 is a dual illustration (3404 and SO-242 records) and an interpretation overlay showing a possible mineralized zone along a prominent fault or dike. Note the white bleached-out alignment within the dark-toned Lower Paleozoic metamorphic host rock. An area of igneous intrusion is postulated on the upper right. Mineralization might also exist along the interface between the postulated igneous intrusive rock and the metamorphic host rock, as well as along the more prominent fault and fractures. Zones of intersection between the major fracture zones might be the most favorable areas for mineralization.

Fig. 5-4 is a dual illustration (3404 and SO-242 records) and an interpretation overlay of an apparent mineralized area in the vicinity of an igneous intrusion and fault zone. Note the high-standing, dark-toned area in the lower part of the photograph, interpreted to be an area of basic intrusive igneous rocks. This area, occurring within the Lower Paleozoic metamorphic host rock, is encircled by arcuate, annular streams. This is indicative of the uplift and resistant nature of the igneous intrusion. Along the northern edge of the igneous body is a prominent fault zone, oriented in a northwest-southeast direction. Within this zone are numerous areas exhibiting a brick-red staining, distinctly visible on the color film but indistinguishable on the black-and-white record. The red staining suggests a high concentration of iron oxide, possibly accompanying a concentration of other heavy metallic minerals. Note the faint red staining within the alluvium of the main river flood plain. The smaller streams obviously carry the heavy minerals to the main river, dropping their load on the near bank as they reach the lower level. Note the snake-like outcrop in the upper right corner of the photograph. This is interpreted to be a linear zone of folded sedimentary rocks, possible of Devonian-Permian age (designated D-P).

HANDLE VIA
TALENT-KEYHOLE
CONTROL SYSTEM ONLY

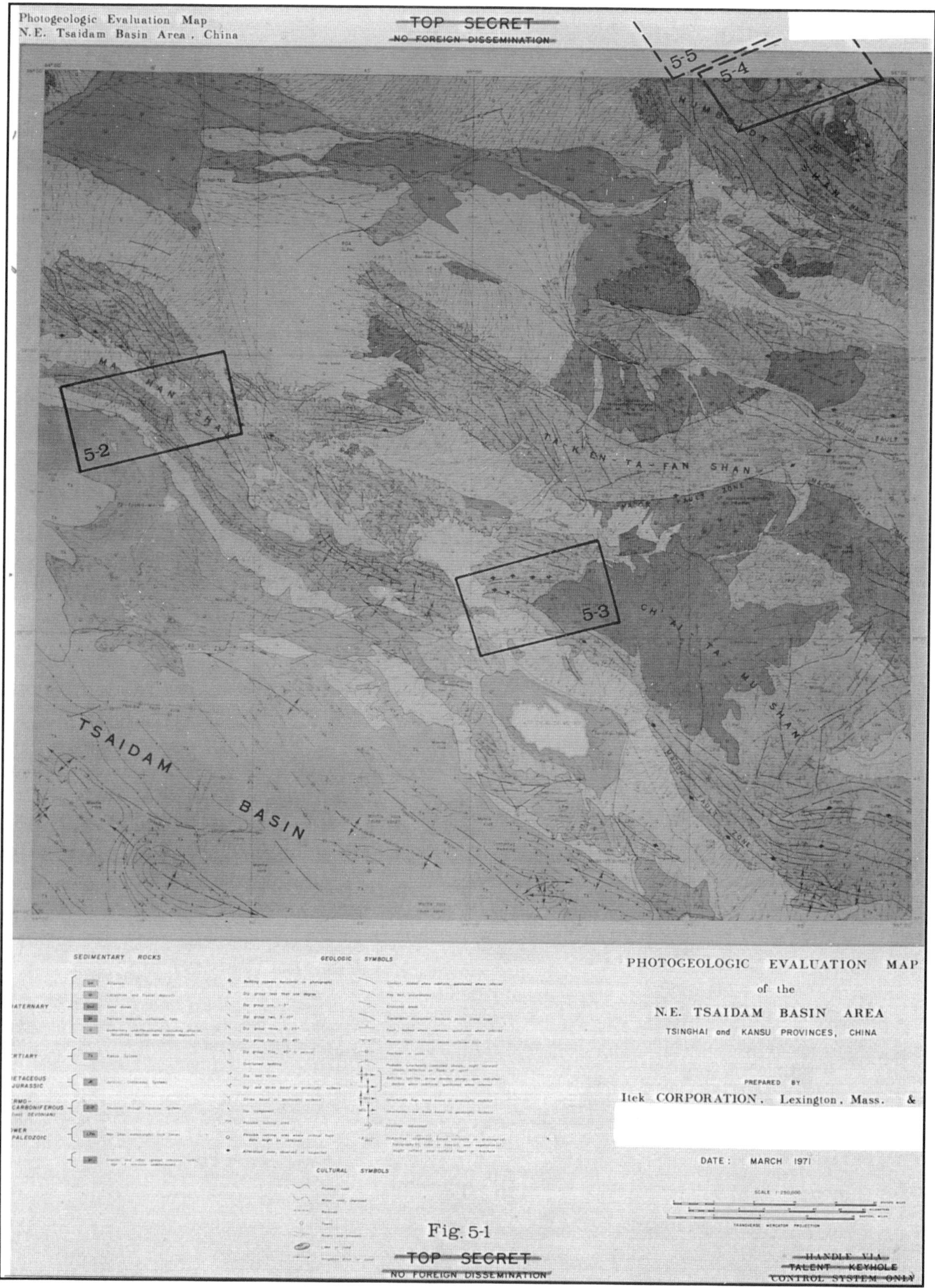

Photogeologic Evaluation Map
N.E. Tsaidam Basin Area, China
TOP SECRET
NO FOREIGN DISSEMINATION
5-5
5-4
5-2
5-3
MA HAN SHAN
TA KEN TA FAN SHAN
CH AI TA MU SHAN
TSAIDAM BASIN
SEDIMENTARY ROCKS
GEOLOGIC SYMBOLS
CULTURAL SYMBOLS
PHOTOGEOLOGIC EVALUATION MAP
of the
N.E. TSAIDAM BASIN AREA
TSINGHAI and KANSU PROVINCES, CHINA
PREPARED BY
Itek CORPORATION, Lexington, Mass. &
DATE: MARCH 1971
TOP SECRET
NO FOREIGN DISSEMINATION
HANDLE VIA
TALENT KEYHOLE
CONTROL SYSTEM ONLY

Fig. 5-1

-NO FOREIGN DISSEMINATION-

Fig. 5-5 is a stereo pair (3404 and SO-242 records) including an interpretation overlay showing an apparent mineralized area in the vicinity of an igneous intrusion and fault zone. This example is located immediately northwest of Fig. 5-4. It lies along the same fault trend as in Fig. 5-4 and exhibits similar igneous activity and mineralization. This might be only part of a regional northwest-trending mineralized zone along the northeastern edge of the Humboldt Shan, in the northeast corner of the map sheet. The stereo pair depicts the strong relief in the mountains bordering the Tsaidam Basin. It also points out the usefulness of stereoscopy for accurate photogeologic mapping.

5.5 MINERAL RESOURCES

The following general statements are made with respect to the possible petroleum and mineral potential of Area B in light of the evidence revealed by the photogeologic study.

5.5.1 Petroleum

The southwest one-third of the subject area has good petroleum potential. Only that part of the area covered by Tertiary Kansu rocks is considered prospective however. The Jurassic-Cretaceous sequence appears to be too strongly deformed to be prospective. Several broad anticlinal folds mapped within the Tertiary Kansu rocks are considered most favorable as traps for the accumulation of hydrocarbons. Those nearer the southwest edge of the mapped area are broader and appear to be less fault controlled than those situated near the basin margin.

The broad topographically low area in the northwest corner of the map sheet should totally discounted for possible petroleum accumulations. This might be an isolated arm of the Tsaidam Basin and might contain fairly thick sequences of Mesozoic and Tertiary rocks beneath the unconsolidated Quaternary materials.

5.5.2 Metallic Minerals

The mountainous region of Area B offers excellent potential for metallic mineral concentrations. Most of the mountain ranges are composed of Lower Paleozoic metamorphic rocks and have undergone repeated and complex deformation. Considerable igneous intrusive activi. apparent in many areas. The margins of most of the ranges are outlined by major fault zones. The most prospective areas are along the major fault zones and at their points of intersection with secondary fault or fracture belts. The contact, or interfaces, between the postulated intrusive igneous rocks and the Lower Paleozoic metamorphic sequence are likewise most prospective.

Of particular interest are: (1) the northwest-trending mineralized fault zone in the northeast corner of the area, as depicted by Figs. 5-4 and 5-5; (2) the possible mineralized zone shown in Fig. 5-3; and (3) the numerous areas of apparent alteration indicated on the photogeologic maps.

5.5.3 Nonmetallic Minerals

Commercial coal beds might exist in a few places within the Jurassic-Cretaceous sequence rimming the basin proper. Salt, potash and gypsum might be found in commercial quantities in the vicinity of the modern interior lake basins.

5.5.4 Other

No doubt other mineral possibilities exist in the area. This study could be improved im-surably with any additional ground truth available, such as, any other previous mining activities, information on the composition of some of the rock suites within the Lower Paleozoic metamorphic sequence, etc.

~~TOP SECRET~~
~~NO FOREIGN DISSEMINATION~~

~~HANDLE VIA~~
~~TALENT KEYHOLE~~
~~CONTROL SYSTEM ONLY~~

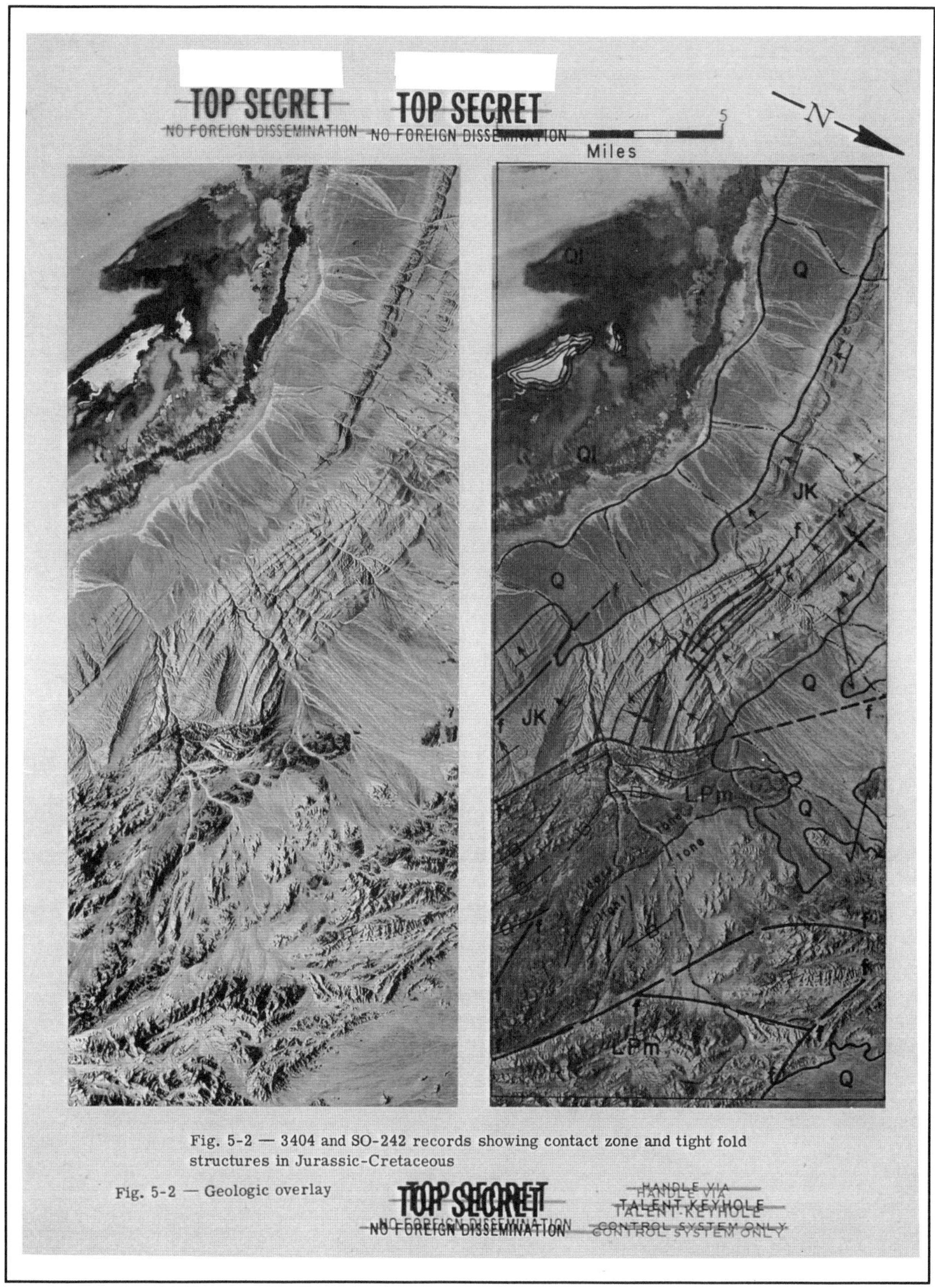

Fig. 5-2 — 3404 and SO-242 records showing contact zone and tight fold structures in Jurassic-Cretaceous

Fig. 5-2 — Geologic overlay

29. *(Continued)*

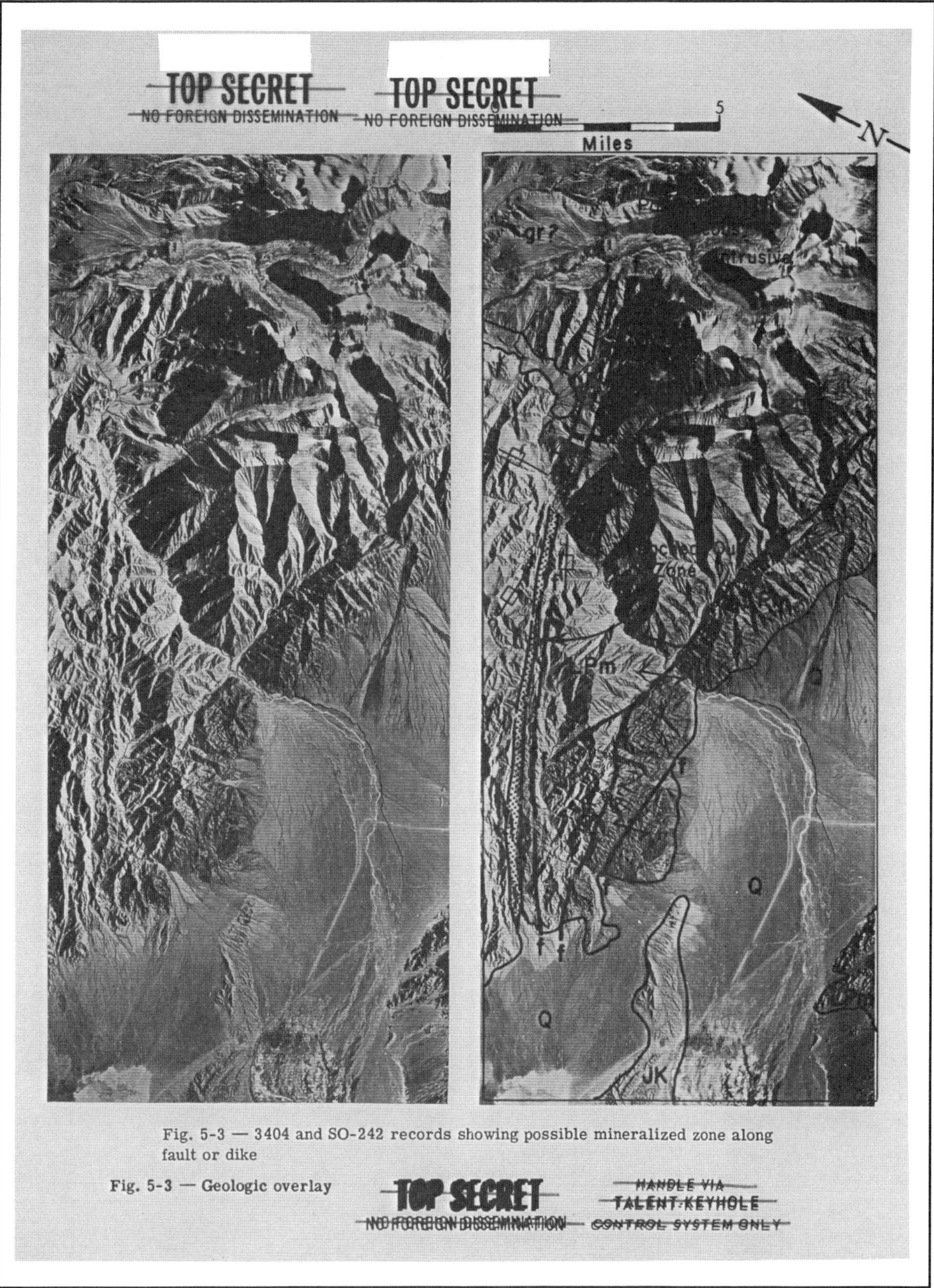

Fig. 5-3 — 3404 and SO-242 records showing possible mineralized zone along fault or dike

Fig. 5-3 — Geologic overlay

29. *(Continued)*

Fig. 5-4 — 3404 and SO-242 records showing apparent mineralization (red stains) in vicinity of igneous intrusion

Fig. 5-4 — Geologic overlay

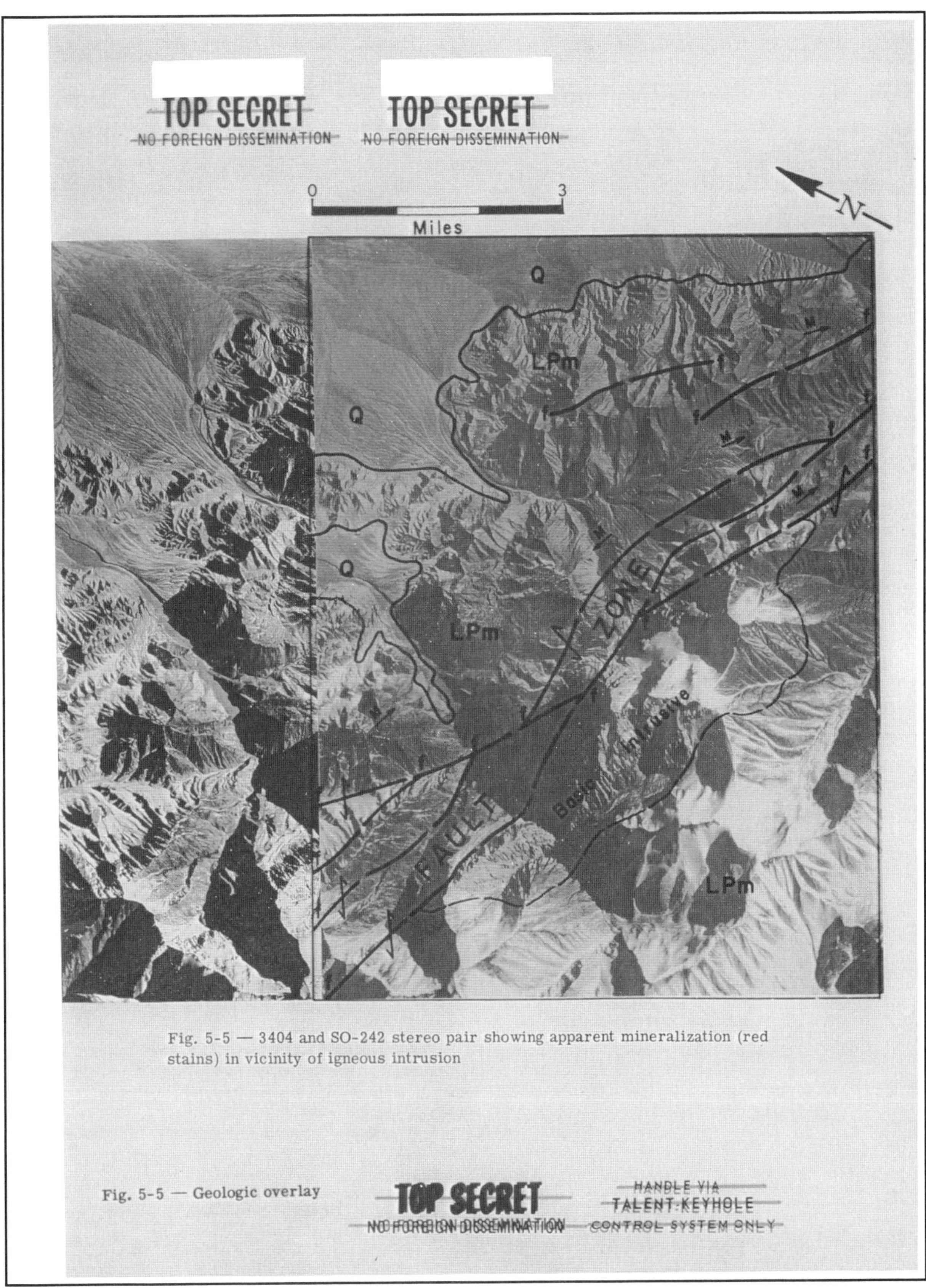

Fig. 5-5 — 3404 and SO-242 stereo pair showing apparent mineralization (red stains) in vicinity of igneous intrusion

Fig. 5-5 — Geologic overlay

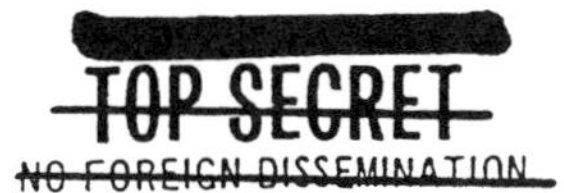

6. AREA C—KAFIRNIGAN-PYANDZH AREA

6.1 GENERAL GEOLOGY

The Kafirnigan-Pyandzh Area falls principally within the Tadzhik Region of the USSR but includes on the south a small part of Afghanistan. The project name relates to the Kafirnigan River, which crosses the western part of the area, and on the south the Pyandzh River, separating Afghanistan from the USSR.

The background research effort was most fruitful for this study. Personnel from OBGI in Washington, D. C. located and provided an excellent geologic reference for the project, i.e., Terrain Atlas, Kafirnigan Area, USSR (C), 1969, sponsored by the Advanced Research Projects Agency, ARPA Order No. 485. The Atlas was produced to provide earth-science data for evaluating the geologic environment in terms of its potential for secret underground nuclear testing. The report is based largely on previously published earth-science data, and its great value was the synthesis and interpretation of the basic information. Although this reference only covers the northwest part of the area, the ground truth it provided proved an excellent guide to the mapping of the area as a whole.

According to the ARPA Report, the subject area lies within the Tadzhik Depression in the eastern part of the vast Scytho-Turanian Platform. It contains a very thick sequence of Mesozoic and Tertiary sedimentary rocks, marine below and continental at the top. The formations were subject to Alpine folding which in this area culminated in late Tertiary time. Linear elongated folds, associated with high-angle reverse faulting were produced, resulting in a rugged linear terrain. The ridges are closely spaced in the north and tend to diverge farther south. This phenomenon, resembling the spreading fingers of a hand, is called the Tadzhik Virgation.

6.2 ROCKS EXPOSED

Geologically this area is quite different from the Tsaidam Basin. The rocks exposed are entirely of the sedimentary type and no igneous or metamorphic (hard rock) areas are exposed. The sedimentary sequence includes rocks of Jurassic, Cretaceous, and Teritary ages overlain in places by various Quaternary deposits. The sequence is characterized by seven individual formations (or units), each of which has its own identifying lithologic characteristics. The stratigraphic sequence, from oldest to youngest, is as follows.

6.2.1 Upper Jurassic Undifferentiated—(Ju)

This is the oldest sequence in the project. It includes gypsum with thin beds of gypsiferous claystone, and local rock-salt beds (20 to 30 meters exposed). Most exposures are associated with major reverse faults and often occur below an irregular boundary marked by dome-like swellings separated by saddles reflecting in the overlying younger strata.

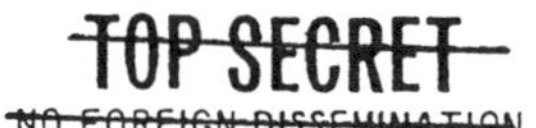

HANDLE VIA
TALENT KEYHOLE
CONTROL SYSTEM ONLY

29. *(Continued)*

6.2.2 Lower Cretaceous Undifferentiated—(K_1)

This is relatively a thick complex of interbedded red claystone and red and gray sandstone, laterally very variable in lithology and in thickness of individual lithologic units. The total thickness remains rather constant however, from 500 to 700 meters. This sequence is commonly exposed below the Upper Cretaceous in the eroded, eastern flanks of the mountain-forming anticlines.

6.2.3 Upper Cretaceous Undifferentiated—(K_2)

This sequence includes largely grayish-claystone commonly interbedded with sandstone or limestone with occasional interbeds of gypsum, capped in places by a thin sequence of interbedded limestone and gypsum. The thickness varies from 400 to 1,300 meters. These beds are generally exposed in the relatively steep east-facing slopes of the mountain ranges, below the Bukhara limestone cap rock.

6.2.4 Bukhara Limestone (Paleocene)—(Tb_1)

This is a hard dense gray limestone with dolomite and gypsum interbeds. This dark-toned, resistant formation is the main ridge-former, capping the crests of nearly all of the anticlinal mountain ranges within the region.

6.2.5 Eocene-Lower and Middle Oligocene Undifferentiated—(Tc)

This sequence is composed primarily of vari-colored marine claystone with occasional beds of limestone, marl, and sandstone. It ranges in thickness from 365 to 500 meters. It is typically light-toned and moderately resistant, oftenforming V-shaped hogbacks along the western flanks of the Bukhara anticlinal ridges.

6.2.6 Bol'dzhuan Formation (Upper Oligocene-Lower Miocene)—(Tbs)

This is a continental facies sequence composed of maroon, wine-red, or brick-red sandstone and siltstone, often including claystone. Its thickness varies greatly from 220 to 1,000 meters. It usually crops out in bands of varying widths, indicating considerable variation in thickness, on the western mountain slopes above the more resistant marine rocks of the Tc unit.

6.2.7 Garauty Formation (Miocene)—(Tg)

This is a continental sandstone and siltstone sequence, generally light brown to tan in color. It varies in thickness from 435 to 1,800 meters and rests on the eroded upper surface of the Bol'dzhuan formation, generally along the gentle western dip-slopes of the mountain ranges.

Most of these rock units are considerably mantled by various Quaternary deposits, the most widespread of which is the Dushanbe and Ilyak Series, a thick loess deposit.

Although the ARPA Report, from which the above descriptions are summarized, only covers the northwest part of the area under consideration, the ground truth it provided proved an excellent guide to the mapping of the entire area. From the lithological descriptions above, it was possible to identify the various formations outside the area of the ARPA Report and map the entire area, probably more accurately than it had ever been done before. Fig. 6-1 is a photographic reduction of the photogeologic map.

29. *(Continued)*

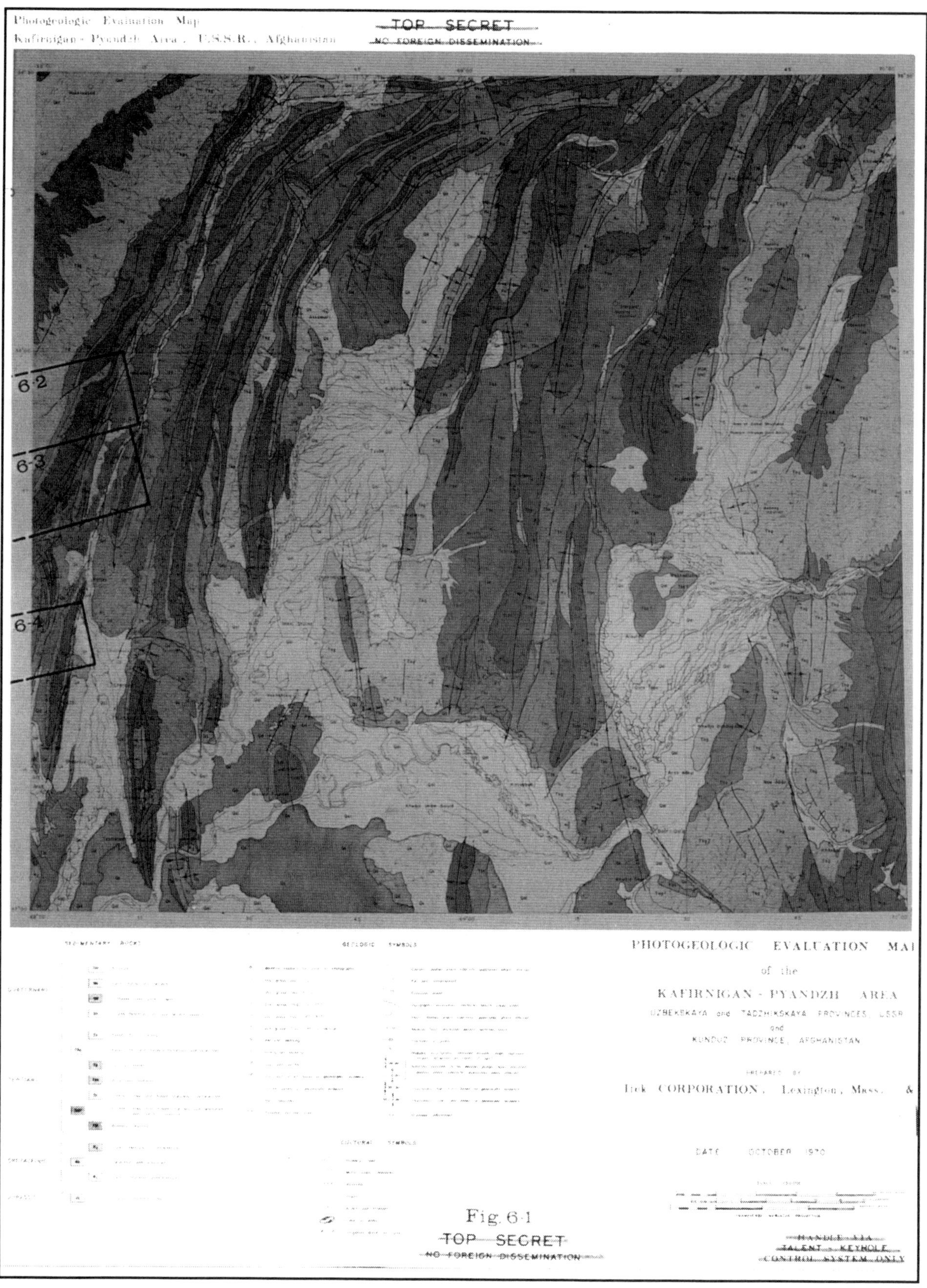

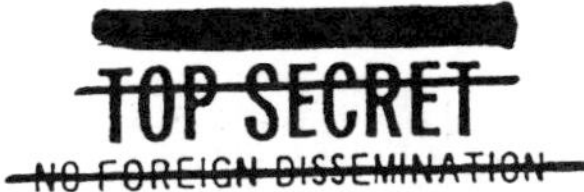

6.3 STRUCTURE

The regional structure of the subject area is relatively simple. In detail it is complicated and only partly understood. The linear, subparallel elongated folds of the region are generally outlined on the east by long, probably high-angle, reverse faults, down-thrown on the east. These faults cut all formations of Tertiary age or older and commonly form rugged fault scarps. A number of normal faults of considerable length were mapped along the west edge of some of the major structures. These linear faulted anticlines are closely spaced in the north but tend to diverge and become more open farther south. In the northern part of the subject area, several east-west trending strike-slip faults were found. These appear to be left-lateral structural features associated with the northern zone of major structural change.

6.4 ILLUSTRATIONS

The following examples depict some of the more important stratigraphic and structural features revealed by the photogeologic evaluation within the subject area. These, more than any others, vividly portray the value of color for photogeologic mapping with the KH-4B System. Their locations are depicted on Fig. 6-1

Fig. 6-2 is a dual illustration and interpretation overlay depicting graphically the value of color for distinguishing between the various rock formations within a given area. The black and white photograph on the left is useful up to a point. The bold mountain-forming Bukhara limestone is easily identified by its topographic prominence, as it forms the "backbone" of most of the linear mountain ranges in the region. Likewise, the Eocene claystone unit, labeled "Tc," is identifiable by its V-shaped hogback ridges. Above these marine units the stratigraphic sequence changes to a continental facies. The change in color reflects this characteristic. Note the deep maroon-red color of the Bol'dzhuan formation and how easily it can be distinguished from the overlying Garauty formation on the SO-242 color imagery. This contact (interface) is virtually indistinguishable on the 3404 record on the left. The deep red signature of the Bol'dzhuan formation proved to be the most reliable mapping marker within the project.

Fig. 6-3 is a dual illustration and an interpretation overlay showing essentially the same part of the stratigraphic section as in Fig. 6-2. Observe the continuity of formational color and topographic characteristics. Note how the prominent backbone of the mountain ridge is formed on the characteristic Bukhara limestone. The west flank is the dip-slope and the east flank is the rugged and highly-faulted obsequent slope. The Lower Cretaceous rocks beneath the Bukhara are relatively easily eroded and do not display recognizable identifying characteristics.

Fig. 6-4 is a stereo-pair and an interpretation overlay depicting an elongated, faulted anticline along the west edge of the study area. The bold Bukhara limestone forms the backbone of the anticlinal mountain range. The V-shaped hogbacks etched by erosion on the "Tc" unit encircle the prominent uplift. The reddish-hued Bol'dzhuan formation is apparent on the west flank, even though it is heavily mantled by Quaternary loess deposits.

This structure typifies the characteristic structural forms found within the area. The linear faulted anticlinal ranges broaden toward the south and become more prospective for the entrapment of hydrocarbons. To the north they become tighter and more highly faulted, thus diminishing their petroleum potential.

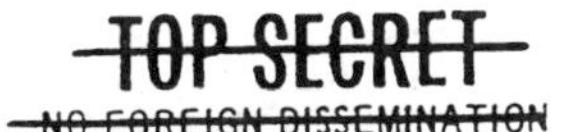

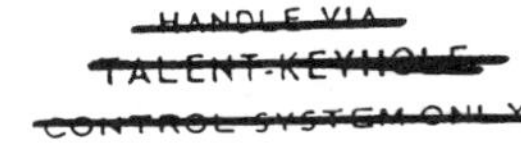

29. *(Continued)*

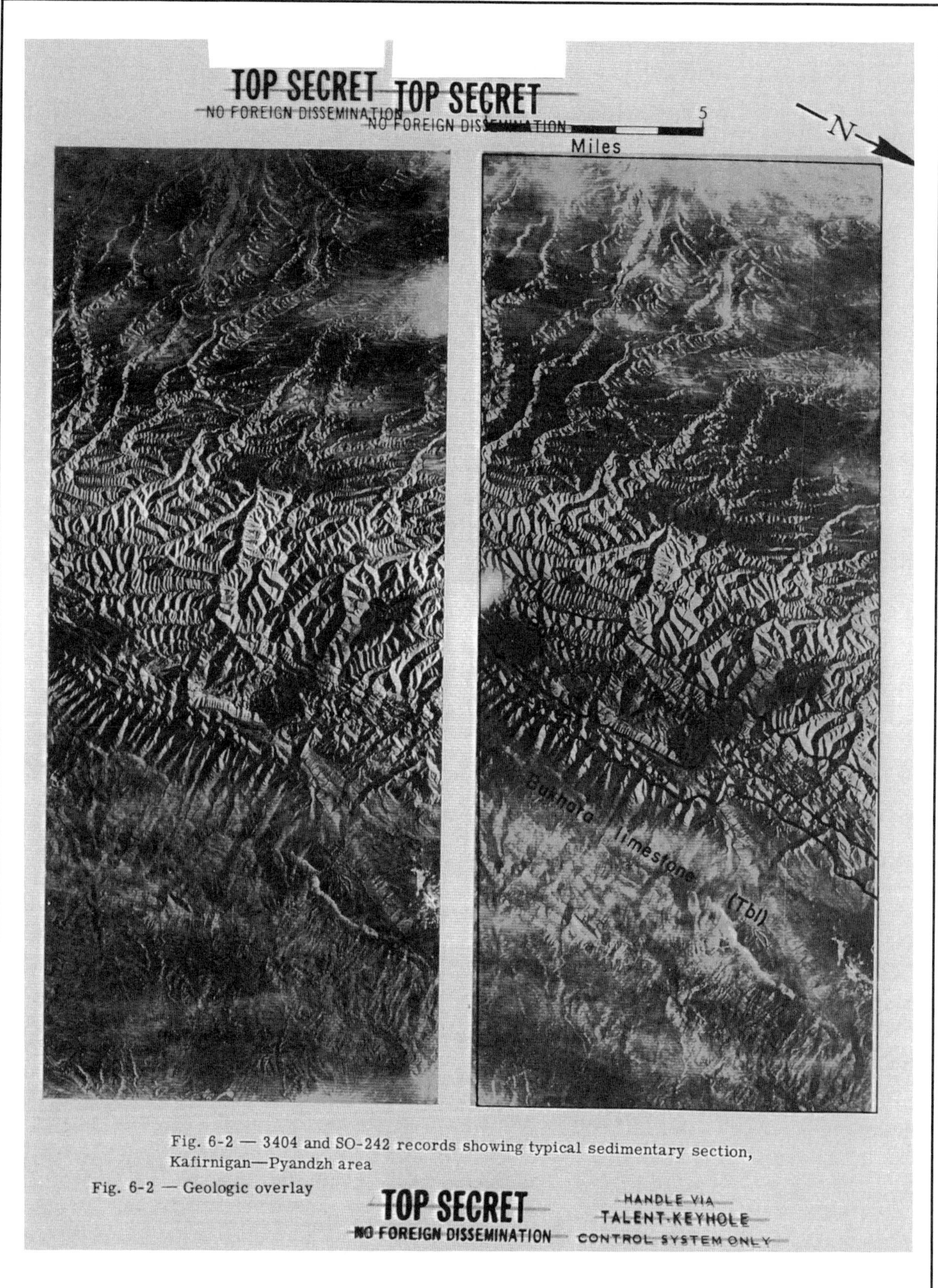

Fig. 6-2 — 3404 and SO-242 records showing typical sedimentary section, Kafirnigan—Pyandzh area

Fig. 6-2 — Geologic overlay

6.5 MINERAL RESOURCES

As a result of the photogeologic mapping together with the information contained in the ARPA Report, the following general statements can be made with respect to the mineral and petroleum potential of the area.

6.5.1 Petroleum and Natural Gas

Oil and gas are being produced from anticlinal structures east and west of the study area. No production data has been obtained for these fields. The numerous elongated anticlinal folds of the region are excellent prospects where closure exists and where faulting is not too severe. Therefore, the southern part of the area is most prospective since the folds broaden in that direction.

6.5.2 Metallic Minerals

The potential for metallic minerals within the region is not known. No igneous or metamorphic rocks have been reported in the area. The greatest potential for metallic mineral concentrations would likely be along the northern margin of the area where the structural deformation is known to be strongest.

6.5.3 Nonmetallic Minerals

Some local bituminous coal deposits are mined north of the area, but the coal potential for most of the region is slight. Sand, gravel, and loess deposits are plentiful from the various Quaternary materials widely distributed across the area. Brick clay is likely abundant from the upper Cenozoic and Quaternary deposits. Lime, marl, dolomite, and building stone is plentiful from the Bukhara and "Tc" formations. Gypsum and rock salt are available from the Jurassic and Cretaceous outcrops as well as the Bukhara limestone.

6.5.4 Other

Although doubtless other mineral possibilities exist for the area, they cannot be realistically appraised without additional ground truth.

HANDLE VIA
TALENT-KEYHOLE
CONTROL SYSTEM ONLY

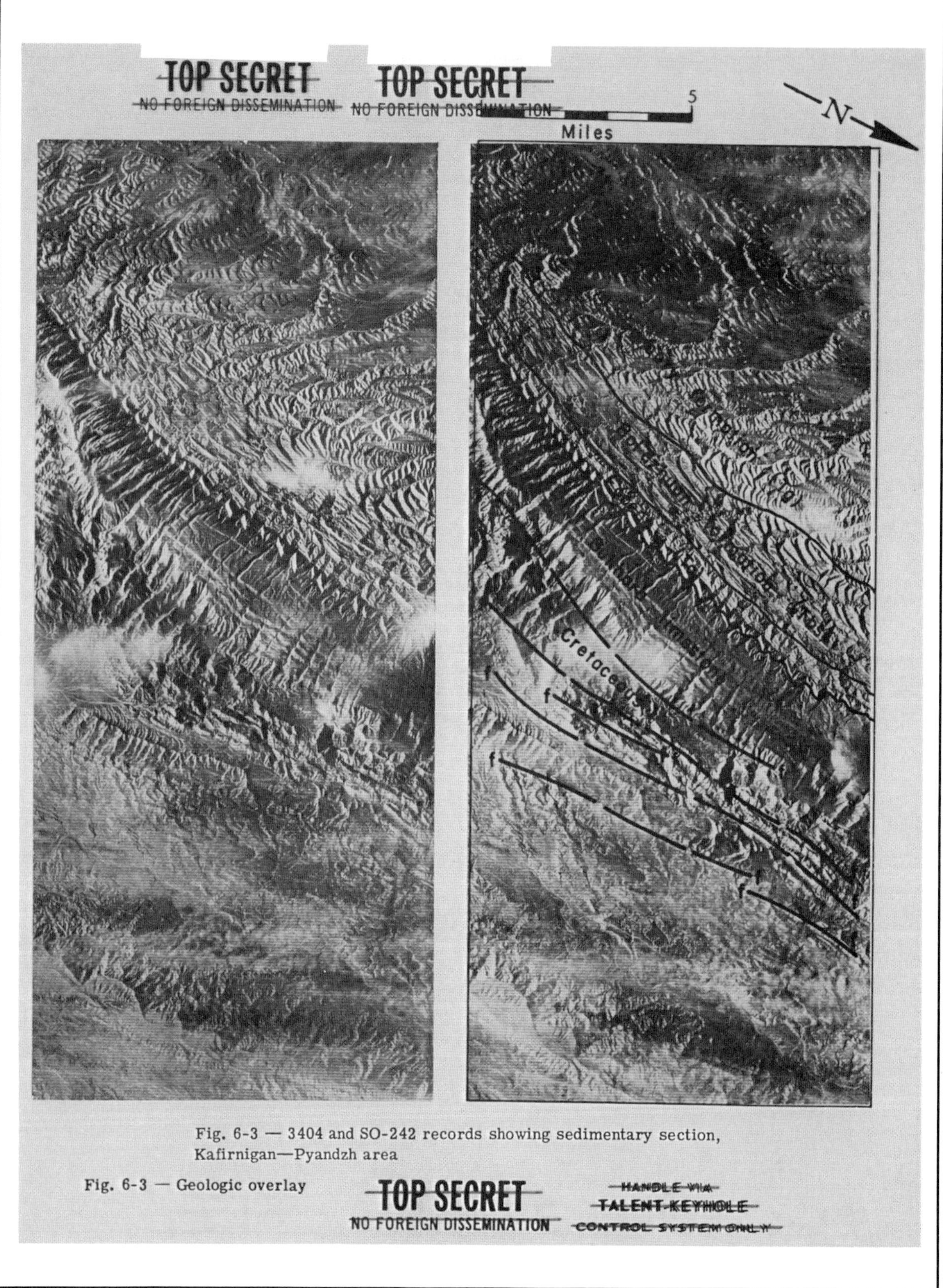

Fig. 6-3 — 3404 and SO-242 records showing sedimentary section, Kafirnigan—Pyandzh area

Fig. 6-3 — Geologic overlay

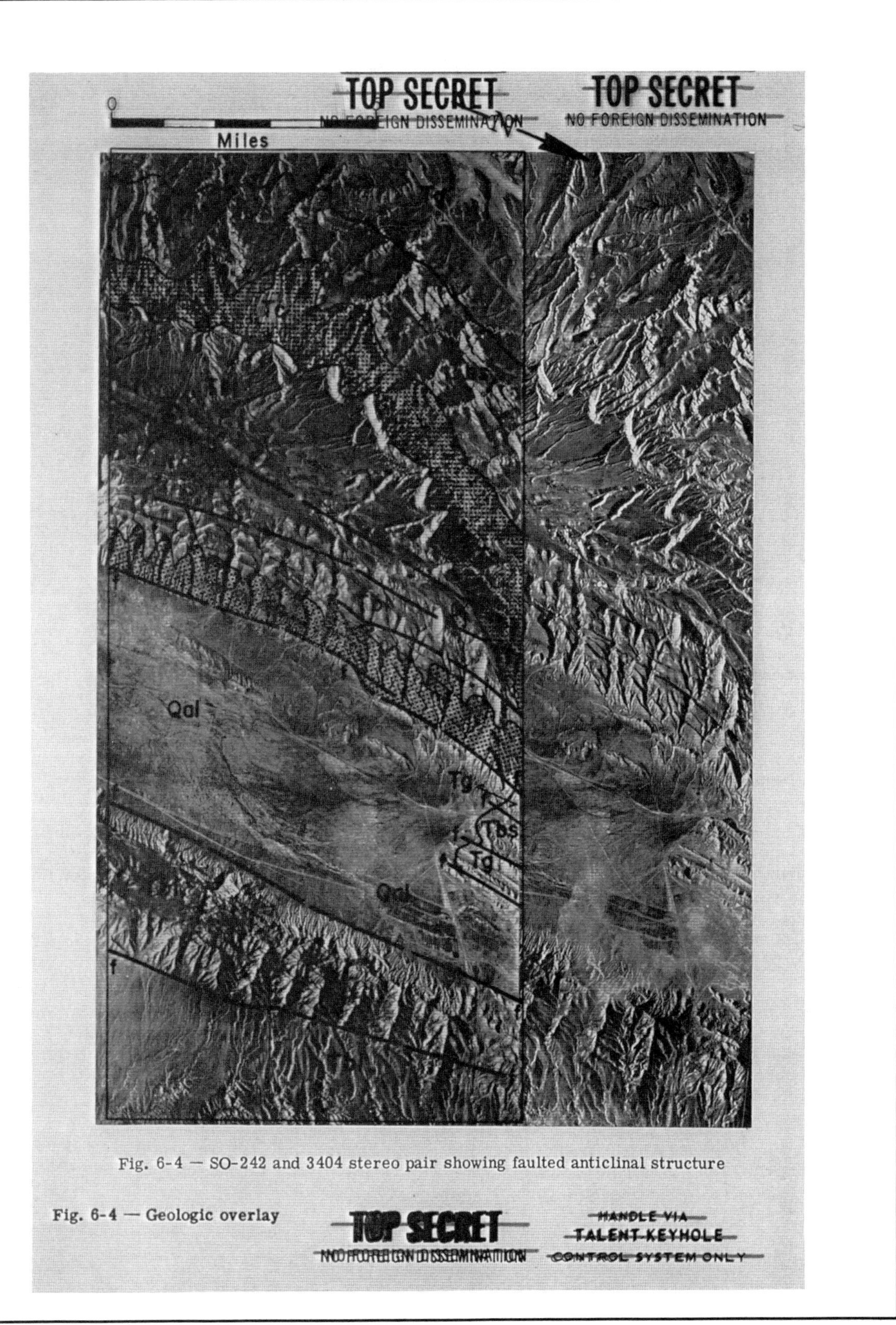

Fig. 6-4 — SO-242 and 3404 stereo pair showing faulted anticlinal structure

Fig. 6-4 — Geologic overlay

7. CONCLUSIONS

The following conclusions have been reached from the foregoing study.

7.1 EFFECTIVE PHOTOGEOLOGIC MAPPING CAN BE ACHIEVED USING THE KH-4B SYSTEM

The results of the three photogeologic mapping projects indicate that the KH-4B System, originally developed for military intelligence purposes, is uniquely well suited to photogeologic mapping. The completed photogeologic maps are immediately useable for assessing the mineral and petroleum potential of the subject areas. They could be presently used, if feasible, to make new mineral discoveries. The quality of the maps, however, could be improved by employing the best aspects of interpretation and compliation learned from these initial studies.

Comparing these studies with conventional photogeologic mapping projects provides some interesting insights into the efficiency of using the KH-4B materials. In total area, these three studies embrace approximately 36,000 square miles. A standard photogeologic study using conventional aerial photography at the basic approximate scale of 1:40,000 would require an experienced photogeologist to expend approximately 3 man-years and would require him to analyze about 3,600 aerial photographs. This is compared with him using approximately 72 PAN images and expending about 3 months to achieve essentially comparable results using the KH-4B System. This is to say that a photogeologic study using the KH-4B material would require from 10 to 15 percent expenditure of time and money as compared to using conventional aerial photography.

For regional photogeologic mapping, the main advantages of the KH-4B System are: (1) overall synoptic view, (2) polar orbit (no inaccessible areas), (3) stereoscopic perspective, (4) vertical (rectified), (5) resolution, and (6) color. The first four of these, while important, are not necessarily indigenous to this system but are typical of other space photography systems. The unique qualities of the KH-4B System, resolution and color, are most important, as discussed below.

7.2 THE VALUE OF COLOR CANNOT BE OVERSTATED

For photogeologic mapping, the use of color photography has distinct advantages over black and white. Color provides: (1) easier differentiation between rock types; (2) more accurate tracing of individual sedimentary beds; (3) more definitive clues as to the exact nature of lithology (rock type), and hence is far more valuable in areas of limited ground truth; (4) better identification of specific formation signatures, and (5) oxidation halos and discoloration zones indicative of possible mineralization.

It is recognized that the use of the SO-242 color film has resulted in reduced resolution from the 3404 operational standard. For photogeologic analysis, however, this loss of resolution is insignificant when compared to the interpretive value gained by color. Resolution of 20 to 30

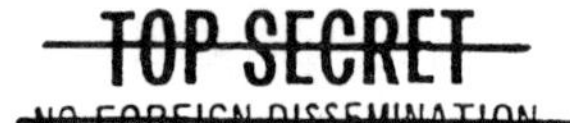

HANDLE VIA
TALENT-KEYHOLE
CONTROL SYSTEM ONLY

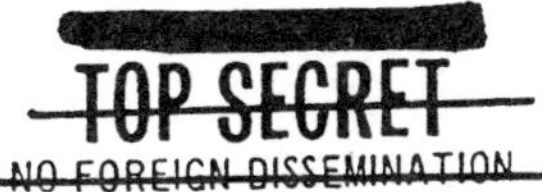

feet is entirely adequate for most regional photogeologic mapping projects, and the SO-242 color film easily meets that standard. By using both the 3404 and SO-242 film in a stereoscopic mode, as was done for this study, the advantages of both are obtained with very little sacrifice of the useful qualities of each type.

7.3 THE KH-4B SYSTEM REPRESENTS AN IMPORTANT BREAKTHROUGH FOR NATIONAL RESOURCES EXPLORATION

The economic and political impact of this cannot be overstated. While the world-wide demand increases dramatically for minerals and fossil fuels (those resources in fixed supply), our ability to locate and harvest these hidden deposits lags far behind.

Experts agree that exploration from space offers a potential breakthrough in large scale explorationtechniques. Virtually every major exploration advance inthe last 20 years has been on-the-ground detectors of one sort or another. These are detailing geophysical tools, whose use is very expensive in relation to area analyzed, and must be used selectively. A prerequisite to their proper and efficient use is a conduct of effective preliminary reconnaissance studies to localize areas of most promise.

Exploration from space provides an enlarged prospective, a previously unattainable synoptic view of the earth. Though the geologists' discipline is a study of the earth, until now he has never seen it. With his vision broadened from this space perspective, he is enabled to search for oil provinces instead of oil fields and mineral districts instead of mineral deposits.

The barrier of inaccessibility has been broken. No area is inaccessible or too remote for the polar-orbiting satellite. Now the entire earth is the geologist's true laboratory. The dramatic oil discovery at Prudhoe Bay, north of the Arctic Circle in Alaska, and the subsequent $900 million investment in adjacent land by oil companies indicate that no areas are too remote for raw materials exploration.

~~TOP SECRET~~
~~NO FOREIGN DISSEMINATION~~

~~HANDLE VIA~~
~~TALENT-KEYHOLE~~
~~CONTROL SYSTEM ONLY~~

Appendix

Appendix: Executive Order 12951, Release of Imagery Acquired by Space-Based National Intelligence Reconnaissance Systems, 22 February 1995

10789

Federal Register
Vol. 60, No. 39
Tuesday, February 28, 1995

Presidential Documents

Title 3—

The President

Executive Order 12951 of February 22, 1995

Release of Imagery Acquired by Space-Based National Intelligence Reconnaissance Systems

By the authority vested in me as President by the Constitution and the laws of the United States of America and in order to release certain scientifically or environmentally useful imagery acquired by space-based national intelligence reconnaissance systems, consistent with the national security, it is hereby ordered as follows:

Section 1. *Public Release of Historical Intelligence Imagery.* Imagery acquired by the space-based national intelligence reconnaissance systems known as the Corona, Argon, and Lanyard missions shall, within 18 months of the date of this order, be declassified and transferred to the National Archives and Records Administration with a copy sent to the United States Geological Survey of the Department of the Interior consistent with procedures approved by the Director of Central Intelligence and the Archivist of the United States. Upon transfer, such imagery shall be deemed declassified and shall be made available to the public.

Sec. 2. *Review for Future Public Release of Intelligence Imagery.* (a) All information that meets the criteria in section 2(b) of this order shall be kept secret in the interests of national defense and foreign policy until deemed otherwise by the Director of Central Intelligence. In consultation with the Secretaries of State and Defense, the Director of Central Intelligence shall establish a comprehensive program for the periodic review of imagery from systems other than the Corona, Argon, and Lanyard missions, with the objective of making available to the public as much imagery as possible consistent with the interests of national defense and foreign policy. For imagery from obsolete broad-area film-return systems other than Corona, Argon, and Lanyard missions, this review shall be completed within 5 years of the date of this order. Review of imagery from any other system that the Director of Central Intelligence deems to be obsolete shall be accomplished according to a timetable established by the Director of Central Intelligence. The Director of Central Intelligence shall report annually to the President on the implementation of this order.

(b) The criteria referred to in section 2(a) of this order consist of the following: imagery acquired by a space-based national intelligence reconnaissance system other than the Corona, Argon, and Lanyard missions.

Sec. 3. *General Provisions.* (a) This order prescribes a comprehensive and exclusive system for the public release of imagery acquired by space-based national intelligence reconnaissance systems. This order is the exclusive Executive order governing the public release of imagery for purposes of section 552(b)(1) of the Freedom of Information Act.

(b) Nothing contained in this order shall create any right or benefit, substantive or procedural, enforceable by any party against the United States, its agencies or instrumentalities, its officers or employees, or any other person.

Sec. 4. *Definition.* As used herein, "imagery" means the product acquired by space-based national intelligence reconnaissance systems that provides a likeness or representation of any natural or man-made feature or related objective or activities and satellite positional data acquired at the same time the likeness or representation was acquired.

William J. Clinton

THE WHITE HOUSE,
February 22, 1995

[FR Doc. 95–5050
Filed 2-24-95; 2:13 pm]
Billing code 3195–01–P